ECOTOXICITY AND HUMAN HEALTH

A Biological Approach to Environmental Remediation

Edited by
Frederick J. de Serres • Arthur D. Bloom

CRC LEWIS PUBLISHERS
Boca Raton New York London Tokyo

Library of Congress Cataloging-in-Publication Data

de Serres, Frederick, J.
Ecotoxicity and human health: a biological approach to environmental remediation / by Frederick J. de Serres and Arthur D. Bloom.
p. cm.
Includes bibliographical references and index.
ISBN 1-56670-141-1
1. Pollution—Environment aspects. 2. Bioremediation. I. Bloom, Arthur D. II. Title.
RA566.D4 1995
615.9′02—dc20 95-15280
CIP

Direct all inquiries to CRC Press, Inc., 2000 Corporate Blvd., N.W., Boca Raton, Florida 33431.

Lewis Publishers is an imprint of CRC Press

No claim to original U.S. Government works
International Standard Book Number 1-56670-141-1
Library of Congress Card Number 95-15280
Printed in the United States of America 1 2 3 4 5 6 7 8 9 0
Printed on acid-free paper

THE EDITORS

Dr. Arthur D. Bloom is a geneticist and pediatrician. He has had a long-standing, active interest in the effects of environmental agents on human health, beginning with his work with the atomic bomb survivors in Hiroshima and Nagasaki, and his later research into the effects of chemicals on reproduction and cancer. Dr. Bloom has served on the full-time faculties at the University of Michigan and at Columbia University, where he was professor of pediatrics and of human genetics and development and director of the division of genetics at Babies Hospital of Columbia-Presbyterian Medical Center in New York. In recent years he has headed the Silvio O. Conte Institute for Environmental Health in Massachusetts and has founded the Comité International de Scientifiques pour la Santé et l'Environnement (the International Committee of Scientists for the Environment and Human Health) in Paris, from where he now works on the environmental problems of Central and Eastern Europe. He has been on the Board of Scientific Counselors of the National Institute of Environmental Health Sciences of the National Institutes of Health and is the author of many scientific papers and books.

Dr. Frederick J. de Serres received his B.S. degree in Biology and Chemistry at Tufts University, Medford, MA, and his M.S. and Ph.D. degrees in Genetics and Botany at Yale University, New Haven, CT. Since 1994, he has been a Guest Researcher in the Toxicology Branch of the Environmental Toxicology Program at the National Institutes of Environmental Health Sciences (NIEHS), Research Triangle Park, NC. From 1986 to 1994, he was Research Director of the Center for Life Sciences and Toxicology in the Chemistry and Life Sciences Unit of the Research Triangle Institute (RTI) in Research Triangle Park, NC. In that position, he managed multidisciplinary research in a comprehensive program of toxicological testing for commercial and government clients. All studies were performed under Good Laboratory Practice guidelines, and included both study design and performance, and submission of final reports. At RTI, he supervised contract work for commercial clients consisting of broad-spectrum toxicological evaluation of individual chemicals as well as product formulations. In connection with this work, Dr. de Serres also has worked with other commercial outfits with an interest in bioremediation technology. From 1972 to 1986, Dr. de Serres was Associate Director for Genetics in the Office of the Director at NIEHS. In this position, he arranged numerous workshops, conferences, symposia, and collaborative studies both nationally and internationally in collaboration with the International Programme on Chemical Safety for the World Health Organization in Geneva, Switzerland. From 1955 to 1972, Dr. de Serres was a member of the staff of the Biology Division of the Oak Ridge National Laboratory, Oak Ridge, TN, where he conducted a research program with *Neurospora crassa* designed to study the induction of specific-locus mutations after treatment with mutagenic environmental agents including environmental chemicals as well as both ionizing and nonionizing radiations. In this connection, he has developed a program to utilize the data generated with this assay in combination with other eukaryotic assays in genetic risk assessment exercises.

PREFACE

The present writing grew out of the perceived need for the development of safe, nontoxic remediation of chemical pollutants in the environment. These problems range from cleanup of oil spills to the removal of pollutants from the environments of the home and especially of the workplace. It is now widely thought that toxic waste sites are producing undesirable biological effects on both humans and the indigenous flora and fauna of our planet. The lack of definitive guidelines for remediation of complex chemical mixtures in such toxic waste dumps as well as of specific chemical pollutants in a safe, nontoxic manner was the driving force for the preparation of this book. Our primary intent is to review the biological aspects of the remediation problem. Our emphasis here is on the new field of **ecotoxicology,** in which we try to assess the impact of chemical toxins on human health as well as on other organisms that comprise the complex ecosystems of land, sea, and air.

The individual chapters of this book are planned to provide a biological approach to environmental remediation, and to provide an extensive characterization of the overall problem areas. In addition, the chapters discuss specific approaches to bioremediation of environmental pollutants that may be made with existing technology in a manner that is both efficient and cost effective.

We anticipate that this book will serve to introduce and to summarize various aspects of bioremediation for the uninitiated as well as for those actively working on the cleanup of the environment both in the United States and abroad. Particularly useful is the listing of Federal and State regulations and the standards that have been developed to regulate the overal process of bioremediation. The dumping of a myriad of complex mixtures of toxic chemicals and radioactive waste in the environment has created problems of enormous magnitude that will require a heavy investment of money, time, and professional expertise. Effective resolution is probably the most difficult challenge ever presented to our professional and legislative communities. It is our intention that this book will help provide a focus as well as a broad perspective for all interested parties.

The work of the study group convened to plan and to write this book was funded by Sunshine Makers, Inc., Huntington Harbor, California, the makers of Simple Green®, and by core support to the Conte Institute by the National Institute of Environmental Health Sciences (NIEHS), Research Triangle Park, North Carolina (contract number N01-ES-75185). We thank Mr. Bruce FaBrizio, CEO and President, and Mr. Milton Krause, Director, Research and Development, Sunshine Makers, Inc. We also thank the staff at Lewis Publishers for their assistance in the production of this new volume in the Conte Institute book series.

Arthur D. Bloom, M.D.
Frederick J. de Serres, Ph.D.
Pittsfield, Massachusetts
1995

CONTRIBUTORS

Yog R. Ahuja, Ph.D.
Professor
Genetics Unit
Mahavir Medical Research Centre
Hyderabad, India

Arthur D. Bloom, Ph.D.
President
Comite International de Scientifiques pour la Sante el l' Environment
Paris, France

Celia Bonaventura, Ph.D.
Director
Marine/Freshwater Biomedical Center
School of the Environment Marine Laboratory
Duke University
Beaufort, North Carolina

Joseph Bonaventura, Ph.D.
Professor
Department of Cell Biology
School of the Environment Marine Laboratory
Duke University
Beaufort, North Carolina

Donovan F. Bodishbaugh
Graduate Student
School of the Environment Marine Laboratory
Duke University
Beaufort, North Carolina

Larry D. Claxton, Ph.D.
Acting Director
Environmental Carcinogenesis Division
Health Effects Laboratory
U.S. EPA
Research Triangle Park, North Carolina

Frederick J. de Serres, Ph.D.
Guest Researcher
Toxicology Branch, Environmental Toxicology Program
National Institute of Environmental Health Sciences
Research Triangle Park, North Carolina

Donald P. Gaver, Ph.D.
Professor
Mathematics and Applied Sciences
Department of Operations Research
Naval Postgraduate School
Monterey, California

S. Elizabeth George, Ph.D.
Research Biologist
Genetic Toxicology Division
U.S. EPA
Research Triangle Park, North Carolina

Virgina S. Houk, Ph.D.
Toxicologist
Environmnental Carcinogenesis Division
U.S. EPA
Research Triangle Park, North Carolina

Barry Johnson, Ph.D.
Assistant Surgeon General
Assistant Administrator
Agency for Toxic Substances and Disease Registry
Atlanta, Georgia

Robert R. Lochmiller, Ph.D.
Professor
Department of Zoology
Oklahoma State University
Stillwater, Oklahoma

William R. Lower, Ph.D.
Senior Researcher
Cancer Research Center
University of Missouri
Columbia, Missouri

Karen McBee, Ph.D.
Associate Professor
Department of Zoology
Oklahoma State University
Stillwater, Oklahoma

Shabeg S. Sandhu, Ph.D.
Research Biologist
Genetic Toxicology Division
Health Effects Research Laboratory
U.S. EPA
Research Triangle Park, North Carolina

Shabeg S. Sandhu, Ph.D.
Research Biologist
National Health and Environmental Effects Research Laboratory
U.S. EPA
Research Triangle Park, North Carolina

Lee R. Shugart, Ph.D.
Senior Researcher
Environmental Sciences Division
Oak Ridge Laboratory
Oak Ridge, Tennessee

Deborah Weimer
Senior Attorney Advisor
Office of the General Counsel
Centers for Disease control and Prevention
Atlanta, Georgia

Charles E. Weinstein, Ph.D.
Professor
Department of Life Sciences
Berkshire Community College and
Conte Institute for Environmental Health
Pittsfield, Massachusetts

CONTENTS

Introduction

Frederick J. de Serres and Arthur Bloom

CONTENTS

Toxic wastes and their removal from the environment present a threat to the public health and to a multiplicity of ecological systems. Public concern for these effects has resulted in state and national legislation intended to remove hazardous wastes from the environment and to regulate the generation, transport, and disposal of currently-produced wastes. The financial burden of waste management on both state and national economies is enormous, with a multi-billion dollar industry having developed in the United States and elsewhere. For prevention of hazardous waste effects on public health and ecological systems, permanent remediation of existing waste sites and an effective program of waste reduction and recycling are required. Bioremediation of hazardous wastes by means of indigenous or introduced microorganisms appears to be a promising method of achieving these goals, and is a focus of this book.

ENVIRONMENTAL FATE AND ECOSYSTEM IMPACT

A widely recognized, but poorly understood relationship exists between the human and ecological effects of environmental pollution. Evidence now exists that both plant and animal populations may be more susceptible to the harmful biological effects of environmental contaminants than humans, and the effects on other species may thereby serve as an early warning sentinel of potential harm to human

1-56670-141-4/96/$0.00+$.50

populations. In addition, concern exists on a global level that toxic chemicals in the environment will produce a loss of biodiversity and other effects that will destabilize and stress our ecosystems. Preservation of ecosystem "integrity" is considered a critical requirement for the protection of human health. It is crucial, therefore, to develop methods of measuring the loss of ecosystem integrity that results from environmental contamination, and to include social and economic costs into an ecosystem approach to remediation.

COMPLEX MIXTURE TOXICITY AND MICROBIOLOGICAL ANALYSES

Environmental situations requiring remediation usually consist of mixtures of multiple toxicants and other chemical pollutants. Most types of remediation efforts, including biological, or bioremediation, produce additional chemical compounds that add to the complexity of evaluating the potential for human health effects. Because the products of bioremediation itself may differ from the original pollutants, the type and distribution of compounds within the different environmental media constitute an evolving situation. With the presence of multiple pollutants in a typical site, remediation may produce additional nonconventional pollutants (NCPs) that cannot be readily identified. However, these NCPs are a potentially important source of hazardous exposure. Preliminary studies suggest that NCPs have important toxicological properties with environmental persistence and mobility. As a result, they represent risk of unknown magnitude. NCPs limit our ability to do risk characterization of remediation sites when only analytical chemistry is used for exposure assessment studies. Incorporation of biological tests provides estimates of potential human toxicant exposure before, during, and after remediation efforts. Indigenous and genetically engineered microorganisms (GEMs) have great potential for use in environmental remediation. In the case of GEMs, it is important to study the potential health effects of the environmentally released organisms themselves prior to their use. At potentially high risk are the bioremediation workers who are exposed both to toxicants and to any genetically engineered microbes; new methods are needed to minimize work-related human health risks.

EVALUATION OF THE GENETIC HAZARDS

The use of biomarkers make it possible to assess the health of the environment and the potential impact of toxic chemical pollutants on the organisms (including man) that comprise various ecosystems. Biomarkers are often divided into those that indicate the level of chemical exposure, those that demonstrate the level of effects resulting from that exposure, and those that indicate the susceptibility of individuals or subpopulations to the environmental pollutants. DNA-damaging events are used to illustrate the use of a biomarker for assessing genotoxic insult. The assessment of the biological hazards of environmental pollutants may require an integration of biological and chemical analyses. A battery of assays with different endpoints is needed to assess the potential health hazards from exposure to industrial wastes. There are a wide variety of such bioassays for genetic damage in various plant species, as well as in aquatic and terrestrial animal species. These have been demonstrated to detect genetic hazards under a variety of natural environments. In the assessment of data derived from the use of such systems, it is important to consider the impact of both intrinsic and extrinsic factors on the variability and the interpretation of the data. Extrapolation from genotoxic assays to *in vivo* human health effects is, of course, the major focus of genetic risk assessment. The difficulties inherent in such extrapolations are by and large yet to be overcome.

WILDLIFE ECOTOXICOLOGY

Much of the work in ecotoxicology has been directed towards the identification and characterization of toxic effects of pollutants in laboratory models, followed by extrapolation to "real-life" situations. In the broad area known as wildlife toxicology efforts are made to determine the effect of exposure on such life forms directly. Almost all of this work has been conducted with wild birds and mammals; very little is known about the toxic effects of pollutants on wild populations of amphibians and reptiles. An argument in support of in situ biomonitoring is that resident animals provide a model system for evaluating the risks associated with "realistic" exposure routes and dosages. Wild animals residing in polluted environments are exposed to a variety of xenobiotics via multiple routes such as simultaneous ingestion and inhalation; these are assessments rarely, if ever, attempted on laboratory animals. In this book, we discuss at length an approach to the cytogenetic monitoring of rodent species exposed in the wild to toxic chemicals.

APPROACHES AND PROCESSES OF ENVIRONMENTAL BIOREMEDIATION

Bioremediation is an approach to dealing with toxic waste problems that relies on the same principles as those that occur naturally for dealing with waste. In a balanced natural system, many biogeochemical waste products are kept from accumulating to disruptive levels by the enzyme activity of indigenous microorganisms, such as bacteria and fungi. In many cases, bioremediation can lead to simple organic constituents via reduction of wastes. In addition, biological systems can be used to accumulate nondegradable wastes, such as heavy metals, possibly leading to recovery of useful products. Microbial sewage treatment and composting systems are well-known examples of how biological processes have been harnessed by man for many generations. New methods of utilizing living organisms in waste reduction are being continually developed. Naturally occurring and/or genetically engineered microorganisms may be used in land treatment, as biofilters and bioreactors for *in situ* treatment of waste. Some risks and limitations of these approaches before bioremediation can be fully exploited are considered here. It will be especially important to determine the exact fates and degradation products of waste in field applications of this technology.

APPLICATIONS AND NEW HORIZONS IN ENVIRONMENTAL BIOREMEDIATION

A variety of mechanisms exist for remediation of toxic chemicals in contaminated soil, ground water, and surface waters. Water-insoluble petroleum wastes in landfills can be emulsified and made more susceptible to microbial action by the addition of chemical solutions containing surfactants. In addition, the population sizes of indigenous bacteria and fungi can be enhanced and their metabolism of xenobiotics stimulated by the addition of chemically balanced fertilizers. Nondegradable wastes in contaminated soil may be concentrated by bioaccumulation; wastes are thereby absorbed and stored in the tissues of the organisms utilized. Biofilters utilize intact organisms or immobilized enzymes to degrade or accumulate waste components of fluid streams in ambient air or ground water. Bioremediation technologies utilizing fungi are an emerging approach with promise for biodegradation of a variety of bacterial-resistant wastes. GEMs can be used in sealed bioreactors to minimize the accidental release of such organisms into the environment. Most current applications of bioremediation are for single-waste products or for well-defined chemical mixtures; unfortunately, most sites in need of remediation are contaminated by many poorly-defined complex chemical mixtures. Biological techniques for remediation of such sites are promising, though the myriad of biological, chemical, and physical interactions must be more thoroughly investigated.

QUANTITATIVE MODELING AND ANALYSIS

Mathematical models have been developed for evaluation of the interaction, the eventual disposition or fate, and the health-related effects of biological and chemical agents that have been released into the environment by man. This book deals in some depth with such modeling. These models provide a tool for exploring the cost and effectiveness of different options for environmental usage and control and may help guide the collection of suitable data to organize and focus the analysis. Modeling connected with ground water, a primary medium of transport pollutants, has a long history that employs tools of physics and engineering, applied mathematics, numerical analysis, and statistics as well as chemistry and biology. Exposure to various chemicals in drinking water provides a need for pharmacokinetic models to describe the transport patterns of accumulations of toxic agents in various body organs. The need to relate the origin and subsequent transport of pollutants to the exposure dosage to humans and other organisms has lead to the development of exposure monitoring and modeling. Dose-response models have been developed to relate dosage levels to response at the organ or host organism level. Major interest focuses on the effects of chemical or biological agents on human beings, providing the need for model-based risk assessment. Direct experimental human response data are seldom available, so it is often necessary to extrapolate from animal data performed at high doses to the low doses usually experienced by the human population. Finally, two case studies are presented here to illustrate the use of mathematical and statistical models to analyze risk to humans exposed to polluted drinking water.

LEGAL AND INTERNATIONAL CONSIDERATIONS

The concern of the public about health effects of toxic waste materials in the environment has been transacted into laws and regulations by numerous legislative bodies. Statutes have been enacted to require premarket testing of materials before they are placed in commerce and to regulate the generation, transport, and disposal of waste. International agreements are in place to control international transportation and disposal of wastes. Specific federal laws in the United States for regulation of toxic materials include the Toxic Substances Control Act; Federal Insecticide, Fungicide, and Rodenticide Act; Resource Conservation and Recovery Act; Comprehensive Environmental Response, Compensation, and Liability Act; and air and water quality regulations, among others. An overview is given in this book of these various laws and regulations through the use of a fictitious case example. The use of this example in the development of a new chemical product illustrates the considerable impact of various federal and state statutes that regulate product testing and waste disposal.

In summary, this book is divided into a discussion of the *in vitro* and *in vivo* biological effects of toxic wastes and on the assessment of such effects, and into a detailed consideration of biological approaches to the remediation itself, with due attention to the complex legal requirements involved in the regulatory processes in the United States, as perhaps a model for Eastern Europe and other parts of the world.

CHAPTER 1

Hazardous Waste: Nature, Extent, Effects, and Societal Responses

Barry L. Johnson

CONTENTS

1-56670-141-4/96/$0.00+$.50

"It is now increasingly evident that more and more diseases stem from the degradation caused by man to his own environment."

H. Nakajima, Secretary General, World Health Organization, 1990

INTRODUCTION

The environment is all that which surrounds us. From the environment we draw the air we breathe, the water we drink, the land upon which we live. The environment is also the house in which we live and the place where we work. It is what nourishes us, nurtures us, and sustains us as a species and those other living things with which we symbiotically share the planet. The quality of the environment will determine how well we exist within it. A healthy environment equates with healthy ecosystems. But because of factors such as industrial development, population growth, technology development, unwise political and economic policies, and wasteful personal lifestyles, the quality and health of the environment have too often been compromised.

This chapter introduces the reader to the problems caused by the generation of wastes that contribute to degradation of the environment. An overview will be given about the huge amounts of hazardous wastes generated and dumped into the environment, the consequences to public health and ecosystems, and resultant societal reactions, including ethical concerns. In particular, discussion will address how the American public and international organizations have responded through social and political actions to the problem of hazardous waste in the environment. Subsequent chapters will elaborate on many of the issues introduced in this chapter.

Included is an introduction to U.S. federal laws, select state laws, and some international agreements pertinent to hazardous waste generation and management. It will be evident that hazardous waste is an international concern. The concept and definition of environmental remediation (i.e., the use of technology to clean up waste and repair environmental degradation) will be introduced, and examples of as well as statistics about current remedial technologies will be given. Emphasis will be given to one particular form of remediation, bioremediation, which is defined as treatment technology that uses biological activity to reduce the concentration or toxicity of contaminants in the environment. The effects on public health of waste and remedial actions to manage and dispose of waste will be interwoven throughout the chapter.

NATURE AND EXTENT OF WASTE GENERATION

Industrial development and modern agricultural practices have both contributed mightily to human advancement and well-being. The former serves as the foundation for modern national economies; the latter is the essential enterprise for producing food and fibers. It can be argued that advances in industrialization and agriculture have had positive benefits on human well-being, including improving public health. For instance, industrialization produced the pharmaceutical industry, which in turn, has produced an abundance of drugs essential to the treatment or prevention of human disease. Moreover, industrialization has led to the development and production of consumer products that make work easier and safer. In a similar context, improvements in agriculture have led to increased food production in many parts of the world, with obvious positive implications for human health.

However, the effects of both industrial development and modern agriculture have led to undeniable adverse effects on environment quality and ecosystem integrity. In this context, one significant adverse effect is the generation of waste. (Waste is defined for the purposes of this chapter as by-products that are deemed useless or worthless by the discarder.) By this definition, chemicals released into the environment are a form of waste. Some effects of environmental contaminants on human health and ecosystems are relatively well understood, others are still being investigated and debated. For example, the association between vehicle exhaust emissions and ambient air pollution is well known, including the deleterious release of lead into the environment through past use of leaded gasoline in the United States and, in turn, its effect on human health by way of lead absorption. Although less well understood, environmental contamination by chlorofluorocarbons has been implicated in the reduction of ozone in the atmosphere, leading

Table 1 Waste Generated per Year per Person

Country	Year	Pounds
United States	1986	1900
Canada	1989	1375
Japan	1988	867
West Germany	1987	824
Britain	1989	785
Spain	1988	708
Sweden	1985	697
France	1989	666
Italy	1989	662

From Plotzman (1992).

ultimately to a predicted increase in the incidence of skin cancer (NRC, 1991a; Rom, 1992).

Some agricultural practices have also contributed to environmental degradation and adverse effects on ecological systems. For example, excessive reliance on chemical fertilizers has contributed to elevated levels of contaminants in groundwater and surface water in some areas of the United States. As another example, the effects on ecological systems of past agricultural, consumer, and public health uses in the United States of DDT are well known, including its effects on reducing the numbers of some wild birds through eggshell thinning and adverse effects on reproductive physiology. A contemporary controversy concerns the effects on the global environment of destruction of rain forests. These examples, and others that could be cited, provide evidence that agricultural practices, if not adequately evaluated for their ecological effects, can cause widespread damage to the environment.

Modern industrial societies produce a multitude of products intended for consumption. Indeed, consumerism by way of free enterprise undergirds national economies in many countries. This production and distribution of consumer goods has brought many benefits to individual consumers; for example, products like video equipment for education and entertainment purposes can improve quality of life. With these positive contributions to individuals and to national economies have come some deleterious effects. One deleterious effect of consumerism is the impact on the environment from increased waste generation. Discarded consumer products soon become waste that requires disposal. In 1988, each person in the United States produced about 4 to 5 pounds of waste per day (PHS, 1990; Plotzman, 1992). This leads to more than 50,000 tons of waste produced each day in Los Angeles County, California. Figures available for several industrial countries shown in Table 1 show a pattern similar to that in the United States—large amounts of waste produced per person per year (Plotzman, 1992).

When disposed of improperly, common household waste can contribute to health problems ranging from noxious smoke from burning trash to creation of breeding grounds for vermin. Household waste adds to the municipal waste stream. In addition, hazardous substances contained in municipal waste can contribute to air, water, and foodchain pollution through releases from landfills or from other forms of disposal. Releases of hazardous substances from hazardous waste sites (HWS) into groundwater can present a significant threat to drinking water supplies. To reduce the amount of household waste will require greater efforts in recycling, improved practices of packaging of products, and changes in personal lifestyles that will contribute to less waste being produced per household.

The point being made is that modern economies are built on both industrialization and agriculture, with the benefits that accrue from both. But environmental degradation, with consequent implications for human health and ecology, has accompanied this development. (Degradation of the environment is used here as an undesirable outcome; in other parts of this monograph degradation of wastes by bioremediation connotes a positive outcome.) In some sense, the old cliché "Haste makes waste" applies here. Hazardous waste left expediently in the environment because it was often inexpensive to do so has now become a major concern of many industrialized countries. What is the basis for this concern and what does it mean for public health and its impact on ecological systems?

PUBLIC PERCEPTION OF HAZARDOUS WASTE

In the 1970s discoveries in the United States of toxic substances in residential communities at Love Canal, in New York; the "Valley of the Drums" site in Kentucky; and Elizabeth City, NJ; among others, riveted the attention of the American public and galvanized legislative action. Hazardous waste and its presence in community environments became a concern to environmentalists, legislators, and public health officials alike. A principal product of the legislative concern was the Comprehensive Environmental Response, Compensation, and Liability Act (CERCLA) of 1980, which was quickly christened and chronicled as the Superfund Act because of the multibillion-dollar size of the Hazardous Substance Respond Fund (which is also sometimes referred to as the Superfund) created by the act to fund the statute's directives of identifying uncontrolled HWS and sites where emergency spills have occurred, protecting public health and environmental media from the effects of releases from identified sites, and cleaning up those sites that merit such action.

At the heart of CERCLA is the philosophy, as evidenced by the statute's enforcement powers given to EPA, to hold accountable those parties responsible for the consequences of hazardous substances released into the environment from HWS. More will be said in subsequent sections of this chapter about this philosophy, its effects, and CERCLA's provisions.

The American public, as expressed in public opinion polls, continues to place a high priority on the need to reduce hazardous waste and to repair existing environmental damage. Moreover, this concern has mobilized community groups and national environmental organizations to oppose the generation, transportation, and disposal (e.g., construction of incinerators) of hazardous waste. This opposition has given rise to the term NIMBY, which stands for Not In My Back Yard. Community groups, in particular, often express NIMBY concerns through local opposition to the placement of landfills, incinerators, and other forms of waste storage or destruction facilities within their locales. NIMBY actions are usually rooted in concerns about public health effects of hazardous waste and economic consequences to homeowners (reduced property values). Local community groups, according to one source, gradually enlarge their sphere of contact beyond local agencies and officials and engage national grassroots organizations. The result is a broadening of environmental awareness and involvement (Szasz, 1994). This kind of grassroots networking is a consequential political force that has had significant impact on national policies on waste reduction, recycling, and the like.

DEFINITIONS OF HAZARDOUS WASTE AND HAZARDOUS SUBSTANCE

Before proceeding further, it is necessary to define what is meant by *hazardous waste,* and *hazardous substance,* because the terms are used throughout this book. In a general sense, hazardous waste is, as the name implies, discarded material that has the potential to do harm. However, because laws and international directives on hazardous waste require more precise definitions, two such definitions follow.

As defined in the Resource Conservation and Recovery Act (see Chapter 10), "Hazardous waste means a solid waste, or combination of solid wastes, that because of its quantity, concentration, or physical, chemical, or infectious characteristics, may (A) cause or significantly contribute to an increase in mortality or an increase in serious, irreversible, or incapacitating reversible illness; or (B) pose a substantial present or potential hazard to human health or the environment when improperly treated, stored, transported, or disposed of, or otherwise managed." (ELI, 1989).

The European Community has taken a more operational approach to defining hazardous waste (ELI, 1992). By their definition, hazardous waste means wastes that have one or more of the following 15 properties: explosive, oxidizing, highly flammable, flammable, irritant, harmful, toxic, carcinogenic, corrosive, infectious, teratogenic, mutagenic, ecotoxic; substances and preparations which release toxic or very toxic gases in contact with water, air, or an acid; substances and preparations capable by any means, after disposal, of yielding another substance, e.g., a leachate, which possess any of the other 14 properties. Using these criteria, lists of hazardous wastes are developed and used for administrative purposes like regulating transport and disposal of individual hazardous wastes.

The term hazardous substance is defined by CERCLA in operational terms as any substance designated by CERCLA itself and other federal statutes (see Chapter 10). At its root, hazardous substance is a legal term but in application refers to a specific pollutant or contaminant. CERCLA defines "pollutant or contaminant" to "include, but not be limited to, any element, substance, compound, or mixture, including disease-causing agents, which after release into the environment and upon exposure, ingestion, inhalation, or assimilation into any organism, either directly from the environment or indirectly by ingestion through food chains, will or may reasonably be anticipated to cause death, disease, behavioral abnormalities, cancer, genetic mutation, physiological malfunctions (including malfunctions in reproduction) or physical deformations, in such organisms or their offspring; except that the term 'pollutant or contaminant' shall not include petroleum ... and shall not include natural gas, liquefied natural gas, or synthetic gas ..." (ELI, 1989). This definition is obviously a term written by attorneys for use by attorneys!

For the purposes of this chapter, hazardous waste and hazardous substance will be used to denote that waste and those substances that possess properties that can cause harm to human health and ecological systems. However, the reader is advised that these terms take on specific meanings under various legal and regulatory acts, like CERCLA.

EXTENT OF THE HAZARDOUS WASTE PROBLEM

Industrialized countries produce huge quantities of hazardous waste. This often results from inefficient or uneconomical means of production and use of energy and materials. In the United States, the "solution" frequently chosen in the past was to dispose of the waste in landfills or waterways (rivers, lakes), or by incineration. The end result is a toxic legacy to the environment. Too often an attitude of "The solution to pollution is dilution" has prevailed. That practice doesn't work, especially for pollutants that are persistent in the environment and therefore bioaccumulate. An attitude of *"The solution to pollution is diminution"* should characterize current and future waste management practices. In a public health and ecological systems perspective, prevention of environmental degradation leads to prevention of adverse health effects because exposure to toxicants does not occur. In public health parlance, this is *primary prevention* at work.

The exact volume of hazardous waste produced globally is unknown. International comparisons are almost impossible because of differences in definition and classification of hazardous wastes from country to country. One source estimates that for a number of western European countries, hazardous waste production is about 5000 tons per billion U.S. dollars of gross domestic product (GDP) (Batstone et al., 1989). On this basis, the United States is estimated to produce 75,000 tons per annum (tpa), and Canada 10,000 tpa. Waste production per GDP was estimated to be 10,000 tons in the former Soviet Union, that in other countries with mature industry at 5000 tons, in newly industrialized countries 2000 tons, and in developing countries 1000 tons.

A separate estimate of hazardous waste production in the United States was developed by the Office of Technology Assessment (OTA). They estimate that 255 to 275 million tons of hazardous waste are produced annually in the United States (OTA, 1983). But by whatever yardstick, the quantities of hazardous wastes generated each year are enormous, and without proper management of its storage or disposal, have the potential for compromising public health and the well-being of ecosystems.

NUMBERS OF HAZARDOUS WASTE SITES AND COSTS

Hazardous waste has often been discarded into the environment and abandoned in landfills and on industrial properties. The U.S. Environmental Protection Agency (EPA) lists 37,987 sites on its inventory of uncontrolled hazardous waste sites (HWS) as of April 1994 (NRC, 1994). Of this number, 1296 are listed on, or proposed for, EPA's National Priorities List (NPL), which lists the sites posing the greatest threats to public health and the environment (EPA, 1994). EPA proposes sites for the NPL by applying a Hazard Ranking System, which, in general terms, is an assessment of the relative public health, environmental, and ecological threat posed by a given site. EPA's placement of a site on the NPL can bring monies from the public trust fund called Superfund for the site's remediation.

Of particular importance are HWS associated with federal facilities. As of August 1994, 150 federal facilities are on the NPL (EPA, 1994). These facilities include sites operated by the Department of Defense (DOD, military bases) and the Department of Energy (DOE, weapons complexes). One source attributes 17,482 contaminated sites at 1855 DOD installations as of September 1990; DOE has identified 3700 sites at 500 facilities in 1991 (NRC, 1994). Some of the DOE sites are current or former nuclear weapons facilities that cover large areas and are complex in terms of wastes produced.

EPA estimates that cleaning up the current NPL sites may cost more than $30 billion (Habicht, 1991). In addition, about 4000 facilities, representing 64,000 solid waste management units, may require cleanup, according to EPA.

A study conducted by University of Tennessee researchers projected that cumulative cleanup costs for all sites from 1990 through 2020, using current remediation practices, will be approximately $750 billion, with plausible lower bound at something less than $500 billion and upper bound at approximately $1 trillion (Russell et al., 1991). Their analysis included both federal and nonfederal sites and covered Superfund NPL sites, state and private sector waste remediation programs, underground storage tanks, and RCRA (see Chapter 10) sites requiring corrective action. Specific to Superfund NPL sites, Russell et al. estimate cumulative costs through year 2020 to range from $106 billion (assuming 2100 nonfederal sites) to $302 billion (assuming 6000 nonfederal sites). Less stringent cleanups would lower these projected costs; greater stringency would increase them. Russell et al. "best guess" of cumulative remediation costs for federal facilities are $240 billion for DOE sites and $30 billion for DOD sites.

A separate set of cost estimates were derived by the Congressional Budget Office (CBO, 1994), which analyzed three scenarios for Superfund costs after 1992. Their estimates are based on assumptions related to the number of *nonfederal NPL* sites. The base-case estimate is $74 billion in discounted, present-worth dollars; the low-case estimate is $42 billion and the high-case estimate is $120 billion. The assumptions of 2300 nonfederal NPL sites in the low case, 4000 in the base case, and 7000 in the high case explain most of the differences in CBO's estimated costs. Annual undiscounted costs in the base-case peak are $9.1 billion in the year 2003 and average $2.9 billion per year through year 2070 (CBO, 1994).

Estimates by the CBO of future Superfund costs are very different from earlier estimates developed by the EPA and the University of Tennessee. The main factors that explain the differences are CBO's broader coverage of costs and use of discounted dollars, different average cleanup costs per site, and different numbers of sites on the NPL. Specifically, the CBO figure includes all future public and private Superfund expenditures, whereas the Tennessee estimate covers public and private costs for study and cleanup at NPL sites, omitting administrative and legal expenses and the costs of screening and removals at non-NPL sites. Furthermore, the CBO estimate is in present-worth dollars; the Tennessee figure is expressed in undiscounted dollars (CBO, 1994). However, by anyone's measure, the cost in money and personnel resources to remediate the legacy of hazardous wastes left in the environment will be huge.

CATEGORIES OF SITES AND CONTAMINANTS

As previously cited, the amount of hazardous waste left in the environment is enormous. To clean up the environment requires knowledge about what has been left in waste sites. Figures 1 and 2 show categories of HWS in the United States and classes of chemicals found at these sites. Figure 1 illustrates that about 40% of the current NPL sites are abandoned, former manufacturing facilities. The next

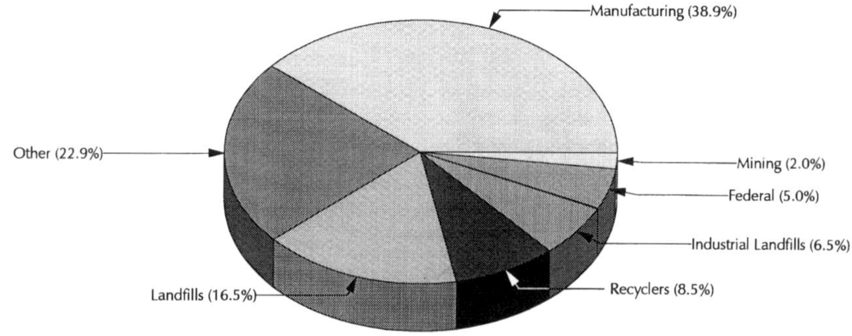

Figure 1 Categories of Superfund sites (OSWER, 1992a).

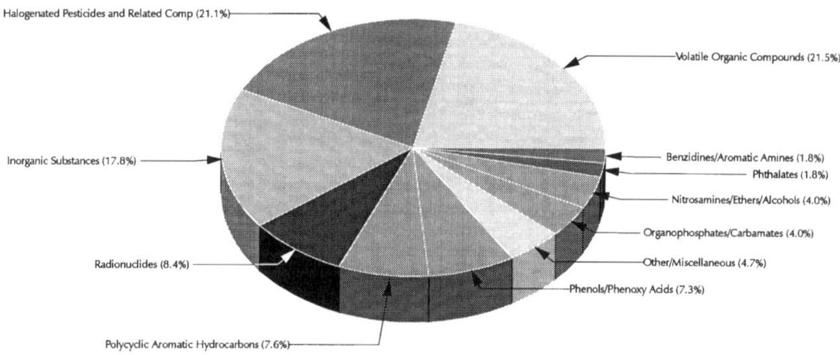

Figure 2 Categories of Superfund substances (ATSDR, 1992c).

largest category of sites is abandoned municipal landfills (OSWER, 1992a). Moreover, there is great diversity among chemical substances left in HWS. The U.S. Agency for Toxic Substances and Disease Registry (ATSDR) lists over 2000 individual chemicals in its national database on NPL and other sites (ATSDR, 1992a). The primary chemical classes for substances in the ATSDR database are shown in Figure 2. The 2000 substances are those identified by samples taken by EPA in their site remediation studies. As noted in the figure, volatile organic compounds, halogenated pesticides, and inorganics constitute about 60% of the substances identified at HWS (ATSDR, 1992a).

ATSDR also has made observations about the kinds of contaminants actually released from NPL sites; the environmental media that are most affected by released contaminants; and the pathways by which people are exposed to contaminants. The top three classes of contaminants of concern derived from contaminants data found in ATSDR's public health assessments* conducted in 1992, and the frequency with which they occurred at sites, were:

- volatile organic compounds (solvents) (87% of sites)
- inorganic compounds (87%)
- halogenated pesticides (50%)

* Public health assessments are a form of hazard evaluation conducted by ATSDR for all NPL sites and sites petitioned of ATSDR (see Chapter 10).

Table 2 Top 25 Superfund Hazardous Substances Found at NPL Sites

Rank	Substance	NPL sites
1	Lead	1674
2	Arsenic	1538
3	Mercury, metallic	1403
4	Vinyl chloride	1338
5	Benzene	1331
6	Cadmium	1286
7	Polychlorinated biphenyls	1256
8	Chloroform	1239
9	Benzo(a)pyrene	1208
10	Trichloroethylene	1191
11	Benzo(b)fluoranthene	1189
12	DDT, p,p'	1188
13	Aroclor 1260	1145
14	Aroclor 1254	1137
15	DDE, p,p'	1119
16	DDD, p,p'	1101
17	Dieldrin	1099
18	Chromium, hexavalent	1091
19	Aroclor 1248	1090
20	Chlordane	1084
21	Tetrachloroethylene	1084
22	Dibenzo(a,h)anthracene	1084
23	Creosote	1067
24	Hexachlorocyclohexane, gamma	1065
25	Aroclor 1242	1057

From ATSDR (1992b).

These classes of contaminants were most often identified in groundwater, followed by subsurface soil, sediment, and surface water.

Using a comprehensive database that contains environmental contamination data on 1300 HWS, ATSDR found that 100 or more different chemicals can be found at a single waste site (ATSDR, 1993a). These chemicals were found in widely varying combinations in water, soil, and air, and some of the combinations may be much more hazardous than any of the individual chemicals. As an example, ATSDR found that soil with levels at health concern for copper, chromium, arsenic, cadmium, or mercury exposures also contained lead in high levels more than 80% of the time. This kind of information is important for conducting site-specific risk assessments and public health assessments and for designing human epidemiological investigations and toxicological studies.

Of course, not all substances released from HWS are equally hazardous to human health and ecological systems. For that reason, ATSDR and EPA have jointly ranked hazardous substances in terms of their threat to human health. Table 2 lists the top 25 substances on a prioritized list of 275 substances, together with the number of NPL sites where each has been found (ATSDR, 1992b). Under provisions of CERCLA, as amended in 1986, placing a substance on the priority list of hazardous substances leads to the development by ATSDR of a Toxicological Profile for each substance and, in turn, a program of directed research to identify and fill data gaps pertaining to a substance's toxicity and health effects. ATSDR's Toxicological Profiles are documents that contain summary information about the toxic and human health properties of substances profiled. Each profile also must contain an analysis of gaps in scientific knowledge about a substance's toxic and physical properties.

The data gaps for priority substances are filled through an applied research program that consists of government-funded projects at universities and federal government laboratories, enforceable consent actions between EPA and private industry through authorities in the Toxic Substances Control Act (Auer, 1994), and research voluntarily conducted by private industry. All these mechanisms to fill key data gaps will ultimately improve chemical and site-specific health risk assessments through use of improved toxicologic science and data.

The wide spectrum of chemical substances at HWS, coupled with factors such as large geographic areas covered by many sites and diverse human populations at potential health risk, makes environmental remediation very difficult and costly.

NUMBERS OF PERSONS AT RISK

The number of persons in the United States at potential health risk from releases of hazardous substances from waste sites is difficult to estimate with any precision, given uncertainties of the frequency and geographic extent of releases. However, using what data are currently available, two estimates exist of numbers of persons at potential risk. Using data from 1134 NPL sites in 1991, EPA estimated that approximately 41 million people reside within 4-mile radii of 1134 NPL sites, and about 3325 persons reside, on the average, within 1-mile radii of NPL sites (NRC, 1991b). This means 1 out of 6 Americans live within 4 miles of an NPL site. A demographics survey of populations around a sample of 1309 NPL sites led ATSDR to estimate that approximately 11 million people live within 1-mile borders (Johnson, 1995). Given the limitations of demographic data available to EPA and ATSDR, the two agencies' estimates are in reasonable agreement. Of course, that doesn't necessarily mean that all these persons are at health risk; that is determined by such factors as patterns of release of substances from sites, degree of exposure of persons in contact with released substances, toxicity of the substances, and a person's individual susceptibility to toxicity.

EFFECTS OF HAZARDOUS WASTE ON HUMAN HEALTH

Scientific literature on the human health effects of hazardous substances in the environment clearly indicates that exposure to some individual substances, or combinations of substances, presents a hazard to human health. The literature base supporting this assertion comes from studies of workers, laboratory animals, and, to a lesser extent, community residents. (In this context, a hazardous substance is one that under specific conditions, will cause premature death, disease, impaired or lost function.) There are three groups of primary concern for the potential health effects of exposure to hazardous waste. One group consists of persons who may be directly exposed to substances released from HWS. The second group consists of waste site cleanup workers and workers at waste disposal facilities. The third group consists of ecological systems that can be affected by hazardous waste. The sections that follow provide a summary of health findings pertinent to each group.

COMMUNITY HEALTH STUDIES

A considerable number of epidemiologic studies have been conducted to investigate health effects in communities around HWS. Many of these studies are negative in the sense that statistically significant increases in adverse health effects were not found (Grisham, 1986; Upton et al., 1989). Although these findings may represent the true health status of the communities due to limited or no exposure to hazardous substances, negative results from epidemiological investigations can also occur from: inadequate cohort sample sizes, inadequate information about exposure to hazardous substances, poor selection of sites studied, or inappropriate health effects studied for the toxicology of the specific substances found at a given HWS. Some of these limitations and confounders have the potential for being palliated through better exposure assessment and by combining sites for study. Regarding aggregating sites, it is possible to combine sites by matching on such factors as common environmental characteristics or by matching sites that have common geographical features (ATSDR, 1992c). The result is a study that includes a larger sample size of persons examined (or their health data if data files are used) than if each site is examined as an independent unit, thereby increasing the statistical power of the study.

Some studies do report adverse health effects in communities or populations exposed to hazardous substances released from waste sites, from improper disposal of hazardous materials, or from permitted release of contaminants by commercial enterprises. These studies have examined many toxic endpoints, including cancer, adverse reproductive outcomes, and toxicity to organ systems (Buffler et al., 1985). The sections that follow provide summary descriptions of findings from various kinds of health investigations of communities and workers potentially impacted by substances released from HWS.

Cancer Investigations

Two sets of investigators have reported an increased frequency of cancers in counties containing HWS. A 1983 study reported that age-adjusted gastrointestinal (GI) cancer mortality rates (all anatomical sites combined) were higher in 20 of New Jersey's 21 counties than national rates. Within specific sex and race groups, cancer mortality rates in the state during the period 1968–1977 significantly* exceeded national rates for cancers of the esophagus, stomach, colon, and rectum. The environmental variables most frequently associated with GI cancer mortality rates were population density, degree of urbanization, and presence of toxic waste disposal sites (Najem et al., 1983). Similarly, a study was conducted of 593 NPL sites in 339 U.S. counties (called HWS counties) of 49 states where analytic evidence showed contaminated ground drinking water as the sole source of water supply (Griffith et al., 1989). For each identified county, age-adjusted, site-specific cancer mortality rates for 13 major cancer sites were extracted from U.S. Cancer Mortality and Trends 1950–1979 for white males and females in the decade 1970–1979. Significant associations between excess deaths and all HWS counties, when

* The word "significantly" is used throughout this chapter to mean statistical significance was found in the data analysis, according to some inferential test used by the investigator.

compared to all non-HWS counties, were shown for cancers of the lung, bladder, esophagus, stomach, large intestine, and rectum for white males; and for cancers of the lung, breast, bladder, stomach, large intestine, and rectum for white females. A study of a leukemia cluster in children living in Woburn, MA, found a confirmed increase in incidence, distributed uniformly over an 11-year period (Cutler et al., 1986). The investigators found no leukemia sufferer who had contact with a HWS, but could not rule out exposure to contaminated water (chloroform, trichloroethylene, tetrachloroethylene) as a factor. None of these studies prove causality between cancer incidence and releases of hazardous substances from waste sites; however, they raise serious questions that should be elaborated through more precise research.

An investigation by the Montana Department of Health and Environmental Sciences and ATSDR of a putative cluster of pancreatic cancer of men yielded some findings of note. This investigation was conducted because area residents were concerned about health effects from contamination emanating from a local railroad refueling and repair station. They were also concerned about an apparent increase in cancer morbidity in the area. The observed number of pancreatic deaths from 1980 through 1989 among white men residing in the area under investigation was elevated compared with the age-adjusted number of deaths expected on the basis of U.S. age- and sex-specific pancreatic cancer morbidity rates (1980–1988). Smoking did not appear to account for this increase. The cluster of pancreatic cancer deaths appeared to be among persons who worked in the railroad industry, indicating that occupation may have been a risk factor. In addition, the addresses of family members seemed to cluster along a railroad right-of-way, suggesting the need to evaluate the role of environmental exposure as a contributing factor in the community's pattern of pancreatic cancer (ATSDR, 1992d).

Reproductive Effects Investigations

A particularly important study examined the association between congenital malformations in children and maternal proximity to HWS in the state of New York (Geschwind et al., 1992). The study consisted of a total of 9313 newborns with congenital malformations, as recorded in the New York State Congenital Malformations Registry, and 17,802 healthy comparison children. A total of 590 HWS in 20 New York counties were selected for analysis of the incidence of malformations. (HWS in New York City were excluded, given what were presumed to be unique sociodemographic and geographic characteristics in comparison to the rest of the state.) A coordinate mapping procedure was used to match all case and comparison geographic coordinates to the HWS coordinates. An "exposure risk index" was created for each respondent within a 1-mile radius of the birth residence. An unconditional linear logistic regression procedure was used to analyze the association between maternal proximity to HWS and congenital malformations. The investigators controlled through statistical analysis for several possible confounding variables, including maternal age, race, education, complications during pregnancy, birth weight, length of gestation, and sex of child. Other possible confounders like smoking and alcohol history, maternal and paternal occupational exposures, and maternal nutritional status were not evaluated due to lack of data. Results indicated that maternal proximity to HWS may carry a small additional risk of approximately 12% of bearing children with congenital malformations. Higher malformation rates were associated with both a higher exposure risk (63% increase in risk) and documentation of off-site chemical leaks (17% increased risk). A follow-up study did not confirm the Geschwind et al. findings, indicating that associations between elevated risks of birth defects may be due to proximity to operating industrial facilities, rather than proximity to HWS (Marshall et al., 1995). Further work will be needed to clarify the two studies.

In a study of a similar investigative intent, a study of birth defects and birth weight was performed in the five-county San Francisco Bay area in California (Shaw et al., 1993). The investigators reviewed case records of congenital malformations and birth weights of all live births and fetal deaths that occurred from 1983 through 1985. A total of 5046 children with congenital malformations were identified from a total of 215,820 live births. Birth weights of 190,400 children were also analyzed. Each mother's census tract of residence at time of delivery was used to categorize exposure potential. Each census tract was categorized into one of three groups, according to the presence or absence of sites of environmental contamination (landfills, chemical dumps, industrial sites, hazardous materials treatment and storage facilities) and whether information was available about potential human exposure to released contaminants. An elevated risk (odds ratio was 1.5) for infants with malformations of the heart and circulatory system was found between census tracts with sites having potential for exposure and census tracts with no sites. No increased risk in other malformations was noted, nor for low birth weight.

In a site-specific study in Tucson, AZ, the proportion of cases of congenital heart disease among live births was significantly greater when parents had first trimester residential exposure to trichloroethylene (TCE) and dichloroethylene in water contaminated by releases from a NPL site than for the comparison group (Goldberg et al., 1990). In another study, a higher prevalence in birth defects was noted among persons living near a thorium waste disposal site in New Jersey when compared to persons living a greater distance from the site (Najem and Voyce, 1990).

In contrast to these findings, very low birth weight, infant and fetal death, prematurity of birth, and congenital malformations were not associated with living in a 1-mile radius of NPL sites during pregnancy (Sosniak et al., 1994). The investigators examined reproductive outcomes reported in a 1988 national database and compared the occurrences of adverse outcomes with mothers' residence during pregnancy, using maternal ZIP codes as a surrogate for exposure to any releases from sites. The NPL database was derived from EPA records for 1990. Possible confounding variables that account for adverse reproductive outcomes, e.g., smoking history, quality of prenatal care, were examined through questionnaires to mothers in the database. The lack of association between maternal proximity to NPL sites nationwide and adverse reproductive outcomes is tempered by lack of actual exposure information during pregnancy in this study, as well as the other reports cited in this section.

Other Adverse Health Outcomes

In addition to studies of cancer rates and adverse reproductive outcomes, several reports address other health effects in community residents putatively impacted by HWS. In brief, the principal health findings include:

1. significantly shorter stature by age for children at the Love Canal, New York, waste site than for comparison children (Paigen et al., 1987);
2. a higher prevalence in liver diseases among persons living near a thorium waste disposal site when compared to persons living a greater distance from the site (Najem and Voyce, 1990);
3. hepatomegaly and elevated liver function tests in residents exposed to leachate from a toxic waste dump (Meyer, 1983);
4. in residents around a Superfund mining site, illness rates for chronic kidney disease, stroke, hypertension, heart disease, skin cancer, and anemia showed statistically significant associations (Neuberger et al., 1990);
5. significantly higher rates of sister-chromatid exchanges in peripheral lymphocytes were found in persons residing near a waste site in Belgium when compared to a reference group (Planard et al. 1994);
6. increased symptoms of systemic lupus erythematosus were found in persons exposed to trichloroethylene and other contaminants in well water contaminated from leachate from a waste site (Kilburn and Warshaw, 1992);
7. tests of neurophysiological and neuropsychological functions administered to persons exposed to well water contaminated with TCE released into groundwater from a HWS showed significant impairment in comparison to a reference community (Kilburn and Warshaw, 1993);
8. a marginal effect of residents' increased use of prescription drugs and the level of presumed exposure to releases of hazardous substances into a community adjacent to a toxic waste site in France (Zmirou et al., 1994).

None of these studies presented actual exposure information, but they raise questions that merit further pursuit.

Symptom and Illness Prevalence Reports

Some investigators have examined the kind and frequency of symptoms of illness reported by persons living near HWS. Increased frequencies of mood, sleep, skin, and respiratory disorders were reported in persons around a Canadian landfill, in comparison with a local community (Hertzman et al., 1987). Results from another study indicated that residents adjacent to an abandoned industrial site complained of odors emanating from the site. Levels of benzene, toluene, TCE, and tetrachloroethylene exceeded urban background, but were below odor thresholds. The residents expressed more self-reported complaints referable to the respiratory system, constitutional ills, sensory irritation, and irregular heartbeat than a comparison group (Ozonoff et al., 1987). Similarly, when compared to a reference community, a health survey of persons near a large NPL site found elevated odds ratios for 23 illness/disease symptoms, and 8 symptoms had odds ratios greater than 1.5: blurred vision, pain in ears, daily cough for more than 1 month, nausea, frequent diarrhea, unsteady gait, and frequent urination (Baker et al., 1988). In a study in Pennsylvania, residents living near an HWS reported significantly increased numbers of skin problems and insomnia than a comparison community, although the investigators reported no "serious diseases" (Logue and Fox, 1986). Another study indicated excess symptom reporting among "exposed" community residents, (i.e., persons living near HWS) compared with "unexposed" persons. Symptom reporting was strongly associated with "environmental worry" (Lipscomb, 1989; Shusterman et al., 1991). At another HWS, an increased prevalence of various self-reported symptoms with onset after moving into the current home was found. After controlling for potential confounders, increased prevalences were found in the target population, when compared with a comparison community, for the following self-reported illnesses diagnosed by a health care provider: eczema and other skin problems; ulcer or other stomach disease, excluding cancer, for males; hay fever or other respiratory allergy for females; and frequent periods of anxiety, nervousness, or depression for some subgroups (Miller, 1994).

In contrast, the results from a community health survey of people living near an HWS in Australia showed no association between an increase in disease rates and proximity to the site; however, symptom prevalence and perceived decline in recent health status correlated with stress and anxiety measures (Dunne et al., 1990). A similar result of elevated reporting of adverse health symptoms in the absence of confirmed chemical releases from a toxic waste site led the investigators to stress the need for symptom and illness prevalence studies that control for subjects' reporting bias (Roht et al., 1985). A similar outcome of lack of elevated illness rates was noted in a New Jersey community adjacent to an NPL site. Structured questionnaires were administered to 676 persons thought most likely to be exposed to releases from an HWS that had documented levels of contaminants released into groundwater and soil. A comparison group consisted of 778 persons from the same community, but at a greater distance from the HWS. After adjusting for various confounding factors like cigarette use and water consumption patterns, no differences between the two groups were found in the respondents' rates of cancer, respiratory disease, liver disease, and skin diseases (Najem et al., 1994).

One set of investigators examined the communities around five HWS in California (Neutra et al., 1991). There was no evidence that rates of cancer and birth defects were elevated. However, the prevalence of subjective health symptoms of illness was elevated in reference to comparison communities. The investigators concluded that toxicologic response and mass psychogenic illness were not valid explanations of the elevated symptom rates. It was observed that excess reporting of symptoms was associated with those persons who complained of odors from the sites and who expressed worry about the effects of environmental chemicals. Based on this observation, the investigators hypothesized that higher symptom rates in persons residing near HWS is due to stress-mediated mechanisms or behavioral sensitization.

In a different kind of health data analysis, but one still based on persons' self-reported health responses, latent adverse health effects in persons exposed to hazardous substances released from NPL sites are suggested by findings from ATSDR's National Exposure Registry (ATSDR, 1993b). Included in the registry are approximately 5000 individuals exposed to TCE in municipal water supplies contaminated by releases from uncontrolled HWS. The 5000 registrants come from 14 sites in Michigan, Indiana, Illinois, and Pennsylvania; 10 of the sites are on the NPL. TCE levels in drinking water ranged from 1 part per billion (ppb) to 19,380 ppm. (The Maximum Contaminant Level for TCE is 5 ppb.) Exposure durations ranged up to 18 years. For essentially all persons on the registry, TCE exposure has ceased due to provision of uncontaminated water. People on the registry reported (through a structured health survey) more health problems when compared to national survey data. Some of these problems appear in only males, some in only females, and some in only specific age groups. The health problems which were reported in excess of those reported by the general population included anemia and other blood disorders, diabetes, hearing impairments, heart disease, hypertension, kidney disease, skin rashes, speech impairments, stroke, and urinary tract disorders. Some of the health problems reported by registrants seem consistent with the toxicology of TCE. However, caution must be exercised when interpreting these health findings until epidemiological health studies are conducted on a sample of the 5000 persons in the TCE registry.

Symptom and illness prevalence studies have a number of limitations and shortcomings. Recall bias can be a problem, i.e., persons' memory of health problems and related events can be faulty. Also, self-reporting of adverse health events has been thought by some to be subject to some degree of biasing for reason of economic gain or attention. On the positive side, this kind of study can be useful in designing hypothesis-generating investigations, for example, investigating a pattern of respiratory health complaints through pulmonary function testing.

CONTAMINANT EXPOSURE INVESTIGATIONS OF COMMUNITIES

For most communities around HWS, few data exist concerning measured human exposure levels to substances putatively released from the sites. There are many reasons for this lack of data. For some sites, releases from the site have been interdicted and ability to measure any past exposures have passed because the substances (or their metabolites) do not remain in body tissues very long. For some sites, there may be on-going exposure, as measured by environmental contamination, but biological measurement methods may not be available to assess actual human uptake.

A small body of literature documents human exposure to hazardous substances released from some waste sites. These are investigations that have focused on measuring exposure to specific hazardous substances in specific human populations thought to be at elevated risk of exposure. As such, the investigators usually did not ascertain information about health outcomes. For example, the mean blood lead level for children living near the California Gulch site in Colorado was almost twice that predicted by a biokinetics uptake model (CHD, 1990). In addition to lead, increased exposure to arsenic has been reported in children living near waste sites in Montana, Washington, and Utah (ATSDR, 1986, 1987, 1988; Binder et al., 1987); elevated serum levels of polychlorinated biphenyls were found in some people at two HWS (Stehr-Green et al., 1988), although not at most sites surveyed; and 1,1,2,2-tetrachloroethylene was noted in the breath of persons whose homes were situated near a former chemical waste dump (Monster and Smolders, 1984).

A study of children in Colorado measured blood lead levels in an area contaminated by releases from a metals refinery, but the levels were low and not related to proximity to the location of the smelter (Gottlieb and Koehler, 1994). This study illustrates the difficulty of relating site-specific HWS releases with body burden levels of toxicants. In this study, removal of lead from gasoline significantly contributed to lowered blood lead levels in young children, perhaps overriding any increase in lead uptake attributable to lead released from the smelter and soil contaminated by the smelter. Investigations that measure, where possible, levels of exposure to hazardous substances in human populations at putative health risk should be increased in number. Exposure data from such investigations are invaluable in terms of prioritizing sites for identifying populations at health risk and for cleanup actions.

Correlative Environmental Contamination Studies

Some of the same substances that are known to be released from HWS have been associated with adverse human health effects when investigated in environmental media through releases other than from HWS. When the specific substances and the levels of environmental contamination are similar to those associated with HWS, then the human health implications may portend possible impacts from HWS.

For example, cancer of the bladder was associated with increased intake of beverages made from tap water that came from chlorinated surface water (Cantor et al., 1987). The risk gradient with intake was restricted to persons with at least a 40-year exposure to chlorinated surface water and was not found among long-term users of nonchlorinated groundwater. Although results from another study of contaminants in public drinking water supplies showed an association between 1,2-dichloroethane and cancers of the colon and rectum, and between nickel and cancers of the bladder and lung, the investigators suggest the chemical contaminants are not causal factors but rather indicators of possible anthropogenic (i.e., related to human activities) contamination of other types (Isacson et al., 1985).

In other studies, (1) when compared to controls, findings from a case-control* study showed the frequency of spontaneous abortion was associated with detectable levels in public drinking water supplies of mercury, high levels of arsenic, potassium, and silica, among other factors (Aschengrau et al., 1989); (2) conduction latency of the blink reflex was lengthened in persons with alleged chronic exposure to TCE in well water when compared with latencies from laboratory controls (Feldman et al., 1988); (3) a study of rates of leukemia and non-Hodgkin's lymphoma (NHL) in 75 towns in New Jersey showed incidence in women of all leukemias and NHL to be associated with levels of TCE in drinking water (Cohn et al., 1994); and (4) an elevated risk of bladder cancer was associated with contaminant levels of perchloroethylene in a municipal drinking water supply (Aschengrau et al., 1994).

LITERATURE REVIEWS OF COMMUNITY HEALTH LITERATURE

In addition to the summary of human health findings presented in the preceding sections, there are at least two comprehensive reviews of the literature on hazardous waste and human health are available. The National Research Council (NRC) conducted a comprehensive review of the published literature on public health implications of hazardous waste (NRC, 1991b). The review concluded, "In spite of the complex limitations of epidemiologic studies of hazardous waste sites, several investigations at specific sites have documented a variety of symptoms of ill health in exposed persons, including low birth weight, cardiac anomalies, headache,

* In a case-control study, persons with disease or organ system impairment ("cases") are compared with individuals without the condition ("controls") with respect to their rate of exposure to the agent of interest (Johnson et al., 1987).

Table 3 ATSDR's Priority Health Conditions for Superfund Communities

Birth defects and reproductive disorders
Cancers (select)
Immune function disorders
Kidney dysfunction
Liver dysfunction
Lung and respiratory illness
Neurotoxic disorders

From Morbidity and Mortality Weekly Report, Centers for Disease Control and Prevention, Atlanta, 41, 72, 1992.

fatigue, and a constellation of neurobehavioral problems. It is less clear whether outcomes with a long delay between exposure and disease also have occurred, because of complex methodological problems in assessing these outcomes. However, some studies have detected excesses of cancer in residents exposed to compounds, such as those that occur at hazardous waste sites." The NRC concluded that the overall impact of hazardous wastes in the American environment is unknown because of limitations in identifying, assessing, or ranking HWS exposures and their potential effects on public health. They recommended additional effort to conduct epidemiologic investigations of communities around HWS. However, an earlier review concluded that little evidence existed to support the proposition that chemical wastes were adversely impacting the health of communities (Grisham, 1986). It is not possible to reconcile these two differing views, other than to note that the two reports were prepared with 5 years between them and by different persons.

In another review of the health effects literature, the following conclusions were reached (Johnson, 1993): "When evaluated in aggregate (i.e., by combining health data from many populations around hazardous waste sites), proximity to sites is associated with a small to moderate increased risk of some kinds of birth defects and, less well documented, some specific kinds of cancers. Health investigations of communities around some individual hazardous waste sites have found increases in the risk of birth defects, neurotoxic disorders, leukemia, cardiovascular abnormalities, respiratory and sensory irritation, and dermatitis. However, no adverse health findings have been reported in many studies. These apparent discrepancies may be due to factors such as limitations inherent to epidemiologic methods, differences in environmental conditions in the communities investigated, and the small numbers of persons evaluated at some sites."

As efforts continue to identify and remediate abandoned HWS, it is prudent to consider each site as a potential source for the release of substances into environmental media that can adversely affect human health. Given this concern, what health problems should health investigators be most concerned about? In response to this question, ATSDR developed a list of seven priority health conditions to (1) assist in evaluating potential health risks to persons living near these sites and (2) determine program and applied human health research activities involving hazardous substances identified at the sites (MMWR, 1992). To arrive at the priority health conditions shown in Table 3, ATSDR evaluated the toxicological and human health effects data for 25 hazardous substances identified as being often released from NPL sites. In addition, ATSDR drew heavily on the findings from approximately 1200 public health assessments that it conducted for NPL and other sites. Each Priority Health Condition is being pursued by federal, state health, and academic researchers in a program of integrated laboratory and epidemiologic endeavors.

COMPARATIVE RISK ASSESSMENT OF HAZARDOUS WASTE SITES

How do HWS rank in comparison to other environmental hazards in terms of risk? For example, do HWS present a greater health risk than indoor air pollutants? These kinds of questions are important in terms of setting public policies on environmental programs and for establishing priority-based budgets. Government agencies have used risk assessment methods to provide a first-cut response to the first question posed here. These methods have been a mix of quantitative risk assessments (i.e., those that use environmental and health data to calculate a numerical estimate of risk) and qualitative risk assessments that draw upon professional judgment and consensus debate. Two illustrations of this comparative risk assessment approach follow.

In 1987 EPA released what has come to be known as its "Unfinished Business" report (EPA, 1987). A group of about 75 career EPA managers ranked 31 environmental problems according to each problem's risk. Four different kinds of risk were considered for each environmental problem: cancer, noncancer health effects, ecological effects, and welfare effects (e.g., materials damage, aesthetic damage). Quantitative cancer risks were developed where possible; other risks were estimated through consensus. EPA concluded that areas of relatively high risk included: indoor radon, indoor air pollution, stratospheric ozone depletion, global warming, nonpoint sources; discharges to estuaries, coastal waters, and oceans; other pesticide risks, accidental releases of toxics, consumer products, and worker exposures. Areas of medium or low risks included: RCRA sites, Superfund sites, underground storage tanks, and municipal nonhazardous waste sites.

Overall, EPA ranked inactive HWS (i.e., Superfund sites), as 17th on a list of their list of 31 environmental problems. They concluded that Superfund sites ranked 8th out of 29 environmental problems in terms of cancer risk, but presented relatively low risk on noncancer health endpoints, and relatively medium risk on ecological effects and welfare effects. The EPA analysis was based on a sample of environmental contamination data from 35 Superfund sites. Six contaminants (TCE, vinyl chloride, arsenic, tetrachloroethylene, benzene, and 1,2-dichloroethane) were selected for risk assessment. These six chemicals were estimated to cause approximately 1000 cases of cancer annually as result of releases from Superfund sites.

The EPA analysis, while reasonable for the data and conditions of circa 1987, is limited because of inadequate human health data, incomplete exposure pathway analyses, and limited number of sites and hazardous substances selected for risk assessment. For example, of the six substances selected by EPA for risk assessment, only one is a metal (arsenic). Further, much of the noncancer health effects data pertaining to HWS have been reported after the EPA comparative risk assessment study was completed.

In addition to EPA, several states also have performed comparative risk assessments of environmental hazards. As one example, Louisiana used risk assessment methods to rank their state's hazards (DEQ, 1991). Risk assessments prepared by a state government task force were reviewed and consensus formed by a panel of industry, environmental, government, and academic representatives. This effort led to ranking Inactive & Abandoned Waste Sites as 10th on a list of 33 environmental problems. It is interesting to observe that Louisiana's ranking of Superfund and similar sites was higher than EPA's, but both rankings place abandoned HWS in the second tier of hazard.

In distinction to the EPA comparative risk assessment just described, two sets of investigators focused only on the health risks presented by hazardous wastes on a national level. In one examination, researchers at Duke University conducted a comprehensive evaluation of the site-specific risk assessments conducted by EPA for each NPL site (Viscusi and Hamilton, 1994). An EPA baseline assessment of the risks at each site is the first step in a process that in time leads to options for managing the risks. The investigators' database consisted of risk assessments for 78 NPL sites. They focused on estimated cancer risk expressed in each risk assessment. The investigators found a number of quite interesting results. For instance, the magnitude of risk presented by Superfund sites exceeded estimated risks for other activities falling under federal cancer regulations. Also, the dominant risks arose from consideration in risk assessments given to future use scenarios that involve alternative uses of land. Viscusi and Hamilton note "Consideration of the risk assessments for Superfund sites indicates, however, that it is not the existing risks that are most significant. ... Indeed, these future risks account for 90 percent of all the risk-weighted pathways for the Superfund sites in our sample."

In contrast to Viscusi and Hamilton's estimate, ATSDR officials have reported a higher level of *current* health hazard associated with HWS (Lichtveld and Johnson, 1994; Johnson, 1995). Under its CERCLA public health authorities, ATSDR conducts a public health assessment of each site placed, or proposed for placement, on the NPL. A public health assessment consists of a qualitative evaluation of environmental contamination data, community health concerns, and state health department data (when available) that can be used to estimate the level of health hazard posed by each site. Each site is put into one of five possible health hazard categories (with percent of 136 sites assessed in 1993 through 1994): Urgent Public Health Hazard (5%), Public Health Hazard (49), Indeterminate Public Health Hazard (34), No Apparent Public Health Hazard (10), or No Public Health Hazard (2). By this ranking scheme, about 54% of sites presented a current public health concern.

HEALTH OF WORKERS

HWS and emergency releases of toxic substances have the potential to cause adverse health effects in community populations. The scientific database supporting this health concern has been summarized in the preceding sections. However, in addition to community groups, other persons also are at possible health risk due to exposure to hazardous substances. In particular, there are safety and health implications for workers whose occupations place them in contact with waste sites or facilities.

Remediation Workers

In addition to concern for the health of persons residing in communities potentially affected by the release of substances from HWS, care must be exercised that the workers who remediate (i.e., clean up) the sites are not harmed by their work. The removal of soil or water contaminated with toxic substances clearly presents the potential for adversely affecting the health of cleanup workers if the necessary precautions are not taken. Work conditions can involve toxic chemicals, biologic hazards, radiation, heavy workloads, and heat or cold stress. Primary protective measures include personal protective equipment, engineering controls, site monitoring equipment, and training and education of workers. Secondary protection measures include periodic medical evaluations of remediation workers and a program of workers' health surveillance. An adequate program of worker health protection requires both primary and secondary measures.

Two major U.S. federal statutes address workplace safety and health problems potentially faced by remediation workers (see Chapter 10). The Occupational Safety and Health Act of 1970 was intended to ensure safe and healthful conditions in the American workplace. The act requires employers to take steps to protect employees from recognized workplace hazards that could cause illness or injury. Numerous safety standards developed by the Occupational Safety and Health Administration (OSHA) to protect the safety and health of workers involved in construction work and general labor are also applicable to activities on HWS (Andrews, 1990). Specific regulations for HWS remediation have been developed and promulgated by OSHA as a mandate in CERCLA, as amended (Moran, 1994). These regulations, which are contained in 29 CFR 1910.120, require the employer of remediation workers to develop and implement a comprehensive safety and health program for workers involved in hazardous waste operations, the details of which can be found in other sources. A significant requirement of the regulations is periodic medical surveillance of HWS workers (Melius, 1990).

As an industry, remediation of HWS is relatively new in the United States, but one that involves large numbers of workers. Given the recency of site-specific work, there is only a very limited scientific database on any potential health effects associated with such work. According to one source, clinical and research findings from clinical centers performing health surveillance examinations on hazardous waste workers have not shown health abnormalities related to the workers' activities (Favata et al., 1990). From a public health perspective, this is a very desirable outcome. However, the investigators caution that the lack of any health problems in hazardous waste workers may be due to the fact that most of the workers examined had performed feasibility (i.e., inspection and assessment) work, not actual remediation of sites. In a study of Italian workers who performed cleanup operations of the Seveso site that had been highly contaminated with dioxins following an industrial release, no meaningful differences in biochemical outcomes were found between remediation workers and a comparison group (Assennato et al., 1989). These gratifying results may be attributable to the safety measures taken by workers during the cleanup operations, but could also be a consequence that sufficient time has not elapsed to evidence any potential health problems.

Waste Disposal Workers

In addition to waste site remediation work, the permitted disposal of hazardous waste by incineration, on-site destruction, or other permitted method of disposal can bring workers into contact with hazardous substances. Without adequate safeguards and work practices, workers' exposure to the wastes could lead to adverse health effects. Although there is not an extensive occupational health literature on the subject, the following studies illustrate some of the hazards to the health of waste disposal workers.

In Sweden, mortality was investigated among 176 male workers employed for at least 1 year between 1920 and 1985 at a municipal waste incinerator (Gustavsson, 1989). There was an excess of deaths from ischemic heart disease that was associated with duration of work. Exposure to combustion products and polycyclic aromatic compounds was common. In a study of 14 incineration workers at a solvents disposal facility in the United States, investigators found severe neurologic damage (myoclonus and severe tremor) in two workers, and all workers had abnormal psychiatric symptoms, but with different diagnoses (Kawamoto, 1992). In Poland, dermatologic examinations were conducted on 393 workers whose work required contact with waste volatile ashes in electric power stations. Eczema was found in 18% of the workers and hypersensitivity to chromium was noted in 21% (Kiec-Swierczynska, 1989). In Japan, an investigation was conducted of workers' exposure to organic solvents at 35 wastewater disposal facilities. The method of disposal was incineration of liquid waste. Workers' exposure to all solvents combined exceeded an acceptable threshold limit value. Exposures were greatest in the incineration areas (Ikatsu et al., 1989). Mexican workers at a landfill were evaluated for uptake of hazardous substances and adverse health effects. Twenty-two workers whose duties required them to handle hazardous wastes were compared to manual laborers from an adjacent community. No protective measures or equipment were used by the landfill workers. The landfill workers were found to have elevated levels of arsenic in urine and hair; other metals were not found elevated in biological samples. No patterns of adverse health effects were found, which is not surprising given the newness of the landfill operations (Diaz-Barriga et al., 1993). In Nigeria, macrocytic anemia and leucopenia were noted in workers who had contact with radioactive waste when compared with workers not similarly exposed (Ogunranti, 1989).

It is apparent that work that brings workers into contact with hazardous wastes has the potential to cause adverse health effects if adequate safeguards are not taken. Work practices should be used that prevent direct contact with hazardous materials and protective equipment must be provided to workers and used. Programs of medical monitoring and health surveillance should also be implemented if chronic health effects are to be prevented.

EFFECTS OF HAZARDOUS WASTE ON ECOSYSTEMS

Hazardous waste in the environment has led, as previously described, to concern about its impact on public health. Indeed, concern for the effects on public health of toxic substances in the environment has been a principal factor in the passage of environmental legislation (Miller, 1992). Less often in legislation has been any stated concern for the health of ecological systems, that is, concern for all the organisms with which humans share the planet. Yet, the same hazardous substances that can affect human health also have the potential, in most instances, to affect the health of plants, fish, fowl, and mammalian populations. What has often been missing from environmental research and legislation has been an ecological perspective on the effects of environmental pollution. This is beginning to change with support for investigations of the environmental fate of toxicants and the effect on ecological systems becoming more evident (see Chapter 2).

It should seem obvious that concern for the environment must encompass all significant effects of global pollution. An illustration of the importance of thinking in terms of ecosystems (of which human health is a component) is that of the Great Lakes basin (Chapter 2). The Great Lakes basin comprises one fifth of the total freshwater on the Earth's surface. It is a most valuable natural resource for both the United States and Canada. Approximately 20% of the U.S. population and 30% of the Canadian population live in the basin. Over two centuries of commercial, recreational, and municipal use of the Great Lakes have led to a deteriorated water quality, including deterioration caused by leachate from HWS and industrial discharges of toxic substances. Although progress has occurred in reducing levels of contaminants in the Great Lakes, a large effort is currently under way in the United States and Canada to further reduce the levels of pollutants in the Great Lakes and contributing waterways. Approximately 1000 chemicals have been identified in the Great Lakes, with 11 of the more persistent toxicants being labeled as "critical Great Lakes pollutants" (Hicks and Katz, 1992). All 11 substances (which include, e.g., polychlorinated biphenyls, mirex, alkylated lead, methylmercury) tend to bioaccumulate in organisms and biomagnify through food chains. Sport fish at the top of the predatory structure in the Great Lakes ecosystem contain the greatest concentration of contaminants. Birds and mammals that feed on the fish, in turn, are exposed to toxic substances that had their origin in the pollution left in the Great Lakes. Humans who consume sport fish from the Great Lakes are similarly at risk of exposure to the contaminants. What are the effects of these exposures?

As described in Chapter 2, a series of studies of fish, reptiles, and small mammals in the Great Lakes basin have revealed a disturbing pattern of adverse effects in the populations studied (Hicks and Katz, 1992; Fulton et al., 1992). Wildlife investigators have associated the presence of Great Lakes pollutants in fish and wildlife with interference of reproduction processes in birds and lake trout, increased congenital abnormalities in birds, and the development of neoplastic tumors in bottom-dwelling fish (e.g., Gilbertson et al., 1976; Maccubbin et al., 1985; Price and Weseloh, 1986). In particular, many of the toxicants found in the Great Lakes appear to cause disruptions of the endocrine system. These disruptions can be profound because the endocrine system plays a crucial role in controlling the extent and pace of an individual's development, including reproductive success and development of offspring. For instance, a study of nesting pairs of bald eagles reintroduced to the north and south shores of Lakes Erie thrived, but the viability of some hatchlings was compromised because of wasting, a condition ascribed to persistent toxic substances (Colborn, 1991).

In addition to studies that report effects of persistent toxicants on ecological systems, several human health investigations seem to associate consumption of Great Lakes sport fish with adverse health effects (Hicks and Katz, 1992). Most of the studies have focused on PCBs in sport fish. In one study, 572 persons who resided along the Lake Michigan shore and ate more than 24 pounds of sport fish per year were compared to 419 persons from the same communities who reported less sport fish consumption. Persons in the upper range of fish consumption were found to have serum PCB levels about four times greater than the comparison group (Humphrey, 1988).

Three investigations have been conducted on the association between consumption of sport fish from the Great Lakes and adverse health effects in human populations. In one study, neonates of mothers who consumed more than 11.8 kg of Great Lakes sport fish over a 6-year period were studied (Fein et al., 1984). The infants showed significant decreased infant birth weights (160–190 g lighter), shorter gestational age (average of 4.9 days), and reduced head circumference (average 0.6 cm). A follow-up study of this cohort of children at 5 months of age indicated that the observed difference in body weights persisted (Jacobson and Jacobson, 1988). Further studies of the cohort at 4 years of age confirmed a continuous deficit in body weight gain when compared to controls. Also, the 4-year-old children had poorer short-term memory (Jacobson et al., 1990).

These studies show that investigators have learned much about the extent and content of contaminants in the Great Lakes, the effects of contaminants on fish and wildlife, and human populations presumed to be at greatest risk of adverse health effects from sport fish consumption. In an ecosystem context, this body of scientific knowledge is unique. In an ecotoxicologic context, the attention of federal, state, and provincial governments to the full range of ecological effects of contamination in the Great Lakes ought to serve as an example of how to handle similar environmental problems in other parts of the globe.

SOCIETAL RESPONSE: LAWS AND INTERNATIONAL CONVENTIONS

Commencing in the 1940s, national legislation has been enacted in the United States bearing on environmental protection and reduction of health risks. Most states also have passed legislation that substantially shapes actions on hazardous substances in the environment. Additionally, recent international conventions that deal with global environmental concerns, including hazardous waste, have been negotiated. Chapter 10 describes in more detail the pertinent legislation and regulations in the United States and covers a key international convention concerning hazardous wastes. The following sections will provide a general introduction to laws and attendant regulations on hazardous wastes.

U.S. FEDERAL LEGISLATION

National governments in the industrialized countries have come to realize the importance of controlling the generation, transportation, storage, and destruction of hazardous wastes. As noted previously in this chapter, the public's concern about the effects of hazardous wastes on public health have been a central expression in national legislation and international treaties.

There are 12 statutes in the United States that form the major portion of the legal basis for federal programs administered by the EPA and other federal agencies (CRS, 1991). They are listed in Table 4 and profiled in Chapter 10. Each federal statute has amendments and attendant regulations developed by EPA or other regulatory agencies. If one takes a broad view of what constitutes "hazardous waste," all 12 statutes, it can be argued, are relevant in some context to hazardous

Table 4 Major Environmental Laws

Federal Insecticide, Fungicide, and Rodenticide Act
Clean Water Act
Clean Air Act
Resource Conservation and Recovery Act
National Environmental Policy Act
Occupational Safety and Health Act
Ocean Dumping Act
Safe Drinking Water Act
Toxic Substances Control Act
Environmental Research and Development Demonstration Act
Comprehensive Environmental Response, Compensation, and Liability Act
Oil Pollution Act

From CRS (1991) and Miller (1992).

wastes; four federal statutes are especially germane for the purposes of hazardous waste management. For the purposes of this introductory chapter, the federal CERCLA statute is particularly relevant.

Superfund Statute

CERCLA, or Superfund, was enacted in 1980 and reauthorized by the U.S. Congress in 1986 (Stoll, 1991). Superfund's basic purpose is to provide funding and enforcement authority for cleaning up waste sites created in the past, and for responding to hazardous substance spills. The statute includes provisions for remediating waste sites, responding to public health concerns, enforcement authority to identify "potentially responsible parties," emergency removal of chemical spills, and a community right-to-know provision.

CERCLA has at its center the social justice philosophy that "polluters pay" for the consequences of their pollution. This philosophy led to CERCLA's establishment of the Hazardous Substance Trust Fund (also sometimes called the Superfund). The fund is financed by petroleum excise taxes, a chemical feedstock tax, a corporate environmental tax, general revenues, and other sources (Probst, 1992). In addition, CERCLA gives broad legal authority to EPA to identify Potentially Responsible Parties (PRPs) and require them to pay for the costs of a site's remediation and other actions related to the effects of hazardous substances released from the site. PRPs include the current and past owners and/or operators of a site, those who arranged for the transportation of hazardous substances to the site, and those who arranged for the treatment or disposal of the substances, and they are subject to retroactive liability (Probst, 1992). This is established under the legal concept of "retroactive joint and several liability" for parties whose wastes have contributed to environmental degradation at the waste sites. The concept means that a company or other accountable entity which long ago had disposed of wastes could now be held responsible for all of a site's remediation costs unless other responsible parties could be identified and costs shared. Needless to say, retroactive joint and several liability has often led to litigation as to who pays what amount of a site's remedial costs.

CERCLA requires reauthorization by Congress every 5 years (unless, as in 1990, Congress chooses to extend the Act without debate). In 1986 extensive revisions were made during reauthorization. The current statute, as amended, and its administration by federal agencies has been criticized for several reasons. Some of these criticisms involve: (1) the small number of sites "cleaned up" and removed from the NPL, (2) the high cost of the program and whether the cost is commensurate with the health and environmental risks, (3) the high cost of "transactional actions" (i.e., legal and administrative costs) that attend the identification of PRPs and recovering cleanup costs from them, and (4) whether the Superfund trust fund is fairly derived.

Congressional hearings held in 1992 addressed the issues of fairness and transaction costs of CERCLA. For example, one source provided testimony that, based on data from 17 Superfund sites, at least 2787 PRPs had been identified for possible recovery by EPA of costs of cleanup. The PRPs, in turn, filed a total of 3236 claims against insurers alleging insurance coverage for their Superfund legal and cleanup costs. All this activity has led to an average of more than five lawsuits per site (Edelstein, 1992). The fraction of Superfund monies spend on actual site cleanups is estimated to be about 30%, with another 19% spent on cleanup studies. About 50% of Superfund costs have gone to transaction and support costs (Christenson, 1992). These cost issues and how the Hazardous Substance Trust Fund is derived, as well as other administrative concerns (Keystone Center, 1994), were debated by Congress in 1994, but failed to reauthorize CERCLA, thereby leaving the 1986 authorities of CERCLA still in force until Congress acts at a future time.

Resource Conservation and Recovery Act (RCRA)

This statute is the federal prescription that hazardous waste be treated, stored, and disposed of so as to minimize the current and future threat to human health and the environment (Case, 1991). RCRA is a regulatory statute designed to provide cradle-to-grave control of hazardous waste by imposing management requirements on generators and transporters of hazardous waste and upon owners and operators of treatment, storage, and disposal facilities. This statute applies principally to active facilities, whereas Superfund applies mainly to inactive sites. RCRA, as amended, represents a significant challenge to the regulated community. In particular, industry is challenged to find new ways to minimize, treat, and dispose of hazardous waste. The use of innovative technologies, like bioremediation, to reduce waste is the subject of active research and development.

Toxic Substances Control Act (TSCA)

This statute gives EPA broad authority to (1) require testing of existing chemicals, those currently in widespread commercial production or use, (2) prevent future chemical risks through premarket screening and regulatory tracking of new chemicals, (3) control unreasonable risk of chemicals, and (4) gather and disseminate information about chemical production, use, and possible adverse effects to human health and the environment (CRS, 1991). TSCA has been amended several times to include legislative directives about how to control specific toxic substances, e.g., asbestos and radon. A large number of substances are considered to be or are classified by law or regulation as being hazardous, and TSCA provides EPA with statutory authority to require industry to develop toxicological and other information about a substance's hazard to health and the environment.

Ocean Dumping Act of 1972

The Ocean Dumping Act of 1972 regulates intentional ocean disposal of materials and authorizes related research. This law prohibits all ocean dumping, except that allowed by permit, in any ocean waters under U.S. jurisdiction, by any U.S. vessel, or by any vessel sailing from U.S. ports. Any dumping of radiological, chemical, and biological warfare agents and any high-level radioactive waste is banned. Permits for dumping any other materials can be issued by EPA where it is determined that the dumping will "not unreasonably degrade or endanger human health, welfare, or amenities, or the marine environment, ecological systems, or economic potentialities." (CRS, 1991). Subsequent amendments to the Ocean Dumping Act prohibited the dumping of municipal sewage sludge or industrial waste.

ILLUSTRATIVE STATE LEGISLATION

State and local governments have also enacted various laws and regulations bearing on toxic substances in the environment. These statutes are intended to meet specific state and local needs or fill gaps in federal statutes. It is not possible to catalog all the state laws pertinent to this chapter. However, two state laws illustrate an important trend toward making information about toxic substances more readily available to the public and the central role in each statute of concern for protecting public health.

Massachusetts Hazardous Waste Management Act

The Massachusetts Hazardous Waste Management Act established a comprehensive framework for protecting public health and the environment from the generation, transportation, and disposal of hazardous wastes. The law provides for the state to develop and enforce standards for the identification and management of hazardous wastes. Of particular note is the requirement that the state conduct public participation programs on matters of policy and programmatic significance. Generators of hazardous wastes are required to maintain records and report to the state the quantity and type of hazardous wastes generated, stored, treated, or disposed of. In turn, the state is required to furnish information annually to local boards of health identifying the hazardous wastes within their communities (DEP, 1979).

California's Proposition 65

A second state statute that has garnered a great amount of attention by environmentalists, the lay public, private industry, and government officials is California's Safe Drinking Water and Toxic Enforcement Act, which is more commonly known as Proposition 65 (which was the number of the state referendum that led to the Act) (Jennings, 1986). Proposition 65, according to news releases of the day, was passed as a referendum on the belief that industry needed to assume a greater role in warning the public about hazardous chemicals and to reduce discharges of toxics into the environment. The Act shifts a burden of proof to industry for knowing the health effects of carcinogens and reproductive toxics. It creates a policy that for known carcinogens and reproductive toxicants, any detectable release into the environment shall be presumed as hazardous, unless evidence can demonstrate otherwise.

There are several key provisions of Proposition 65 that pertain to hazardous waste. They include: (1) requires the Governor of California to publish a list of carcinogens and reproductive toxic chemicals, (2) prohibits discharges of listed chemicals that might enter sources of drinking water, (3) requires warnings about public exposure to listed chemicals, (4) increases criminal penalties for violations of the state's Hazardous Waste Law, and (5) allows citizens to initiate suits and collect a portion of fines for violations of Proposition 65. These provisions, and others in the Act, have been controversial since the Act's passage into law. Without passing judgment on the merits of the Act, it seems accurate to observe that placing a greater burden on industry for toxicological testing, information garnering, and dissemination to the public is a theme common to several federal and state statutes of recent years. Moreover, the notion of citizen "right-to-know" about toxic releases into the environment is being promoted by international environmental and health bodies, as will be discussed in the next section.

UNITED NATIONS ACTIONS

Basel Convention

Another action that affects waste management issues on an international scale is the Basel Convention (ILM, 1989). Negotiated under the auspices of the United Nations Environment Program (UNEP), the Basel Convention was completed in March 1989. Over 50 countries have currently signed the Convention, including the United States, Mexico, and Canada. The Convention entered into force on May 5, 1992, 90 days after 20 signatory countries ratified it. Australia was the 20th signatory. The U.S. Senate ratified the Basel agreement on August 11, 1992 (Anon., 1992c). Ratification signals the party's readiness to fully implement the Convention, including having the necessary legislative authorities to enforce its terms. The Convention covers transboundary movements of hazardous waste and household waste ash, and prohibits movement between nonparties, unless these countries have a separate agreement that ensures sound waste management. The Basel Convention specifies that government-to-government notice must be given and consent obtained before waste exports proceed. It also sets an "environmentally sound management" standard as the basis for all movements. If the exporting government has reason to believe waste will not be managed in the importing country in an environmentally sound manner, then the exporting country must not allow a shipment to proceed, even if the importing country has agreed to accept it.

United Nations Declarations

The UNEP drafted a declaration of 27 principles for encouraging environmentally responsible development (Anon., 1992b). Two of the principles seem especially appropriate for the subject of this chapter. Principle 10 states, "Environmental issues are best handled with the participation of all concerned citizens, at the relevant level. At the national level, each individual shall have appropriate access to information concerning the environment that is held by public authorities, including information on hazardous materials and activities in their communities, and the opportunity to participate in decision-making processes. States shall facilitate and encourage public awareness and participation by making information widely available. Effective access to judicial and administrative proceedings, including redress and remedy, shall be provided." This principle reinforces the importance of the Toxic Releases Inventory section of CERCLA, as described in Chapter 10. There seems to be a global trend to require the reporting, collation, and dissemination of information to the public of information about hazardous substances released into the environment.

Principle 14 says that, "States should effectively cooperate to discourage or prevent the relocation and transfer to other states of any activities and substances that cause severe environmental degradation or are found to be harmful to human health" (Anon., 1992b). However, international cooperation will be needed to effectively implement this principle in order to prevent waste generation and environmental degradation in developing countries as the result of industries being relocated from industrialized countries (LaDou, 1991).

Rio Conference

The member states of the United Nations met in Rio de Janeiro, Brazil, in 1992 to conduct the U.N. Conference on Environment and Development. On June 14, 1992, the conference approved Agenda 21, which is a comprehensive statement of concerns, findings, and recommended actions on specific environmental areas (UNEP, 1992). Agenda 21 is a 580-page document that is organized into 40 chapters, commencing with a Preamble, which states "Humanity stands at a defining moment in history. We are confronted with a perpetuation of disparities between and within nations, a worsening of poverty, hunger, ill health and illiteracy, and the continuing *deterioration of the ecosystems* (emphasis added) on which we depend for our well-being." Agenda 21 is a very significant agreement among the nations of the world. It reflects a global consensus and political commitment at the highest level on development and environmental cooperation.

Several chapters of Agenda 21 relate to the purposes of this book; however, two are directly relevant. They are Chapter 20, "Environmentally Sound Management of Hazardous Wastes Including Prevention of Illegal International Traffic in Hazardous Wastes" and Chapter 16, "Environmentally Sound Management of Biotechnology." Chapter 20 opens by stating: "Effective control of the generation, storage, treatment, recycling and reuse, transport, recovery and disposal of hazardous wastes is of paramount importance for proper health, environmental protection and natural resource management, and sustainable development. This will require the active cooperation and participation of the international community." Further, "Prevention of the generation of hazardous wastes and the rehabilitation of contaminated sites are the key elements, and both require knowledge, experienced people, facilities, financial resources and technical and scientific capacities." In a similar voice of concern, Chapter 16 states: "Despite increasing effort to prevent waste accumulation and to promote recycling, the amount of environmental damage caused by excessive consumption, the quantities of waste generated and the degree of unsustainable land use appear likely to continue growing" (UNEP, Chap. 16, p. 12). These statements from Agenda 21 clearly commit the U.N. members to a cradle-to-grave approach to reducing the volume of hazardous wastes and the need to take an ecologic perspective, and sustainable development approach, on the impact of hazardous wastes.

Chapter 20 establishes four overall targets: (1) Preventing or minimizing the generation of hazardous wastes as part of an overall integrated production approach; eliminating or reducing to a minimum transboundary movements of hazardous wastes, consistent with the environmentally sound and efficient management of those wastes; and ensuring that environmentally sound hazardous waste management options are pursued to the maximum extent possible within the country of origin; (2) Ratification of the Basel Convention on the Control of Transboundary Movements of Hazardous Wastes and Their Disposal and the expeditious elaboration of related protocols; (3) Ratification and full implementation of regional agreements bearing on transboundary shipments of hazardous wastes; and (4) Elimination of the export of hazardous wastes to countries that prohibit the import of such wastes (UNEP, Chap. 20, p. 3). Chapter 20 details a series of recommended policies and actions that support each overall target.

Chapter 16 of Agenda 21 pertains to biotechnology. Agenda 21 defines biotechnology as "a set of enabling techniques for bringing about specific man-made changes in deoxyribonucleic acid (DNA), or genetic material, in plants, animals and microbial systems, leading to useful products and technologies" (UNEP, Chap. 16, p. 2). Specific mention is made of the use of biotechnology in program areas to: increase the availability of food, feed, and renewable raw materials; improve human health; and enhance protection of the environment. Within the area of enforcing protection of the environment, is the objective "to prevent, halt and reverse environmental degradation through the appropriate use of biotechnology in conjunction with other technologies, while supporting safety procedures as an integral component of the program" (UNEP, Chap. 16, p. 12). This objective clearly indicates international support for bioremediation as a means to reduce hazardous wastes and as a method for waste site remediation.

The implementation of Agenda 21, and other agreements from the Rio conference, by the nations of the world will determine the future fate of ecological systems, environmental quality, and human health of peoples at risk from environmental hazards. However ambitious and promising agreements such as Agenda 21 seem, little will be achieved if national governments do not commit to actions to support individual recommendations in the agreements and to efforts to collaborate across national borders.

EUROPEAN COMMUNITY REGULATIONS AND DIRECTIVES

The European Community (EC) was created in 1957 by the Treaty of Rome. The European Union (EU) was created by the Treaty of Maastricht and came into existence as an umbrella organization in November 1993 (see Chapter 10). The EU now consists of membership from 15 nations: Austria, Belgium, Denmark, Finland, France, Germany, Greece, Italy, Luxembourg, the Netherlands, Portugal, Republic of Ireland, Spain, Sweden, and the United Kingdom. The European Economic Community became the European Community. The EC is therefore a structural member of the EU.

The two most common forms of EC legislation are regulations and directives. Regulations become law throughout the EC as of their effective date, generally enforceable in each member state. Directives, in general, are not directly and generally enforceable in member states (Smith and Hunter, 1992). Directives set out goals for the member states to achieve through national legislation. The EC has adopted a substantial body of environmental legislation, covering a range of environmental concerns and issues. Enacted legislation includes conducting environmental impact statements of projects, air and water pollution, solid and hazardous wastes, noise pollution, radioactive waste, conservation of wild fauna and flora, urban waste treatment, freedom of information on the environment, expanded waste regulations, and the establishment of the European Environment Agency. Pending EC legislation of relevance to waste problems include (1) a packaging waste proposal and a plastic waste proposal, (2) regulation of the shipment of waste that would require advance government consent prior to shipment, (3) restructuring of air and water pollution control frameworks to achieve integrated pollution control, and (4) civil liability proposals for environmental pollution analogous to those in CERCLA (Smith and Hunter, 1992).

REMEDIATION OF HAZARDOUS WASTES*

In a public health sense, the prescription to prevent adverse health effects attributable to HWS is to "remediate" the site before releases of contaminants occur. Remediation means to correct or fix a deficiency—in this case, correcting the environmental contamination legacy left in the environment by abandoned wastes. To the public, remediation means "cleaning up" a contaminated site. To the public health professional, remediation means dissociating the public from exposure to hazardous substances that can portend adverse health effects.

Previously employed options for disposal of hazardous waste include: underground injection, landfills, natural storage areas (caves, shale beds), secure landfills, and ocean dumping (Dominguez, 1986). All these options have led to problems of environmental pollution, which in turn has led to national and international laws and regulations regarding waste disposal. Waste sites, in time, disintegrate and materials contained in them leach into groundwater, contaminate surface water and soil, and some can volatilize as air toxicants.

SUPERFUND CLEANUP PROCESS

Cleanup of hazardous wastes left in the environment is the central intent of CERCLA, as amended. Wastes that need removal from the environment can be grouped into two broad categories: those requiring emergency cleanup and those

* The description given here of site remediation pertains to federal actions and authorities under CERCLA, as amended. There are similar programs of remediation of abandoned HWS being conducted by individual states. Their actions may differ from those described for federal agencies.

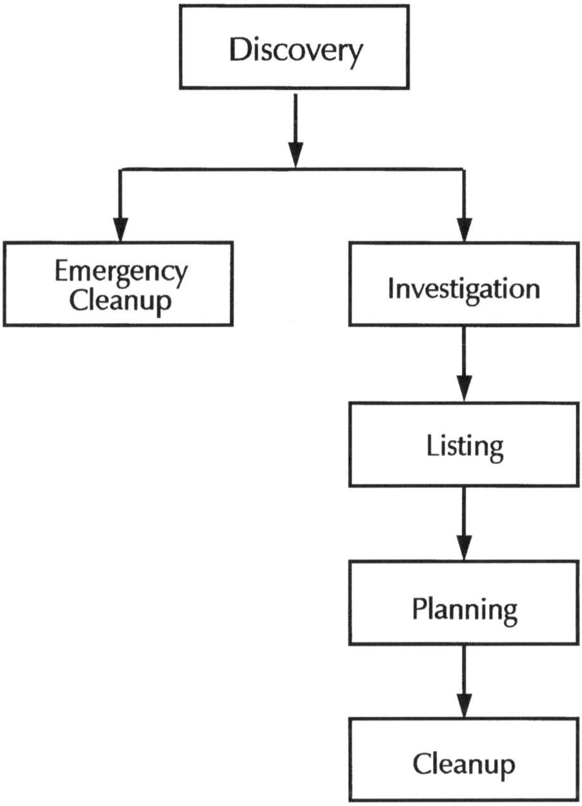

Figure 3 EPA's Superfund process from OSWER (1992b).

found at abandoned waste sites for which longer-term actions (called remediations) are needed (OSWER, 1992b, p. 6). (However, some abandoned sites do sometimes require emergency removal actions.) EPA is the lead federal agency responsible for administering CERCLA, as amended. The process that ultimately leads to cleanup of sites necessarily begins with their discovery. EPA's Superfund process is shown conceptually in Figure 3. It is important to explain some details of the process.

Site Discovery

The identification of potential HWS occurs through several mechanisms. CERCLA of 1980 required owners or operators of hazardous waste facilities to notify EPA of their existence. EPA was, in turn, mandated to notify each state of the sites reported. The result of this reporting is the Comprehensive Environmental Response, Compensation, and Liability Information System (CERCLIS), which is an EPA list of identified HWS and facilities. As of March 31, 1992, there were 35,984 sites on the CERCLIS (OSWER, 1992c, p. 4). Sites are added to the list in a variety of ways. The most frequent method is through site discovery efforts undertaken by state officials, who in turn notify EPA. In addition, federal agencies, local officials, business, and citizens can report to EPA potential waste sites of concern.

Emergency Removals and Site Investigation

Occasionally during the transport of hazardous materials or operations of industrial processes that use toxic substances, spills or releases occur that are a threat to public health and the environment. In other situations, the discovery of abandoned hazardous materials like cylinders of gases may present an imminent threat to public health and safety. When such events occur, e.g., a train derailment that releases chemicals into the environment, Superfund teams from EPA manage the response, in collaboration with other federal, state, and local agencies. (If the spill occurs on a waterway, the Coast Guard is the lead agency.) Most emergencies are cleaned up quickly, leading to removal of the hazardous substances through remedial actions like soil removal. The Superfund removal program therefore responds to short-term emergencies that could threaten public health. By law, such removals can take up to a year to complete and can cost no more than $2 million, though exemptions to these criteria can be granted by EPA. As of March 31, 1992, 1916 emergency removals had been completed (OSWER, 1992c, p. 7).

Sites that do not require emergency cleanups usually are investigated first by local or state officials, leading to a referral to EPA if the site's conditions seems to warrant listing on the Superfund list. According to EPA, only the worst HWS require federal cleanup; the rest are the responsibility of state or local governments, private industry, or citizens (OSWER, 1992b, p. 6).

Site Listing

HWS that are identified as potentially in need of remediation must be ranked for their hazard. The ranking occurs through application of EPA's Hazard Ranking System (HRS), which is a method to numerically score each site according to threat to public health and the environment. Each site receives a numerical score based on the likelihood that people will be exposed to hazardous substances on or off the site. Sites that score high enough (currently a score of 28.50) are placed on the NPL, which is significant because only NPL sites are eligible for long-term cleanup under federal funding through CERCLA. As of December 1992, there are 1275 sites on the NPL (ASTSWMO, 1994), with EPA adding 50 to 100 additional sites annually. EPA also can place sites on the NPL that score less than 28.50 on the HRS if (1) ATSDR has issued a public health advisory on a site, (2) EPA determines that the site poses a significant threat to public health, and (3) EPA anticipates that a remedial cleanup of the site would be more cost effective than a removal action (ASTSWMO, 1994, p. 9).

Planning

Preparing for a long-term cleanup of a NPL site is a complicated and lengthy effort by EPA. A detailed study of the site, called a Remedial Investigation/ Feasibility Study, is conducted by EPA. Wherever possible, EPA negotiates with

PRPs to conduct these studies, while the ultimate responsibility for selecting specific remedial options is the responsibility of EPA.

Site Cleanup

The actual cleanup of a site is the culmination of the Superfund remediation process. Remedial actions at a particular site may take years, depending on such factors as size of the site and the specific cleanup methods chosen. Total cleanup costs currently average $31.6 million per site, according to EPA figures (Christenson, 1992); of this amount, remedial action costs per site average $25 million (OSWER, 1992a, p. 6). NPL sites that are remediated are removed from the NPL; 155 have been removed as of December 31, 1992 (ASTSWMO, 1994). As previously noted, states and territories also have a major program of site remediation.

The current average NPL site takes 7 to 10 years to work through the Superfund process, from discovery to start of cleanup. Actual cleanup can take years—decades if groundwater must be treated (ASTSWMO, 1994, p. 6).

Remediation of a HWS is a costly and complex undertaking, dependent on such factors as the size of the site, the extent of the contamination, the technology selected for the remediation, and the extent of legal and other transactional costs. There are various forms of remediation currently being used by government and private industry. The most common technologies used currently are listed in Table 5 (Russell et al., 1991). It is noteworthy that several kinds of remedial actions and technologies are available; the actual choice of remedial actions taken at a particular site will depend on local conditions to a considerable extent.

The selection of remediation technologies varies according to several factors, including the nature and extent of a site's contamination. Listed in Table 6 are percentages of various technologies used by EPA for hazardous waste remediations. The high use of institutional controls is noteworthy, normally denoting those actions taken as interim measures prior to implementing more permanent remediations.

For protection of public health and minimization of effects on ecological systems, permanent remediation is preferred. That is, removal, or marked diminution, of contact between persons and substances from HWS will lessen the likelihood of adverse health and ecological consequences. The key is permanent destruction of residual wastes. Some investigators believe current remedial practices do not lead to permanent reductions in health risk. One study states: "However, the usual practice is an exercise in transferring pollution from one environmental medium to another. At 68 percent of hazardous waste sites with contaminated groundwater, pumping and treating for aquifer restoration is the selected remedy. However, treatment most often consists of air-stripping the volatile organics from groundwater, thereby releasing the organics into the air. An evaluation of the Superfund program concluded that only 19 percent of source remedies effectively and permanently cleaned up hazardous waste sites. Most remedies provide only for containment or partial treatment, or they utilize a technology of questionable effectiveness and permanence. These remedies are designed to minimize short-term exposure, rather than to treat the waste and thereby permanently eliminate the hazards" (Travis and Hester, 1991).

Table 5 Remedial Action Categories

1. Institutional controls
   a. Monitoring
   b. Access restrictions
   c. Alternate water supply
   d. Deed restrictions
2. Containment
   a. Multilayered RCRA cap
   b. Soil cap
   c. Slurry well
   d. Surface well diversion by grading and dikes
3. Water collection/treatment/discharge
4. Soil/sediment removal, low-intensity treatment, and site restoration
   a. Soil washing
   b. Low temperature thermal treatment
   c. Bioreclamation (e.g., landfarming, windrowing)
   d. Stabilization (e.g., dewatering, solidification)
5. Soil/sediment removal, high-intensity treatment, ash disposal, and site restoration
   a. On-site incineration
   b. Off-site incineration
6. *In situ* treatment
   a. Soil flushing
   b. Vitrification
   c. Aquifer zone biological treatment
   d. Vadose zone vacuum extraction
   e. In-place soil biological treatment
   f. In-place sludge stabilization
   g. Flaring
7. Soil/sediment removal and landfilling
   a. Off-site landfill
   b. On-site landfill
8. Water collection/discharge to existing facility

From Russell et al., *Hazardous Waste Remediation: The Task Ahead,* University of Tennessee Press, Knoxville, TN, 1991. With permission.

Table 6 Distribution of Remediation Technologies and Percent Usage

Rank	Technology	%
1	Institutional controls	66
2	Containment	36
3	Water pump/treatment/discharge	47
4	Low-intensity treatment	22
5	High-intensity treatment	18
6	*In situ* treatment	13
7	Landfilling	11
8	Water collection/discharge	7

From Russell et al., *Hazardous Waste Remediation: The Task Ahead,* University of Tennessee Press, Knoxville, TN, 1991. With permission.

Given the large cost of remediating HWS and the nature of assigning costs to PRPs, it is not surprising to find that remedy selection by EPA and individual companies (where voluntary cleanup is undertaken) is a source of differing opinions. A study by the U.S. General Accounting Office found a number of alleged problems with completeness and consistency of cleanup decisions at both EPA- and responsible-party-managed sites. Problems noted were inadequate cleanup goals and insufficient justification for selecting a particular remedial action (GAO,

1992). Remedy selection and cleanup goals will remain subjects of active debate as the Superfund program matures.

STATE AND TERRITORIAL PROGRAMS

In addition to the federal Superfund program, states and U.S. territories have under way a major national program of removal and remedial actions at non-NPL sites that parallel those of the federal CERCLA. According to one source that acquired data from 39 states and one territory, as of December 31, 1992; 3527 sites have at had at least one completed state removal, 4834 sites were in a state's version of the remedial investigation/feasibility study phase, 2689 sites have completed construction through state remedial processes, and 11,000 sites were described as still active in some part of their state remedial process (ASTSWMO, 1994). This illustrates a substantial commitment on the part of states and territories to removing toxic wastes from the environment.

BIOREMEDIATION

Bioremediation, which is defined for the purposes of this text as being treatment technology that uses biological activity to reduce the concentration or toxicity of contaminants in the environment, is one of the methods of remediating HWS that is receiving increased attention. EPA defines bioremediation as the process that uses microorganisms to transform harmful substances to nontoxic compounds (ORD, 1991, p. 1). Bioremediation currently uses only naturally occurring microorganisms, such as bacteria, fungi, or yeast, to degrade harmful chemicals into less toxic or nontoxic substances. It is likely that genetically altered microorganisms will be used in the future as a means of bioremediation. The development of altered organisms is an active area of research. The actual employment of engineered microorganisms to remediate hazardous waste will require review and approval by government regulatory agencies (Day, 1991).

Bioremediation of hazardous wastes offers a number of attractive options (ORD, 1991, p. 7; Orzech et al., 1991), including the following:

- It is an ecologically sound, "natural" process, using natural organisms rather than engineered technologies, which may be less acceptable to the public.
- The target chemicals are actually destroyed, or changed into less toxic substances, instead of merely transferred from one environmental medium to another. Permanent destruction can occur, in some instances.
- The remedial costs are usually less than for other technologies. One source notes that incineration of toxic wastes often costs 250 to \$500 per ton of waste, whereas biological methods can cost as little as 40 to \$70 per ton (Vervalin, 1989). Another source estimates the costs of treating waste by bioremediation average 50 to \$80 per ton, vs. 250 to \$650 per ton for incineration and \$200 per ton or more for landfill disposal (Anon., 1990a). Another source estimates that costs to bioremediate soil range from 10 to \$180 per cubic yard, and bioremediation costs are estimated to be two thirds less than incineration for large volumes of waste (Orzech et al., 1991, p. 10).

- On-site remediation is achieved, obviating the need to transport hazardous wastes off-site.
- Public acceptance of a "natural" method like bioremediation to clean up hazardous waste may be greater than for methods perceived to be more hazardous (e.g., incineration).

As with any technology intended to remediate something as complex as a typical HWS, there are limitations and some disadvantages to bioremediation. Some of the disadvantages (ORD, 1991, p. 7; Orzech et al., 1991) include:

- Bioremediation often takes longer than other remedial actions (e.g., excavation and removal of soil or incineration).
- There is the difficult problem of scaling up a specific bioremediation action of sufficient potency and size to remediate a large geographic area.
- Many toxic substances are not currently amenable to biodegradation, awaiting the development of engineered organisms specific to the resistant substances.
- Few state officials have experience in coordinating projects in which bioremediation is selected or even seriously considered as the remediation method, thereby leading to a natural reluctance to implement bioremediation methods.

The practice of bioremediation of toxic wastes dates from about 1975. EPA statistics through March 1991 indicate that bioremediation projects are being considered, planned, or implemented in more than 140 sites, including projects at 24 sites on the Superfund NPL (Orzech et al., 1991, p. 9; Austin, 1990). EPA's total excludes many sites being cleaned up under state statutes. Industry sources aver that hundreds of successful full-scale applications have been completed, though no data are available. The total value of the U.S. bioremediation market—including assessment services, treatments, and inocula/nutrient sales—is forecast to grow from $100 million to more than $500 million by the year 2000 (Anon., 1990a).

There are many hazardous substances that are known to be degraded by bacteria. A partial list of degradable chemicals, according to one source, is given in Table 7 (Bell and Tremaine, 1993). In this same table are notations as to the rank of each substance on the 1992 Superfund priority list of hazardous substances and whether an ATSDR Toxicological Profile is available for each substance. It is encouraging to observe from Table 2 that several of the priority Superfund hazardous substances are biodegradable. However, the design of a bioremediation plan for a particular waste site is a complex undertaking, one that requires knowledge of the site's contaminants, levels of contamination, and existing microbial conditions (what microbes are available?, should they be augmented?, or will biostimulation suffice?) (Bell and Tremaine, 1993).

Examples of Bioremediation

Oil Spills — The use of bioremediation methods to cleanup oil spills on beaches is already a reality. Two examples are the *Exxon Valdez* oil spill, March 1989, in

Table 7 Select Chemicals Degraded by Bacteria

Chemical	Rank	In a tox profile?
Acetone	134	Yes
Anthracene	236	Yes
Atrazine	—	No
Benzo(a)anthracene	38	Yes
Benzene	5	Yes
Benzoate (benzoic acid)	249	No
Benzothiophene	—	No
Biphenyl	—	No
Chlorobenzene	70	Yes
Chlorophenol	—	No
Creosote	23	Yes
Creosol	—	Yes
Para-cresol	209	
Ortho-cresol	273	
Dibenzothiophene	—	No
Dichlorophenol (2,4-)	248	Yes
Dodecane	—	No
Ethylene bromide	—	No
Fluoranthene	88	Yes
Haloaromatics	—	No
Hexachlorobenzene	86	Yes
Jet fuels	—	Yes
Linuron	—	No
Methanol	—	No
Methyl ethyl ketone	162	Yes
Missile fuels	—	No
Naphthalene	67	Yes
Naphthol	—	No
Nitroglycerin	—	No
PCBs	7	Yes
Phenanthrene	181	Yes
Phenol	101	Yes
Quinoline	—	No
Tetrachloroethylene	21	Yes
Toluene	37	Yes
Trichloroethylene	10	Yes
Vinyl chloride	4	Yes
Xylene	72	Yes

Adapted from Bell and Tremaine, *Soils,* August/September, 42, 1993. With permission.

Prince William Sound, Alaska, and the June 1990 *Mega Borg* oil spill in the Gulf of Mexico. In Alaska, beaches were sprayed with commercially available slow-release formulations, an oleophilic (i.e., attracted to oil) fertilizer, and water-soluble formulations (Pritchard and Costa, 1991); stimulating naturally occurring bacteria by providing them with nitrogen and phosphorus nutrients. The organisms fed on the oil that had seeped nearly 2 ft into the gravel and sand. Within 3 weeks, the enhanced biological degradation of the oil led to areas nearly free of oil up to a depth of 12 in. (Anon., 1990b). The effectiveness of the bioremediation is shown in Figure 4, which shows the change in weight ratio of octadecane and phytane (a chemical measure of biodegradation) over time following a single application of oleophilic fertilizer (Pritchard and Costa, 1991). Moreover, studies of the mutagenicity of the spilled crude oil and its weathered counterpart collected from

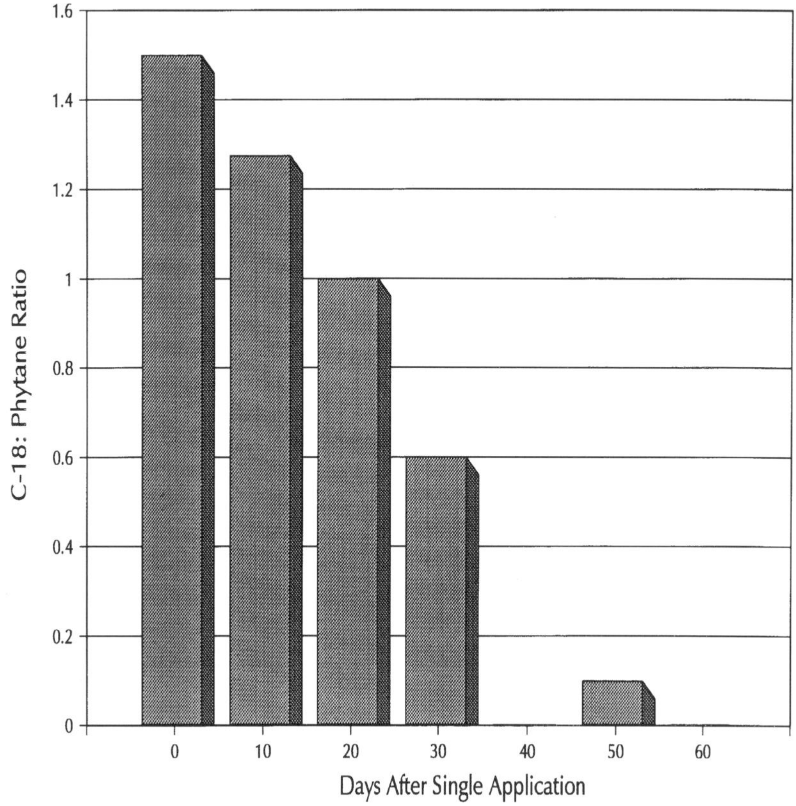

Figure 4 Effects of fertilizer application at Prince William Sound from Pritchard and Costa (1991).

oil-impacted water were found to be weakly mutagenic. The mutagenicity of samples collected from the contaminated beaches declined over time for both enhanced and natural bioremediation (Claxton et al., 1991).

Bioremediation of floating oil spills has also been attempted. The *Mega Borg* spill in the Gulf of Mexico led to seeding a 40-acre oil slick with commercially available mixtures of hydrocarbon-degrading bacteria mixed with fertilizer. No quantitative studies were performed, but it was thought that the microbial preparation accelerated the disappearance of the oil, in that visual inspection indicated enhanced dispersion of the oil slick. An attempt to conduct a controlled application of microbes to the oil slick was later conducted, but led to inconclusive results (Omenn, 1992). Further evaluation of biodegradation of floating oil slicks will be needed to assess its effectiveness (Anon., 1990b).

Another potential use of bioremediation is to clean up contaminated soil around automobile service centers. This is a global environmental problem of great magnitude. One source cites the use of indigenous populations of microorganisms to metabolize oil wastes to carbon dioxide and water. As an example, gasoline-contaminated soil with an average concentration of 1333 mg/kg was reduced to 2.8 mg/kg within 33 days using *ex situ* bioremediation (de Serres, 1994).

Waste Sites — As of 1990, EPA had selected bioremediation as the method of choice to remediate 24 NPL sites. The same source indicates private industry claims over 200 successful remediations of such sites have occurred (Austin, 1990). Two examples illustrate what appear to be successful remediations of chemically contaminated sites. At a Superfund site on the Texas Gulf Coast, a treatment facility was constructed consisting of a lined soil treatment area with a leachate collection system, a water and nutrient distribution system, an emission control system, a microbiological management system, a greenhouse-type enclosure, and support facilities. Approximately 200 yd^3 of soil containing volatile organics were treated in the facility. The activity of indigenous microorganisms was stimulated with inorganic nutrients (nitrogen and phosphorous). Greater than 99% removal of the volatile compounds was achieved after 94 days. The reduction of phenanthrene, the major contaminant, was reduced by 84% (Ross et al., 1988). In a second example, a small company in Iowa found that three underground storage tanks had leaked toxic wood preservative pentachlorophenol (PCP) into the surrounding soil. A commercial firm remediated the site by introducing microbes of the genus *Flavobacterium* into the polluted areas. Within a year, PCP levels had dropped by two thirds, with a thorough cleanup estimated to take 2 to 3 years. The bioremediation was found to be quite cost effective, in comparison with other options like excavation and then placement in landfills, or incineration (Chollar, 1991).

Bioremediation of polluted environmental media appears promising, based on a limited amount of experience. Unfortunately, very few details of field applications have been reported in scientific literature (Luthy, 1991). Controlled scientific studies and field applications will eventually clarify the nature and degree of effectiveness of specific kinds of bioremediation for different environmental pollution problems.

Biotechnology Regulations

The use of biologically based methods to remediate HWS has been, to date, confined to the use of naturally-occurring microorganisms. Genetically engineered organisms that might be targeted to the breakdown of specific toxicants have not been introduced into waste sites for remediation, though this application is likely to occur in the future. The use of engineered microorganisms for site remediation will require adherence to federal regulations on biotechnology. Bioengineered organisms are regulated under specific federal statutes, which are described in brief in this section. Additional details are found in Chapter 10.

The products of biotechnology have been described as a "low probability, high consequence risk"; where, "while there is only a small possibility that damage could occur, the damage that could occur is great" (HCST, 1984). Since the mid-1980s, policymakers and the courts have been grappling with the issues of regulating these products under the existing legal/regulatory framework and also developing new requirements.

The products of biotechnology fall within the scope of relevant laws primarily by their intended use (OSTP, 1985). For purposes of the TSCA, new intergeneric microorganisms used for commercial purposes are subject to the premanufacture notice requirements for new chemical substances. Additional regulations have been proposed or considered including a significant new use rule for pathogenic microorganisms, a rule exempting small scale research and development, and other reporting provisions.

Other major pollution control statutes may be relevant in two contexts. One, products of biotechnology could be considered pollutants under these statutes, and thus subject to emissions limitations, handling, and cleanup requirements. Second, as addressed in this chapter, microorganisms may be used as pollution control techniques, and must meet the various technology-based or performance-based requirements under relevant statutes, primarily, the Clean Air Act; Clean Water Act; Oil Pollution Act; CERCLA; and the RCRA.

Releases to the environment for any use by the federal government or with federal money are potentially covered by the National Environmental Policy Act, and may require preparation of Environmental Assessments, and Environmental Impact Statements. Additional regulatory requirements on biotechnology are described in Chapter 10.

SOCIETAL RESPONSE: ENVIRONMENTAL ETHICS

Is it ethical to pollute the environment? If it is, under what conditions? And if it isn't, under what framework of ethical principles? What are the environmental ethics of national economies based on tenets of consumerism? What are the ethical dimensions of national and international policies regarding the environment? On an individual basis, is it immoral to purchase consumer products that are not "environmentally benign" or, better, "environmentally friendly?" These are examples of serious and socially consequential questions that are being given increasing attention by ethicists who are turning their attention to environmental concerns under the rubric of environmental ethics, or ecoethics.

ECOETHICS

Before turning to what constitutes environmental ethics, it is useful to clarify terms. As noted by one source, even though the words "ethics," "values," and "morals" are often used interchangeably (and sometimes appropriately so), there are important practical differences in the words' meanings. "*Ethics* traditionally refers to the systematic framework of thought and analysis that deals with questions of right and wrong and the nature of the good and proper life" (Miller, A., 1991, p. 32). An ethical act is an action that is consistent within an ethical framework. *Values* refers more to a quality considered inherently worthwhile or desirable. *Morals* generally refers to that natural working-out of a personally affirmed ethical or value system. Morals become the conscious and visible basis for personal conduct and action. Morality is therefore concerned more with how a person acts than with the system that provides the framework for action (Miller, A., 1991). How are these concepts of ethics, values, and morality integrated into what is being called environmental ethics, or ecoethics?

According to one source, "Many people date the rise of 'environmental ethics' as an academic study from a famous 1967 article in the journal *Science* by the American historian of medieval science, Lynn White, Jr. In this piece White accused some aspects of Christianity of fostering an attitude toward nature as an object for exploitation and manipulation. This debate illuminated a number of hidden assumptions in medieval and modern history about the images of the human and the natural" (Timmerman, 1990, p. 214). Environmental ethics is concerned with the framework in which humankind relates to the natural environment. Although the intellectual underpinning of environmental ethics is being provided by philosophers—ecoethics is now considered a subdiscipline in ethics—these concepts will ultimately shape the laws and policies that pertain to the environment.

Ecoethics forces each of us to confront why we do things and the consequences of our actions on the environment. If one accepts the proposition, given the subject of this book, that industrial, agricultural, consumer, and personal practices have resulted in pollution of the environment and adverse effects on human and ecological systems; what accounts for the degraded state of environmental affairs in which we find ourselves? Lutzenberger (1991) posits that an understanding to this question requires an examination of the basic philosophy of modern civilization. He asserts we do what we do because of the dogmas, premises, and postulates on which national and global economic activities are based. In particular, philosophical and religious thinking that has not integrated ecological concepts into their framework is seen by some as a major reason for current environmental problems that include deforestation, waste mismanagement, ozone depletion, and global warming (Lutzenberger, 1991).

Environmental ethics attempts to look at the world in a holistic view and through a different point of view, one that builds around a framework of the planet as an entity which must be shared by all human and natural occupants. In 1975 James Lovelock, a British atmospheric chemist, and Lynn Margulis, an American microbiologist, proposed that Earth be viewed as one enormous, complex ecosystem they called Gaia, after the Greek goddess of the Earth, and humans constitute cells in a tissue of this supraorganism (Lutzenberger, 1991; Miller, A., 1991). This has become known as The Gaia Hypothesis—Earth as a single, self-regulating organism—and has developed a very large following, especially within the activist and "deep ecology" segments of the environmental movement. (The Norwegian philosopher Arne Naess in the 1970s separated ecology into two divisions: "shallow" and "deep." In this scheme, shallow ecology advocates a continuation of current environmental and political paths with only certain revisions to modern lifestyles and ways of life. Deep ecology advocates deep and fundamental changes in human relationships with nature [Timmerman, 1990; Miller, A., 1991, p. 18]). It is interesting to note that Margulis takes issue with how the Gaia hypothesis has been popularized by others. She affirms the notion that the Gaia hypothesis refers to Earth not as an individual organism, but an ecosystem (Margulis, 1993). Regardless of this controversy, it seems important to retain the essence of the Gaia hypothesis: Earth is a complex ecosystem and we as humans are but members of this system. We must function within the ecosystem and understand our relationships within it.

There are a number of scientific questions related to ethical concerns that have contemporary political impact. One subject having increasing impact on U.S. national environmental policies is called environmental justice. Other terms that predate this term are "environmental racism" and "environmental equity" (Sexton et al., 1994). Environmental justice, as the term suggests, addresses the question of whether environmental risk factors are shared equally by all segments of a society. More specifically, are minority groups and persons of color subjected to disproportionate environmental risks as the result of racist policies and actions by government and private sector agents? In consideration of this question, the EPA, following a comprehensive review of published literature, concluded, "Racial minority and low-income populations experience higher than average exposures to selected air pollutants, hazardous waste facilities, contaminated fish and agricultural pesticides in the workplace" (EPA, 1992). Proximity of minorities to HWS and facilities is a therefore a matter of considerable ethical, societal, and political concern.

MINORITIES AND HAZARDOUS WASTE SITES

As a matter of environmental equity, are hazardous waste treatment, storage, and disposal facilities (TSDFs) and abandoned HWS found in minority communities more often than elsewhere? In 1983 a General Accounting Office study found that three of four hazardous waste facilities investigated in the southern United States were located in primarily African-American communities (GAO, 1983). Two of the facilities were commercial TSDFs; the other two were landfills (GAO, 1983).

Drawing upon the GAO report, the United Church of Christ's Commission for Social Justice (CSJ) conducted two studies to determine the racial and socioeconomic characteristics of Americans living in residential areas surrounding commercial TSDFs and uncontrolled toxic waste sites (CSJ, 1987).

The first CSJ study sought to determine whether the variables of race and socioeconomic status played a significant role in the location of commercial TSDFs. The methodologic approach compared characteristics of geographic areas thought by the investigators to be relevant to the location of commercial hazardous waste facilities. The study compared five national data: (1) minority percentage of the population, (2) mean household income, (3) mean value of owner-occupied homes, (4) number of uncontrolled toxic waste sites per 1000 persons, and (5) pounds of hazardous waste generated per person. Racial classifications were taken from the 1980 U.S. Census, and data on TSDFs were obtained from EPA databases.

Minority percentage of the population was used to measure racial composition of communities. Mean household income and mean value of owner-occupied homes were included in the analysis to determine whether socioeconomic factors were more important than race in the location of commercial facilities. Existence of uncontrolled waste sites was evaluated to see whether underlying historic or geographic factors are associated with the siting of commercial hazardous waste facilities in ways that were not accounted for by other variables in the analysis.

Results of discriminant analysis showed that the factor minority percentage of the population in relation to the presence of commercial hazardous waste facilities was significant. The percentage of community residents that belong to a racial and ethnic group was a stronger predictor of the level of commercial hazardous waste activity than was household income, the value of homes, the number of uncontrolled toxic waste sites, or the estimated amount of hazardous wastes generated by industry. A key finding was that in ZIP codes with one commercial TSDF operating in 1986, the percent minority population, on average, was twice that of areas that did not contain a TSDF.

The second study by CSJ was descriptive in nature. Its primary purposes were (1) to quantify the number of racial and ethnic persons who lived in residential areas where uncontrolled toxic waste sites were located and (2) to make comparisons between the extent to which uncontrolled waste sites were located among different racial populations. U.S. Census data for 1980 and EPA's CERCLIS data were used for analysis. At the time of the study, CERCLIS contained information on 18,164 uncontrolled toxic waste sites in the U.S. Residential 5-digit ZIP code areas were used to define "communities." Aggregate population and uncontrolled waste site counts for those metropolitan areas with the largest African-American and Hispanic populations were calculated through use of 3-digit ZIP codes from a base of approximately 36,000 5-digit ZIP codes for the nation.

The descriptive study found the presence of uncontrolled waste sites to be "highly pervasive." More than half of the population in the United States lived in residential ZIP code areas with one or more uncontrolled toxic waste sites. Moreover, three out of every five African-American and Hispanic-Americans lived in communities with uncontrolled toxic waste sites, which amounts to more than 15 million African-Americans and 8 million Hispanics. Also, the investigators estimate 2 million Asian/Pacific Islanders and 700,000 Native Americans live in such communities.

On a smaller scale, investigators examined the demographics of persons residing near 16 commercial TSDFs in three counties in the Detroit area (Mohai and Bryant, 1992). They reported that only 3% of all Caucasians in the three-county area lived within 1 mile of a TSDF, compared with 11% of the minority population.

In contrast to the findings from the GAO, CSJ, and Detroit studies, a comprehensive examination of the demographics of a national sample of TSDFs found no disproportionate representation of minority groups in proximity to operating, commercial TSDFs (Anderton et al., 1994). These researchers, in distinction to the CSJ study, used census track data, not ZIP codes, as the geographic unit of analysis. This is presumably a more accurate, less coarse, index of populations in proximity to TSDFs.

In another study, *National Law Journal* staff analyzed the outcomes of all environmental lawsuits filed in federal courts over a 7-year period (Lavelle and Coyle, 1992). It was found that the average fine imposed for violating federal toxic waste laws in Caucasian residential areas was $335,566, which was more than six times the average fine imposed in minority residential areas. The disparity under the toxic waste law occurs by race alone, not income; the average penalty in areas with areas with the lowest median incomes was only 3% greater than the average penalty in areas with the highest median incomes. Study results also found that abandoned HWS in minority communities took 20% longer to be placed on the Superfund NPL than those sites in Caucasian communities.

The *National Law Journal* study and other studies cited here raise serious questions about whether racism plays a role in the placement of TSDFs and implementing actions to remediate sites. Bullard (1990, 1994), who has conducted comprehensive evaluations of communities in proximity to waste sites and operating facilities, concluded "... there is mounting empirical evidence that people-of-color and low-income communities suffer disproportionately from facility siting decisions involving municipal landfills, incinerators, and hazardous-waste disposal facilities" (Bullard, 1990, p. 139).

As a matter of environmental ethics and social policy, any inequities across cultural and racial groups related to exposure to environmental hazards must be prevented. In 1994 President Clinton issued Executive Order 12898, which directs each federal agency to "... make achieving environmental justice part of its mission by identifying and addressing, as appropriate, disproportionately high and adverse human health or environmental effects of its programs, policies, and activities on minority populations and low-income populations" (Clinton, 1994). Many state agencies are adopting similar directives.

GENETIC MONITORING AND SCREENING

A second significant area of ethical concern of relevance to ecoethics pertains to genetic screening and biological markers in environmental health investigations (Ashford et al., 1990). A distinction is made between genetic screening and genetic monitoring. Each action brings with it certain ethical issues for persons (e.g., workers exposed in the workplace or persons exposed to releases from waste sites) who are exposed to toxic substances. Genetic monitoring includes the periodic testing of biological samples from persons possibly exposed to certain substances (such as known carcinogens) that cause changes in chromosomes or DNA. Generally, genetic monitoring is conducted in an attempt to determine if environmental exposures of a specific population (e.g., a community adjacent to a HWS) to particular toxic substances cause changes in genetic material in statistically significant numbers above background levels (Ashford et al., 1990, p. 9). In the future, genetic monitoring may also serve as an early warning indicator that exposure to certain toxicants has occurred in a *population* of concern.

Advances in molecular biology make it possible to measure an individual's exposure, assess susceptibility, or determine health reaction to some specific environmental hazards. (Chapter 4 details biological markers of DNA damage.) The major ethical issue in genetic monitoring is what to tell persons with "abnormal" marker results about their disease risk, the purposes for which the information is used, and how society should treat persons with "abnormal" findings (NRC, 1991b).

Results from genetic monitoring released before the physiologic significance of a biologic marker is thoroughly assessed by scientists, particularly through laboratory studies, could cause unnecessary public alarm and emotional distress to individuals whose results are being reported. Several recommendations have been made on how to advance scientific development and application of new biological markers, and to be sensitive to attendant ethical issues. One committee recommends that when a biologic marker is included in a study, it must be evaluated against established batteries of tests, and a separate, statistically valid estimation of the new marker must be conducted (NRC, 1991b, p. 247). The committee also concluded that the evaluation process to find new markers should be conducted anonymously, with informed consent of the subjects and coding of specimens to delete identification of all study subjects.

Genetic screening is distinguished from genetic monitoring on the bases of the intent of the genetic testing and in what actions occur subsequent to testing. In an environmental context, genetic monitoring's primary purpose is to better understand, through collection of human genetic data, associations between exposure to environmental hazards and health effects. In distinction, genetic screening focuses on the collection of an individual's genetic data in order to take administrative actions intended to protect the individual, or employer, from health effects that may be related to abnormal genetic findings for the individual. In an occupational health context, in distinction to genetic monitoring, genetic screening has been the subject of removal of the worker before exposure and is preventive *for that person only* (Ashford et al., 1990, p. 160). The primary ethical issue is whether a person's genetic constitution should be the basis for excluding a person from a job or workplace. On the other hand, is it ethical on the part of an employer to place an employee in a position of potential health harm if the person is known to be at risk because of genetic predisposing factors? As is evident, the ethical issues of genetic screening are complicated and still being debated by ethicists and health officials alike.

Biomedical research on human subjects in the United States now commonly requires review and approval by an Institutional Review Board. These boards are composed of scientists, ethicists, medical experts, and lay persons who must review and approval all investigative protocols that involve human experimentation. Researchers are required to provide "informed consent" to study participants, i.e., an understandable explanation of the purposes, conditions, and impacts of the study. It is a matter of ethics that a study's investigators obtain informed consent from study participants prior to administering any investigative procedures (Goldsmith, 1986). Informed consent is a form of the right-to-know ethic (i.e., an individual's right to have information about environmental conditions that could affect one's health and well-being) that is increasingly appearing in national environmental legislation and international directives. Right-to-know requirements, in a legislative context, are discussed in Chapter 10.

Ecoethics, articulated as a separate framework within philosophy like the Gaia hypothesis or the "deep ecology" thought, or in individual expressions of concern for particular environmental problems (e.g., the ethics of polluted water supplies), ultimately contributes to attempts to legislate or regulate actions that are intended to control how humankind interrelates with the environment.

SUMMARY AND CONCLUSIONS

Hazardous wastes left in the environment through consequences of industrialization and agricultural practices present a threat to public health, environmental quality, and ecological systems. Adverse health effects have been noted in some communities in proximity to HWS, though many sites investigated have not shown excess morbidity or mortality. In the aggregate HWS as a collection of point sources of environmental degradation appear associated with a small to moderate increase in adverse reproductive outcomes and possibly some cancers. Environmental contamination data confirm that HWS pollute groundwater and soil. Moreover, many HWS are found in communities of low income and where persons of color reside, which raises issues of environmental justice. Public concern for these effects has resulted in national and state legislation and international agreements intended to remove hazardous wastes from the environment and regulate the generation, transport, and disposal of currently produced wastes.

Reflection on the experience gained from actions under the U.S. Superfund Law and future directions evident in U.N. and EC statements leads to several conclusions:

- The amount of uncontrolled hazardous waste left or being placed in the environment exceeds earlier predictions of the amount. Because of huge costs associated with remediation of hazardous wastes, national programs will attempt to minimize wastes and to prioritize how and what hazardous wastes to remove from the environment.
- Waste management programs will increasingly recognize and incorporate an international perspective. This will lead to national governments requiring multinational companies to manage hazardous wastes in a societally acceptable manner.
- Increasingly, national and local waste management programs will require waste generators to share information with the public in "right-to-know" programs.
- The current philosophy of "the polluter pays" for the consequences of environmental degradation will be revised in ways to make sharing of costs more equitable.
- Concern by legislators, policy makers, and regulators for the ecological effects of hazardous waste and other environmental hazards will become more frequently expressed and made evident in environmental legislation.

In closing, the ethics of environmental protection have developed as concern for protecting ecological systems has increased. Because of legal and ethical concerns pertaining to the management of all wastes, the financial burden of waste management on national economies is enormous. Permanent remediation of abandoned waste sites and effective programs of waste reduction and recycling are required to prevent hazardous waste's detrimental effects on public health, ecological systems, and environmental quality. Bioremediation of hazardous wastes appears a promising method to help achieve these prevention goals.

ACKNOWLEDGMENTS

The author gratefully acknowledges the assistance of Louise Lewis and Kim Jenkins, information specialists, and Maureen Lichtveld, M.D., Mark Bashor, Ph.D., and Charles Xintaras, Sc.D., manuscript reviewers, in the preparation of this chapter.

REFERENCES

Anderton, D. L., Anderson, A. B., Oakes, J. M., and Fraser, M. R., Environmental equity: The demographics of dumping, *Demography,* 31, 229, 1994.

Andrews, L. P., *Worker Protection During Hazardous Waste Remediation,* Van Nostrand Reinhold, New York, 1990, 6.

Anon., Bioremediation market to reach $500 million by end of decade, *J. Air Waste Manage. Assoc.,* 40, 961,1990a.

Anon., A tiny solution to a big problem, *Chemecology,* 19, 2, 1990b.

Anon., Cleanup costs could mount to $750 billion, *Hazardous Materials Control Resources Institute Focus,* 8, 1, 1992a.

Anon., Draft of environmental rules: Global partnership, *New York Times,* April 4, 1992b.
Anon., Senate ratifies Basel Convention on transboundary shipments of waste, *Environ. Rep.,* August 21, 1992c.
Aschengrau, A., Zierler, S., and Cohen, A., Quality of community drinking water and the occurrence of spontaneous abortion, *Arch. Environ. Health,* 44, 283, 1989.
Aschengrau, A., Ozonoff, D., Paulu, C., Coogan, P., Vezina, R., et al., Cancer risk and tetrachloroethylene-contaminated drinking water in Massachusetts, *Arch. Environ. Health,* 48, 284, 1993.
Ashford, N., Spadafor, C., Hattis, D., and Caldart, C., *Monitoring the Worker for Exposure and Disease,* Johns Hopkins University Press, Baltimore, 1990.
Assennato, G., Cannatelli, P., Emmett, P., Ghezzi, I., and Merlo, F., Medical monitoring of dioxin cleanup workers, *Am. Ind. Hyg. Assoc. J.,* 50, 586, 1989.
Association of State and Territorial Solid Waste Management Officials (ASTSWMO), A Report on State/Territory Non-NPL Hazardous Waste Site Cleanup Efforts for the Period 1980–1992, U.S. Environmental Protection Agency, Washington, D.C., 1994.
ATSDR, East Helena, Montana Child Lead Study, U.S. Department of Health and Human Services, Agency for Toxic Substances and Disease Registry, Atlanta, 1986.
ATSDR, Arsenic Exposure Pathway Study—Commencement Bay, Tacoma, Washington, U.S. Department of Health and Human Services, Agency for Toxic Substances and Disease Registry, Atlanta, 1987.
ATSDR, The Silver Creek Mine Tailings Exposure Study, U.S. Department of Health and Human Services, Agency for Toxic Substances and Disease Registry, Atlanta, 1988.
ATSDR, Personal Communication, Dr. S. Susten, U.S. Department of Health and Human Services, Agency for Toxic Substances and Disease Registry, Atlanta, 1992a.
ATSDR, Notice of revised priority list of hazardous substances, *Fed. Regist.,* 57, 48801, 1992b.
ATSDR, The Feasibility and Value of Performing Multisite Epidemiologic Studies for Superfund Sites, U.S. Department of Health and Human Services, Agency for Toxic Substances and Disease Registry, Atlanta, 1992c.
ATSDR, Investigation of a Cluster of Pancreatic Cancer Deaths, Livingston and Park County, Montana, U.S. Department of Health and Human Services, Atlanta, 1992d.
ATSDR, Biennial Report to Congress, U.S. Department of Health and Human Services, Agency for Toxic Substances and Disease Registry, Atlanta, 1993a.
ATSDR, National Exposure Registry: Trichloroethylene Subregistry Baseline Technical Report, U.S. Department of Health and Human Services, Agency for Toxic Substances and Disease Registry, Atlanta, 1993b.
Auer, C., Notice of opportunity to initiate negotiations for TSCA Section 4 enforceable consent agreements, *Fed. Regist.,* 59, 49934, 1994.
Austin, T., Bio bonanza?, *Civ. Eng. NY,* 60, 49, 1990.
Baker, D., Greenland, S., Mendlein, J., and Harmon, P., A health study of two communities near the Stringfellow waste disposal site, *Arch. Environ. Health,* 43, 325, 1988.
Batstone, R., Smith, J. E., Jr., and Wilson, D., *The Safe Disposal of Hazardous Wastes: The Special Needs and Problems of Developing Countries,* Vol. 1, The World Bank, Washington, D.C., 1989, 5.

Bell, P. and Tremaine, S., Debate continues between biostimulation, bioaugmentation, *Soils,* August/September, 42, 1993.

Binder, S., Forney, D., Kaye, W., and Paschal, D., Arsenic exposure in children living near a former copper smelter, *Bull. Environ. Contam. Toxicol.,* 39, 114, 1987.

Buffler, P., Crane, M., and Key, M., Possibilities of detecting health effects by studies of populations exposed to chemicals from waste disposal sites, *Environ. Health Perspect.,* 62, 423, 1985.

Bullard, R. D., *Dumping in Dixie,* Westview Press, Boulder, CO, 1990.

Bullard, R. D., Ed., Unequal Protection: Environmental Justice and Communities of Color, Sierra Club, San Francisco, 1994.

Cantor, K., Hoover, R., Hartge, P., et al., Bladder cancer, drinking water source, and tap water consumption: A case-control study, *J. Natl. Cancer Inst.,* 79, 1269, 1987.

Case, D. R., Resource Conservation and Recovery Act (RCRA), in *Environmental Law Handbook,* Arbuckle, J. G., et al., Eds., Government Institutes, Inc., Rockville, MD, 1991, 406.

Chollar, S., The poison eaters, *Discover,* 11, 76, 1991.

Christenson, L., Statement of the Superfund Landfill Solutions Group, Committee on Ways and Means, House of Representatives, U.S. Congress, Washington, D.C., August 12, 1992.

Claxton, L. D., Houk, V. S., Williams, R., and Kremer, F., Effect of bioremediation on the mutagenicity of oil spilled in Prince William Sound, Alaska, *Chemosphere,* 23, 643, 1991.

Clinton, W. J., Federal actions to address environmental justice in minority populations and low-income populations, *Fed. Regist.,* 59, 32, 7629, 1994.

Cohn, P., Bove, F., Klotz, J., Berkowitz, M., and Fagliano, J., Drinking water contamination and the incidence of leukemia and non-Hodgkin's lymphoma, in *Proc. Int. Congress on Public Health and Hazardous Waste,* Andrews, J., Frumpkin, H., Johnson, B., Mehlman, M., Xintaras, C., and Bucsela, J., Eds., Princeton Scientific Publishing, Princeton, NJ, 742, 1994.

Colborn, T., Epidemiology of Great Lakes eagles, *J. Toxicol. Environ. Health,* 33, 395, 1991.

Colorado Health Department (CHD), Leadville Metals Exposure Study, State of Colorado, Denver, 1990.

Commission on Social Justice (CSJ), Toxic Waste and Race in the United States, United Church of Christ, New York, 1987.

Congressional Budget Office (CBO), The total costs of cleaning up nonfederal Superfund sites, U.S. Government Printing Office, Washington, D.C., 1994.

Congressional Research Service (CRS), Summaries of Environmental Laws Administered by the Environmental Protection Agency, Library of Congress, Washington, D.C., 1991.

Cutler, J., Parker, G., Rosen, S., Prenney, B., Healey, R., and Caldwell, G., Childhood leukemia in Woburn, Massachusetts, *Public Health Rep.,* 101, 201, 1986.

Day, S., Federal regulations: How they impact research and commercialization of biological treatment, in *Environmental Biotechnology for Waste Treatment,* Sayler, G., Fox, R., and Blackburn, J., Eds., Plenum Press, New York, 1991.

Department of Environmental Protection (DEP), Hazardous Waste Management Act, Chapter 21C, State of Massachusetts, Boston, 1979.

Department of Environmental Quality (DEQ), Leap to 2000, Louisiana's Environmental Action Plan, State of Louisiana, Baton Rouge, 1991.

de Serres, F. J., New technology for bioremediation of contaminated soil in auto service stations, *Prev. Contam.,* 3, 14, 1995.

Diaz-Barriga, F., Santos, M., Yanez, L., Cuellar, J., Ostrosky-Wegman, P., Montero, R., Perez, A., Ruiz, F., Garcia, A., and Gomez, H., Biological monitoring of workers at a recently opened hazardous waste disposal site, *J. Exposure Anal. Environ. Epidemiol.,* 3 (Suppl. 1), 63, 1993.

Dominguez, G., Hazardous waste management, in *The Law of Toxics and Toxic Substances,* Dominguez, G. and Bartlett, K., Eds., CRC Press, Boca Raton, FL, 1986, 4.

Dunne, M., Burnett, P., Lawton, J., and Raphael, B., The health effects of chemical waste in an urban community, *Med. J. Aust.* 152, 592, 1990.

Edelstein, J., Statement of American International Group to the Subcommittee on Oversight, House Ways and Means Committee, House of Representatives, U.S. Congress, Washington, D.C., August 12, 1992.

Environmental Law Institute (ELI), *Environmental Law Deskbook,* Environmental Law Institute, Washington, D.C., 1989, 166, 221.

Environmental Law Institute (ELI), *European Community Deskbook,* Environmental Law Institute, Washington, D.C., 1992, 357.

EPA, Unfinished Business: A Comparative Assessment of Environmental Problems, U.S. Environmental Protection Agency, Washington, D.C., 1987.

EPA, Environmental Equity: Reducing Risk for All Communities, Vol. 1, U.S. Environmental Protection Agency, Washington, D.C., 1992, 3.

EPA, National Priorities List for uncontrolled hazardous waste sites, *Fed. Regist.,* August 23, 1994.

Favata, E., Barnhart, S., Bresnitz, E., and Campbell, V., Clinical experiences: Development of a medical surveillance protocol for hazardous waste workers, *J. Occup. Med.,* 5, 117, 1990.

Fein, C., Jacobson, J., Jacobson, S., Schwartz, P., and Dowler, J., Prenatal exposure to polychlorinated biphenyls: Effects on birth size and gestational age, *J. Pediatr.,* 105, 315, 1984.

Feldman, R., Chirico-Post, J., and Proctor, S., Blink reflex latency after exposure to Trichloroethylene in well water, *Arch. Environ. Health,* 43, 143, 1988.

Fulton, E., Durnil, G., Welch, R., Cleveland, H., Lanthier, C., and Goodwin, R., Sixth Biennial Report on Great Lakes Water Quality, International Joint Commission, Washington, D.C., 1992, 15.

GAO, Siting of Hazardous Waste Landfills and Their Correlation with Racial and Economic Status of Surrounding Communities, General Accounting Office, Washington, D.C., 1983.

GAO, Superfund: Problems with the Completeness and Consistency of Site Cleanup Plans, General Accounting Office, Washington, D.C., 1992, 3.

Geschwind, S., Stolwijk, J., Bracken, M., Fitzgerald, E., Stark, A., Olsen, C., and Melius, J., Risk of congenital malformations associated with proximity to hazardous waste sites, *Am. J. Epidemiol.,* 135, 1197, 1992.

Gilbertson, M., Morris, R., and Hunter, R., Abnormal chicks and PCB residue levels in eggs of colonial waterbirds on the Lower Great Lakes, *Auk,* 93, 434, 1976.

Goldberg, S., Lebowitz, M., Graver, E., and Hicks, S., An association of human congenital cardiac malformations and drinking water contaminants, *J. Am. College Cardiol.,* 16, 155, 1990.

Goldsmith, J., *Environmental Epidemiology: Epidemiological Investigation of Community Environmental Health Problems,* CRC Press, Boca Raton, FL, 1986.

Gottlieb, K. and Koehler, J., Blood lead levels in children from lower socioeconomic communities in Denver, Colorado, *Arch. Environ. Health,* 49, 260, 1994.

Griffith, J., Duncan, R., Riggan, W., and Pellom, A., Cancer mortality in U.S. counties with hazardous waste sites and ground water pollution, *Arch. Environ. Health,* 44, 69, 1989.

Grisham, J., Ed., *Health Aspects of the Disposal of Waste Chemicals,* Pergamon Press, New York, 1986.

Gustavsson, P., Mortality among workers at a municipal waste incinerator, *Am. J. Ind. Med.,* 15, 245, 1989.

Habicht, H., Plans, problems, and promises for the years ahead, *Hazard. Mater. Control,* 60, 3, 1991.

Hertzman, C., Hayes, M., Singer, J., and Highland, J., Upper Ottawa Street landfill site health study, *Environ. Health Perspect.,* 75, 173, 1987.

Hicks, H. and Katz, L., Impact on Public Health of Persistent Toxic Substances in the Great Lakes Region, Agency for Toxic Substances and Disease Registry, Atlanta, 1992.

House Committee on Science and Technology (HCST), The Environmental Implications of Genetic Engineering, U.S. House of Representatives, U.S. Congress, Washington, D.C., 1984.

Humphrey, H., Chemical contaminants in the Great Lakes: the human health impact, in *Toxic Contaminants and Ecosystem Health: A Great Lakes Focus,* Evans, M., Ed., John Wiley & Sons, New York, 1988, 153.

Ikatsu, H., Nakajima, T, Okino, T., and Murayama, N., Health care of workers engaged in waste water treatment, *Sangyo Igaku,* 31, 355, 1989.

International Legal Materials (ILM), United Nations document IG.80/3, Final act and text of the Basel Convention, *Am. Soc. Int. Law,* 28, 2, 1989.

Isacson, P., Bean, J., Splinter, R., Olson, D., and Kohler, J., Drinking water and cancer incidence in Iowa, *Am. J. Epidemiol.,* 121, 856, 1985.

Jacobson, J. and Jacobson, S., New methodologies for assessing the effects of prenatal toxic exposure on cognitive functioning in humans, in *Toxic Contaminants and Ecosystem Health: A Great Lakes Focus,* Evans, M., Ed., John Wiley & Sons, New York, 1988, 373.

Jacobson, J., Jacobson, S., and Humphrey, H., Effects of in utero exposure to polychlorinated biphenyls and related contaminants on cognitive functioning in young children, *J. Pediatr.,* 116, 38, 1990.

Jennings, B., Analysis of Proposition 65 — Safe Drinking Water and Toxic Enforcement Act, Senate Office of Research, State of California, Sacramento, 1986.

Johnson, B., Testimony, Subcommittee on Superfund, Recycling, and Solid Waste Management, Senate, U.S. Congress, Washington, D.C., 1993.

Johnson, B., Nature, extent, and impact of Superfund hazardous waste sites, *Chemosphere,* in press, 1995.

Johnson, B., Baker, E., El Batawi, M., Gilioli, R., Hänninen, H., Seppälainen, A., and Xintaras, C., Eds., *Prevention of Neurotoxic Illness in Working Populations,* Wiley, New York, 112, 1987.

Kawamoto, M., *The Caldwell Group,* National Institute for Occupational Safety and Health, Centers for Disease Control and Prevention, Cincinnati, 1992.

Keystone Center, Final Consensus Report of the National Commission on Superfund, The Keystone Center, Keystone, CO, 1994.

Kiec-Swierczynska, M., Waste volatile ashes, a new source of hypersensitivity to metals, *Przegl. Dermatol.,* 76, 410, 1989.

Kilburn, K. and Warshaw, R., Effects on neurobehavioral performance of chronic exposure to chemically contaminated well water, *Toxicol. Ind. Health,* 9, 391, 1993.

Kilburn, K. H. and Warshaw, R. H., Prevalence of symptoms of systemic lupus erythematosus (SLE) and of fluorescent antinuclear antibodies associated with chronic exposure to trichloroethylene and other chemicals in well water, *Environ. Res.,* 57, 1, 1992.

LaDou, J., Deadly migration, *Technol. Rev.,* July, 1991, 47.

Lavelle, M. and Coyle, M., Unequal protection: The racial divide in environmental law, *Natl. Law J.,* September 21, 1992.

Lichtveld, M. Y. and Johnson, B. L., Public health implications of hazardous waste sites in the United States, in *Proc. Int. Congr. Public Health and Hazardous Waste,* Andrews, J., Frumpkin, H., Johnson, B., Mehlman, M., Xintaras, C., and Bucsela, J., Eds., Princeton Scientific Publishing, Princeton, NJ, 1994, 14.

Lipscomb, J., The Epidemiology of Symptoms Reported by Persons Living Near Hazardous Waste Sites, Ph.D. Dissertation, University of California, Berkeley, 1989.

Logue, J. N. and Fox, J. M., Residential health study of families living near the Drake Chemical Superfund site in Lock Haven, Pennsylvania, *Arch. Environ. Health,* 41, 222, 1986.

Lutzenberger, J., Environmental ethics, *Ecodecision,* 1, 81, 1991.

Luthy, R. G., Bioremediation—promises and problems (editorial), *Res. J. Water Pollut. Control Fed.,* 63, 99, 1991.

Maccubbin, A., Black, P., Tryeciak, L., and Black, J., Evidence for polynuclear aromatic hydrocarbons in the diet of bottom-feeding fish, *Bull. Environ. Contam. Toxicol.,* 34, 876, 1985.

Margulis, L., Gaia in science, *Science,* 259, 745, 1993.

Marshall, E., Gensburg, M., Geary, N., Deres, D., and Cayo, M., Maternal residential exposure to hazardous waste sites and the risk of central nervous system and musculoskeletal birth defects, Presented at the International Congress on Hazardous Waste: Human and Ecologic Health, Atlanta, 1995 (abstract).

Melius, J., OSHA standard for medical surveillance of hazardous waste workers, *J. Occup. Med.,* 5, 143, 1990.

Meyer, C., Liver dysfunction in residents exposed to leachate from a toxic waste dump, *Environ. Health Perspect.,* 48, 9, 1983.

Miller, A., *Gaia Connections: An Introduction to Ecology, Ecoethics, Economics,* Rowman & Littlefield, Savage, MD, 1991.

Miller, J., Environmental law and the science of ecology, in *Environmental and Occupational Medicine,* 2nd ed., Rom, W., Ed., Little, Brown, Boston, 1992, 1307.

Miller, S. S., Southbend Subdivision Health Outcomes Study, Harris County, Texas, Agency for Toxic Substances and Disease Registry, Atlanta, 1994.

Mohai, P. and Bryant, B., Environmental racism: Reviewing the evidence, in *Race and the Incidence of Environmental Hazards: A Time for Discourse,* Bryant, B. and Mohai, P., Eds., Westview, Boulder, CO, 1992, 163.

Monster, A. and Smolders, J., Tetrachloroethylene in exhaled air of persons living near pollution sources, *Int. Arch. Occup. Environ. Health,* 53, 331, 1984.

Moran, J., Safety and health issues arising at Superfund sites, in *Proc. Int. Congr. Public Health and Hazardous Waste,* Andrews, J., Frumpkin, H., Johnson, B., Mehlman, M., Xintaras, C., and Bucsela, J., Eds., Princeton Scientific Publishing, Princeton, NJ, 1994, 70.

Morbidity and Mortality Weekly Report (MMWR), Hazardous Waste Sites: Priority Health Conditions and Research Strategies — United States, Centers for Disease Control and Prevention, Atlanta, 41, 72, 1992.

Najem, G., Strunck, T., and Feuerman, M., Health effects of a Superfund hazardous chemical waste disposal site, *Am. J. Prev. Med.,* 10, 151, 1994.

Najem, G., Thind, I., Lavenhar, M., and Louria, D., Gastrointestinal cancer mortality in New Jersey counties, and the relationship with environmental variables, *Int. J. Epidemiol.,* 12, 276, 1983.

Najem, G. and Voyce, L., Health effects of a thorium waste disposal site, *Am. J. Public Health,* 80, 478, 1990.

Nakajima, H., A wounded planet, *World Health,* January–February, 1990, 3.

National Research Council (NRC), *Rethinking the Ozone Problem in Urban and Regional Air Pollution,* National Academy Press, Washington, D.C., 1991a, 19.

National Research Council (NRC), *Environmental Epidemiology: Public Health and Hazardous Wastes,* Vol. 1, National Academy Press, Washington, D.C., 1991b, 20, 114.

National Research Council (NRC), *Ranking Hazardous-Waste Sites,* National Academy Press, Washington, D.C., 1994, 29, 37.

Neuberger, J., Mulhall, M., Pomatto, M., Sheverbush, J., and Hassanein, R., A heavy metal mining Superfund site, *Sci. Total Environ.,* 94, 261, 1990.

Neutra, R., Lipscomb, J., Satin, K., and Shusterman, D., Hypotheses to explain the higher symptom rates observed around hazardous waste sites, *Environ. Health Perspect.,* 94, 31, 1991.

Office of Research and Development (ORD), Understanding Bioremediation: A Guidebook for Citizens, Environmental Protection Agency, Washington, D.C., 1991.

Office of Science and Technology Policy (OSTP), *Fed. Regist.,* 50, 47174, 1985.

Office of Solid Waste and Emergency Response (OSWER), The Challenge of Superfund, Report 9200.5-008A, Environmental Protection Agency, Washington, D.C., 1992a.

Office of Solid Waste and Emergency Response (OSWER), EPA Superfund Progress: Spring 1992, Report 9200.1-12, Environmental Protection Agency, Washington, D.C., 1992b.

Office of Solid Waste and Emergency Response (OSWER), Superfund Progress — Aficionado's Version, Report PB92-963267, Environmental Protection Agency, Washington, D.C., 1992c.

Office of Technology Assessment (OTA), Technologies and Management Strategies for Hazardous Waste Control, U.S. Government Printing Office, Washington, D.C., 1983.

Ogunranti, J., Hematological indices in Nigerians exposed to radioactive waste, *Lancet,* 16, 667, 1989.

Omenn, G., Environmental biotechnology: Biotechnology solutions for hazardous chemical wastes and oil spill cleanup, *J. Clean Technol. Environ. Sci.,* 2, 47, 1992.

Orzech, M., Solyst, J., and Thompson, P., States' Use of Bioremediation: Advantages, Constraints, and Strategies, National Governors' Association, Washington, D.C., 1991.

Ozonoff, D., Coltzen, M., Cupples, A. et al., Health problems reported by residents of a neighborhood contaminated by a hazardous waste facility, *Am. J. Indus. Health,* 11, 581, 1987.

Paigen, B., Goldman, L., Magnant, M., Highland, J., and Steegmann, A., Growth of children living near the hazardous waste site, Love Canal, *Human Biol.,* 59, 489, 1987.

Planard, C., Ros, Y., Bazzoni, D., Pays, A., Ottogali, M., et al., Cytogenetic analysis of peripheral lymphocytes in a human population exposed to alleged environmental pollution by a chemical waste dump. Paper presented at the International Congress on Public Health and Hazardous Waste, Agency for Toxic Substances and Disease Registry, Atlanta, May 1993 (abstract).

Plotzman, F., A nation's recycling law puts businesses on the spot, *New York Times,* p. K1, July 12, 1992.

Price, I. and Weseloh, D., Increased numbers and productivity of double-crested cormorants, *Phalacrocorax Auritis,* on Lake Ontario, *Can. Field Nat.,* 100, 474, 1986.

Pritchard, P. H. and Costa, C. F., EPA's Alaska oil spill bioremediation project, *Environ. Sci. Technol.,* 25, 372, 1991.

Probst, K., Testimony, Subcommittee on Oversight, Ways and Means Committee, House of Representatives, U.S. Congress, Washington, D.C., August 12, 1992.

Public Health Service (PHS), Healthy People 2000, U.S. Department of Health and Human Services, Washington, D.C., 1990, 325.

Roht, L.H., Vernon, S.W., Weir, F.W., Pier, S.M., Sullivan, P., and Reed, L.J., Community exposure to hazardous waste disposal sites: Assessing reporting bias, *Am. J. Epidemiol.,* 122, 418, 1985.

Rom, W.N., Chlorofluorocarbons and destruction of the ozone layer, in *Environmental and Occupational Medicine,* 2nd ed., Rom, W., Ed., Little, Brown, Boston, 1992, 1299.

Ross, D., Stroo, H. F., and Bourquin, A. W., Bioremediation of hazardous waste sites in the USA: Case histories, in *Contaminated Soil '88,* Wolf, K., van den Brink, W. J., and Colon, F. J., Eds., Kluwer Academic Publishers, Norwell, MA, 1988, 717.

Russell, M., Colglazier, E. M., and English, M.R., *Hazardous Waste Remediation: The Task Ahead,* University of Tennessee Press, Knoxville, 1991, A-3.10.

Sexton, K., Olden, K., and Johnson, B., Environmental justice: The central role of research in establishing a credible scientific foundation for informed decision making, *Toxicol. Indust. Health,* 9, 685, 1993.

Shaw, G., Schulman, J., Frisch, J., Cummins, S., and Harris, J., Congenital malformations and birthweight in areas with potential environmental contamination, *Arch. Environ. Health,* 47, 147, 1993.

Shusterman, D., Lipscomb, J., Neutra, R., and Satin, K., Symptom prevalence and odor-worry interaction near hazardous waste sites, *Environ. Health Perspect.,* 94, 25, 1991.

Smith, T. T. and Hunter, R. D., The European Community environmental legal system, in *European Community Deskbook,* Environmental Law Institute, Washington, D.C., 1992, 3.

Sosniak, W., Kaye, W., and Gomez, T., Data linkage to explore the risk of low birthweight associated with maternal proximity to hazardous waste sites from the National Priorities List, *Arch. Environ. Health,* 49, 251, 1994.

Stehr-Green, P., Burse, V., and Welty, E., Human exposure to polychlorinated biphenyls at toxic waste sites: Investigations in the United States, *Arch. Environ. Health,* 43, 420, 1988.

Stoll, R. G., Comprehensive Environmental Response, Compensation, and Liability Act, in Arbuckle, J. G., Ed., *Environmental Law Handbook,* Government Institutes, Inc., Rockville, MD, 1991, 471.

Szasz, A., *EcoPopulism: Toxic Waste and the Movement for Environmental Justice,* University of Minnesota Press, Minneapolis, 1994.

Timmerman, P., Grounds for concern: environmental ethics in the face of global change, in *Planet Under Stress,* Mungall, C. and McLaren, D., Eds., Oxford University Press, London, 1990, 211.

Travis, C. C. and Hester, S. T., Global chemical pollution, *Environ. Sci. Technol.,* 25, 815, 1991.

United Nations Environment Program (UNEP), U.N. Conference on Environment and Development: Agenda 21, United Nations, New York, 1992.

Upton, A., Kneip, T., and Toniolo, P., Public health aspects of toxic chemical disposal sites, *Annu. Rev. Public Health,* 10, 1, 1989.

Vervalin, C. H., Bioremediation on the move, *Hydrocarbon Process.,* 68, 50, 1989.

Viscusi, W. K. and Hamilton, J. T., Superfund and real risks, *The American Enterprise,* March/April, 1994, 37.

Zmirou, D., Deloraine, A., Saviuc, P., Tillier, C., Boucharlat, A., Maury, N., Short-term health effects of an industrial toxic waste landfill: A retrospective follow-up study in Montchanin, France, *Arch. Environ. Health*, 49, 228, 1994.

CHAPTER 2

Ecotoxicology: Environmental Fate and Ecosystem Impact

Charles E. Weinstein

CONTENTS

OVERVIEW

The purpose of this chapter is to describe the major factors affecting the environmental fate and ecosystem impact of toxic chemicals and to demonstrate their significance for human health. It draws on examples from the Great Lakes ecosystem on the United States–Canada border, which has been the focus of remediation efforts for over 20 years.

1-56670-141-4/96/$0.00+$.50

THE RELATIONSHIP BETWEEN HUMAN AND ECOLOGICAL HEALTH

There is a widely recognized but poorly understood relationship between the human and ecological health effects of environmental pollution. In recent years, it has become more obvious that assessing the impacts of chemical toxins on human health requires an interdisciplinary approach. This approach weaves together investigating the environmental fate of these toxins, their effects on ecological systems (ecotoxicology), and their significance for human health. Ecological assessments are ultimately concerned with protecting the structure and function of ecosystems. Human health risk assessments have tried to address the hazardous waste problem as described in Chapter 1, but they have tended to ignore the damage done to ecosystems, especially where the presence of chemical pollutants poses no immediate human health risk. However, there is evidence that animal populations may be more susceptible to harmful health effects of environmental contaminants than humans, thereby serving as an early warning sentinel of potential harm to human populations (NRC, 1991). Similar harmful effects on human health may be delayed and may be difficult to predict. Concerns about toxic chemicals, therefore, should focus *not only* on their short- and long-term effects on human health but also on their potential for damaging ecosystems. This damage may include a reduction in the "services" that ecosystems provide in sustaining civilization, such as water and air purification, flood control, nutrient cycling, soil formation, and climate stabilization. Environmental pollution causes ecological changes on a global scale that could result in significant human suffering (McCally and Cassell, 1990).

There are no widely accepted criteria for determining the damage to ecosystems. However, there is strong support for the maintenance of ecosystem "integrity," the proper functioning of the physical, chemical, and biological processes that support all organisms. Certainly the real and perceived threats from acid deposition have been sufficient to focus significant resources on controlling the damage being done to regional forest ecosystems and other related ecosystems, such as streams and lakes (Pitelka and Raynal, 1989). Contamination of the marine environment with complex chemical mixtures has caused an increase in certain fish diseases (e.g., hepatic neoplasms) and raised significant concern regarding the use of fish as food (Murchelano, 1990; Dawe, 1990).

The Great Lakes ecosystem of the United States and Canada, the world's largest supply of freshwater, includes the drainage basins of Lakes Superior, Michigan, Huron, Erie, and Ontario. It has suffered significant damage from over 1000 chemicals of potential concern that have been released into its waters. These lakes act as reservoirs for pollutants from the land and air while serving as a source of multigenerational human exposure for the nearly 40 million people living in this region. For example, the consumption of fish, which is the major source of human exposure to polychlorinated biphenyls (PCBs) in the Great Lakes region, has been banned in some regions despite uncertainty about the long-term outcome for human health (Humphrey, 1988). The major problem continues to be the exposure and impact of persistent toxic chemicals throughout this complex ecosystem (International Joint Commission [IJC], 1992).

There is concern on a global level that tens of thousands of toxic chemicals continue to contribute to the loss of biodiversity and other effects that destabilize

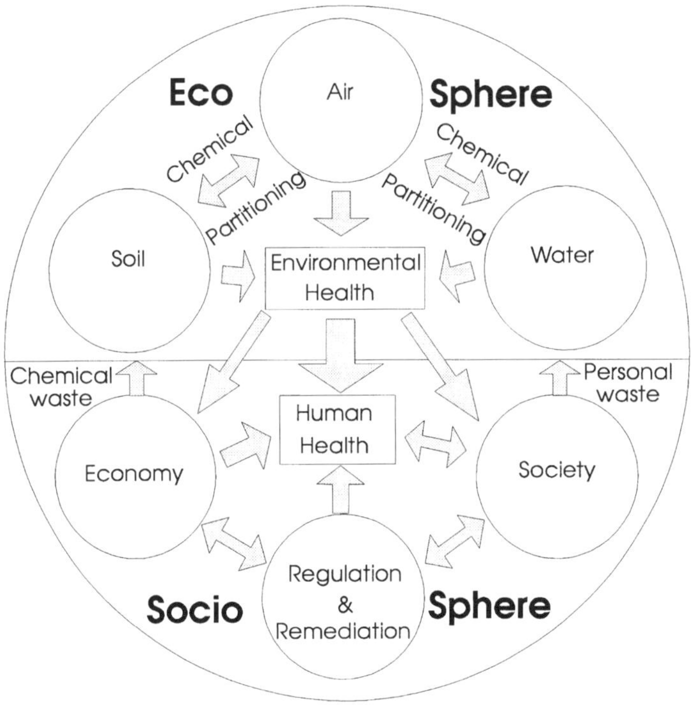

Figure 1 Interrelationship of environmental and human health.

and stress ecosystems (Barrett and Rosenberg, 1981; Rasmussen, 1984; Draggan et al., 1987; Bascietto et al., 1990). The proposition has been advanced that all human economic and social interests are best served by preserving biodiversity on a global scale (Woodwell, 1987). As Figure 1 illustrates, society and the environment interact in a number of ways to affect environmental and human health. A biological approach to remediation of damaging effects would utilize biological criteria and monitoring techniques, such as those discussed in subsequent chapters, and consider preservation of ecosystem "integrity" as a critical requirement for the protection of human health.

ENVIRONMENTAL FATE AND EXPOSURE ASSESSMENT

Estimating the amount of contaminant to which a population is exposed is a difficult task because of the need to consider complex physical, chemical, and biological processes that affect the concentration of contaminants in environmental media (air, water, soil, and biota). These processes change these contaminants and move them among the various media (Figure 2). Analysis of environmental risks has not typically incorporated complex environmental fate processes (Paustenbach, 1989a), and initial risk estimates typically assume a constant exposure to a toxic chemical, particularly when based solely on preliminary monitoring data. The

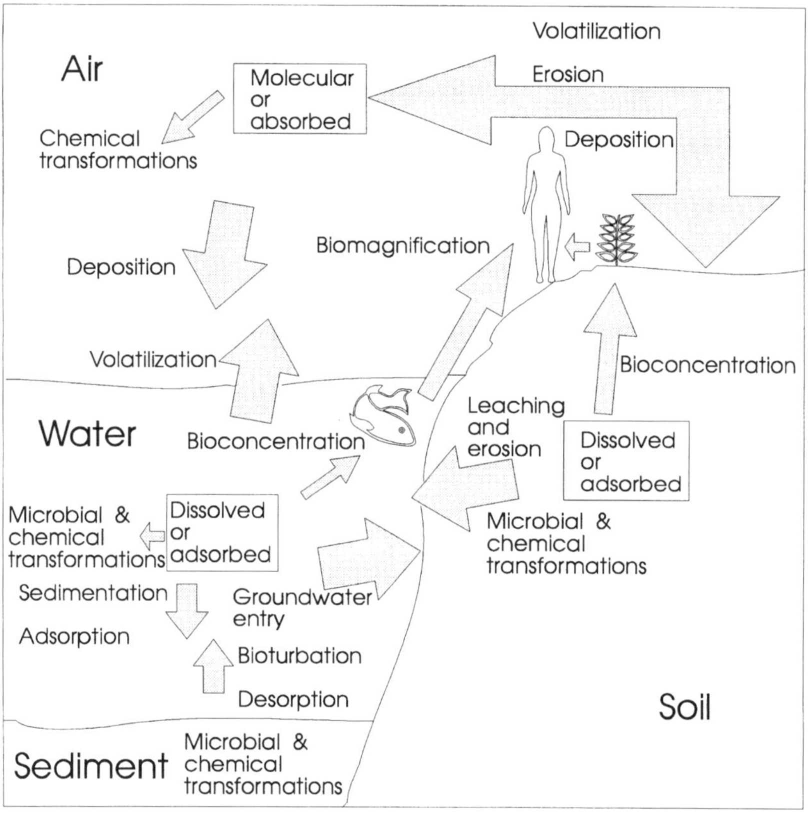

Figure 2 Fate processes affecting environmental contaminants.

assumption that contaminant levels remain constant throughout the period of exposure is, in many cases, a misrepresentation of the actual conditions (USEPA, 1988). Compilations of data on the environmental fate of various chemicals exists (Lyman et al., 1982; Howard, 1989) and are becoming more readily available for use in exposure assessment. In the use of modeling to predict the accumulation of pollutants in soil, for example, the behavior and longevity of each chemical must be known (Stevens and Swackhamer, 1989). Qualitative and quantitative consideration of the environmental fate of contaminants can have a significant impact on the selection of chemicals of concern, determination of significant exposure pathways, and the feasibility and potential impacts of remediation strategies.

PHYSICAL AND CHEMICAL BEHAVIOR

The distribution or partitioning of a pollutant among the various environmental media is the major consideration in predicting its environmental fate, including its availability for biological processes. Partitioning is based on the physical and chemical properties of a contaminant and the physical, chemical, and biological characteristics of the environment into which it is released. In the case of PCBs in the Great Lakes, it is recognized that physical processes are primarily responsible for their environmental fate, especially for the more persistent congeners (different chemical forms of PCB) (Mackay et al., 1983). The properties of a chemical, such as its solubility in water or lipid (e.g., fatty tissues), can be very useful in rapidly predicting its behavior during an initial analysis when field data is not available. Techniques exist for estimating these physical and chemical properties from other characteristics of compounds (Lyman et al., 1982). In appropriate cases information describing the behavior of a structurally related compound under similar conditions may be used in analyzing the environmental fate of compounds for which this information is lacking (Zamuda, 1989). When used in conjunction with site-specific environmental parameters such as rainfall amounts, soil characteristics, and subsurface hydrology, it may provide a useful description of the complex processes that affect the environmental fate of toxic chemicals under field conditions.

The octanol-water partition coefficient (K_{ow}) is the ratio of a chemical's concentration in octanol (representing lipid, "fat", in general) compared to water, and measures a chemical's affinity for the lipid portions of an organism's tissues. It correlates well with the bioconcentration factor (BCF) for chemicals (which measures their concentration in organisms relative to that of the environment) and the adsorption coefficient (K_{oc}) (a constant that measures its ability to attach to particulate organic matter). Compounds with a K_{ow} of 10^5–10^6 are said to be "lipophilic" and showed the greatest bioconcentration in benthic (bottom dwelling) invertebrates from Lake Ontario sediments (Oliver and Charlton, 1984). These invertebrates are important in the diets of aquatic life such as fish and waterfowl, which have even higher tissue concentrations of some persistent toxicants. Due to concern over human exposure to persistent chemicals through the food chain, K_{ow} is considered a required property to be determined for new or problematic chemicals (Lyman et al., 1982). The accumulation of a chemical within an organism, however, is a time-dependent process that is influenced not only by the chemical's K_{ow}, but also by its assimilation, metabolism, and excretion by a particular species.

A chemical's ability to attach to small particles of organic matter rather than remain dissolved in water is described by the partitioning constant, K_{oc}. It is commonly used in modeling a contaminant's mobility and transport by water and is an important parameter affecting biological processes, such as bioconcentration and biodegradation (breakdown by living organisms). In general, large insoluble molecules have the greatest tendency to exist adsorbed to particulate matter (Sims et al., 1990). For example, the fate of persistent toxic organic chemicals in the Great Lakes is affected by their adsorption-desorption in sediments and by the behavior of those sediments (Allan, 1989). Resuspension of previously buried contaminated sediments by the activity of benthic invertebrates or humans recycles them into the food chain and creates special problems in areas of concern in the Great Lakes. However, adsorption may delay or prevent the degradation of some compounds in soils by making them unavailable to the effects of microorganisms. In addition, adsorbed pollutants in soils may become airborne with particles as a result of natural erosive processes or human activities at a site. However, since these processes are complex and highly variable, the estimation of a long-term average erosion rate is difficult (USEPA, 1988).

The mobility of contaminants in soils or sediments may be affected by a variety of factors in addition to K_{oc}. Retardation delays the migration of contaminants and involves interactions with the soil matrix that are reversible (USEPA, 1988). The movement of polar compounds (with positively and negatively charged ends) or positive ions (e.g., heavy metal cations such as mercury) may be retarded by their attraction to negatively charged particles associated with the soil matrix. Retardation may also occur when insoluble compounds adsorb onto organic material in the soil matrix, as in the case of some polyaromatic hydrocarbons (PAHs). However, toxic organic compounds, such as TCDD (2,3,7,8-tetrachloro-dibenzo-*p*-dioxin) and PAHs (e.g., benzo[a]pyrene), may attach to *dissolved* organic material in soils and be transported with soil water (McCarthy and Zachara, 1989). Contaminants may also desorb (the reverse of adsorb) and become even more mobile when converted to a more soluble form by other chemicals that bind to them and release them from sediments or through biological transformations, especially by microorganisms. In this manner the leaching of persistent organic chemicals from landfills and industrial sites along rivers feeding the lower Great Lakes continues to be a major source of their contamination (Allan, 1989). Desorption of PCBs may occur within Great Lakes sediments liberating them for exposure to aquatic biota (DiToro and Horzempa, 1983) and more rapid exposure to human populations.

A contaminant's water solubility influences its fate and transport in all environmental media and is especially relevant to exposure via aquatic pathways. Soluble compounds in water or soil are more available for chemical and biological transformations and are subject to the complex forces affecting the movement of water. Less soluble metal cations, such as aluminum, may enter solution at lower pH as a result of leaching from soils (as with acid rain) and be available for uptake by plants and aquatic organisms. In the Great Lakes, the annual overturns of water determine the amount of oxygen reaching deep water sediments. This in turn affects the oxidation or reduction of metals at the sediment-water interface into forms which may be more or less soluble (Konasewich, 1979). It is clear that we must gain a better understanding of the complex processes that affect a chemical's solubility in natural ecosystems.

The volatilization of chemicals from soil and water to air may result in a significant reduction in contaminant exposure in soil or water but an increase in atmospheric concentrations that may be of concern. Both perchloroethylene (tetrachloroethylene) and less chlorinated PCBs are volatile and are released from Great Lakes waters; however, the dynamics of their release and possible diffusion back into surface waters is not well understood (Allan, 1989). After volatilization of initial contamination in soil, volatiles may continue to separate over time from "heavy end" petrochemical mixtures (Potter, 1989) and recontaminate sites that have been "cleaned" (Brett et al., 1989). The fate of petrochemicals is complex because they are complex mixtures whose composition may change over time as they are subject to the effects of physical, chemical, and biological processes.

Releases of organic compounds, acids, and heavy metals into the atmosphere from power plants, incinerators, industry, transportation, pesticide spraying, and land and water disposal techniques make a major contribution to the global hazards from all environmental media. Once in the atmosphere, a number of transport phenomena may occur. These involve such processes as wet (during rain or snow) and dry deposition, gaseous diffusion, and horizontal and vertical mixing (Connell and Miller, 1984). Dry deposition of airborne PCBs made a more significant contribution than wet deposition in all of the Great Lakes (Eisenreich et al., 1980), and studies suggest that a number of organic compounds such as PCBs (Hallet,

1989) and benzo[a]pyrene (Thomann, 1989) are input to the Great Lakes primarily from the air. Toxaphene, an organochlorine insecticide, also entered the Great Lakes ecosystem from the air after being transported a thousand miles from the southern United States, its region of primary use. Size and physical form, as aerosol or adsorbed to particulates, affect dispersion and subsequent removal or transformation processes.

Among the least understood factors affecting the fate of environmental toxics are chemical processes that result in changes in their molecular structure. The nature of the intermediate and final products of such processes must be carefully analyzed, because they may be more or less toxic than the original chemical contaminant. A chemical's half-life is an estimate of the time required for half of the original contaminant to be transformed by both chemical and biological processes. The half-life in a particular environmental medium should be included in the exposure assessment for pollutants whose chemical transformations are known or can be reliably estimated (USEPA, 1988). Failure to include consideration of half-life can result in exposure estimates that are orders of magnitude greater than those obtained when reasonable half-life estimates are considered (Paustenbach, 1989). The International Joint Commission on Great Lakes Water Quality (IJC, 1992) utilizes a half-life of greater than 8 weeks in any medium to classify toxic substances as persistent. Estimates of half-life may be experimentally determined based on the overall disappearance of the contaminant as a result of one or more chemical transformation processes. (Biological transformations will be discussed in a later section of this chapter.)

Direct photolysis in the atmosphere or at the water or soil surface may significantly reduce contaminant levels for organic chemicals that decompose after absorbing UV-Vis wavelengths, such as TCDD and some pesticides (Moilanen et al., 1975). It may also occur as an indirect process through the transfer of energy from some secondary substance that initially absorbs the light energy (Conway, 1982). However, due to the lack of exposure to sunlight for the major portion of contaminants in soil and the reduced levels of UV radiation in water, it is of less significance in soil and water than biological and other chemical degradation processes.

Other purely chemical transformation processes, such as hydrolysis and oxidation/reduction reactions, have been studied in atmospheric and aquatic systems, but these may be difficult to distinguish from biological processes when they occur in soils. Hydrolysis involves splitting of larger molecules and is dependent on pH and temperature. Laboratory-derived rate constants for hydrolysis in soil may be used in predictive models, but they may be unreliable unless corrected for local environmental conditions (Bonazountas, 1989). Oxidation reactions occur when chemicals interact with highly reactive forms of oxygen naturally found in all media. In air, secondary atmospheric pollutants, such as photochemical smog, may form as a result of both photolytic and oxidative reactions. These chemical transformations may make a significant contribution to half-life under the dynamic conditions of natural ecosystems and may yield end-products that are more or less toxic than the original compounds.

The bioavailability of a chemical may be defined as the percentage of a chemical in an environmental medium (e.g., water or soil) that is absorbed following exposure via a particular route (e.g., dermal contact or ingestion). It is highly variable, depends greatly on the medium in which it is present, and is influenced by a number of physical, chemical, and biological factors. Environmental conditions such as temperature, salinity, pH, and the presence of and binding to organic material can influence bioavailability and have a significant effect on the accumulation of chemicals in living tissue. The adsorption and desorption of chemical pollutants previously discussed is a key determinant of bioavailability in aquatic sediments and soils. Bioturbation, the physical disturbance of upper sediment layers in lakes, is caused by the activities of benthic invertebrates. It increases bioavailability of insoluble compounds such as some PAHs (e.g., benzo[a]pyrene) and may enhance oxidative microbial processes, such as mercury methylation in the Great Lakes (Nalepa and Landrum, 1988). The resulting organo-mercury compounds are well known to cause birth defects and neurologic diseases. Rapid uptake of soluble toxic organic compounds by zoo- and phytoplankton in the Great Lakes acts as a "biological sponge" to keep these chemicals in the food chain (Evans and McNaught, 1988). In humans, uptake of pollutants from soil particles via the lungs, skin, or gastrointestinal tract may range over an order of magnitude from their pure form at or near 100% to a highly adsorbed form at 10% depending on such factors as the soil type, time following contamination (ageing), presence of co-contaminants, and dose received. Reliable information on absorption efficiency for a compound may be available in the literature and may be applied as an adjustment to the estimated exposed dose if exposure conditions are reasonably similar (USEPA, 1989a).

These factors affecting the physical and chemical behavior of environmental contaminants may act as "buffering" mechanisms in ecosystems to limit the bioavailability of pollutants and delay biological effects until these mechanisms are altered or their capacity overloaded. Therefore, due to the soils/sediments capacity to store/immobilize some toxic chemicals in "chemical sinks," chemical loading may occur long before the effects are observed (as with acid rain and aluminum leaching). Despite reduced input of persistent toxicants (e.g., some pesticides and PCBs) into the Great Lakes, their release from sediments over many years may serve as a continuing source of exposure to humans and other biota. Slow alterations of ecosystems may ultimately lead to an increase in their vulnerability, and the resulting adverse effects may occur over a relatively short time period as a "chemical time bomb" (Stigliani et al., 1991).

BIOLOGICAL INTERACTIONS WITH ENVIRONMENTAL CONTAMINANTS

Since the food chain is an important source of exposure to a large variety of lipophilic organic chemicals (with high K_{ow}) and heavy metals, the evaluation of this exposure pathway should receive considerable attention in both ecological and human health studies. Observed PCB concentrations in Lake Michigan lake trout (*Salvelinus namaycush*) were primarily due to their transfer through the food chain by a number of the processes previously described (Thomann and Connolly, 1984). The BCF is often used to measure accumulation of contaminants in organisms and is considered to be the *ratio of the contaminant concentration in tissues to its concentration in the environment* (e.g., water for aquatic organisms, soil for plants). Water quality criteria to protect human health from contaminated fish and shellfish are developed through use of the BCF along with other toxicological and epidemiological data (USEPA, 1989b). Connell (1990) has summarized the characteristics of organic compounds that tend to bioconcentrate, which include: (1) persistence in the soil in the order of years; (2) a chemical structure with few functional groups and a high proportion of stable bonds (e.g., C–C, C–H, or C–halogen); (3) molecular weight of between 100 and 600; (4) a very low degree of ionization and water solubility; and (5) a log K_{ow} of between 2 and 6. BCFs for many organic chemicals have been determined for a number of aquatic and terrestrial species and range up to six orders of magnitude. This means that a water concentration of 10 parts per trillion (which is below the detection limit of many sampling techniques) could result in aquatic organisms with tissue concentrations in parts per million (which may exceed standards for consumption). A general BCF for PCBs in biota in Lake Michigan has been estimated at 40,000 due to the slow turnover time for water in the lake of 90 years (Peterle, 1991). BCFs can be predicted using correlations between K_{ow}, K_{oc}, water solubility, and known bioconcentrations in terrestrial and aquatic organisms (Kenaga, 1980; Travis and Arms 1988).

The term biomagnification has been used extensively to refer to the presence of higher concentrations of a toxic chemical in the tissues of organisms at higher trophic levels of the food web. Nalepa and Landrum (1988) suggest that this process may account for higher levels of PCBs (than would be expected through direct uptake from the water) in young fish and waterfowl in the Great Lakes. They consume large amounts of benthic invertebrates in their diet after these invertebrates have bioconcentrated PCBs from ingestion of contaminated sediments. In an analysis of the effects of human activity on the mercury budget of a new reservoir, Potter and Standiford (1975) found that mercury concentrations increased up the food chain in attached plants (3000 times), fish lower on the food chain (9000 times) and larger, predatory game fish (31,000 to 43,000 times). Using an "ecosystem" approach, they predicted that power plant development in the area would significantly increase mercury levels in game fish over the 500 ppb FDA standard. The phenomenon of biomagnification is complex and is affected by food and water intake, direct absorption from the environment, and the time since release of the contaminant into the environment (Boudou and Ribeyre 1989). Under some conditions, excretion and metabolic transformation of chemical pollutants may eliminate contaminants at lower trophic levels, resulting in lower rather than higher concentrations in higher trophic levels.

In terrestrial ecosystems, plant uptake from soils of lipophilic organics may result in significant bioconcentration in the roots but their accumulation in aboveground foliage is highly variable. Despite the low solubility of TCDD in water, there may be sufficient annual flow of water to the roots of plants to result in significant contamination (Zapponi and Lupi, 1989). Plants may, however, bioconcentrate these organics in aboveground foliage when they volatilize from the soil (Connell, 1990) or deposit as contaminated soil or dust (Fries, 1987). Chlorinated hydrocarbons and other types of pesticides that are more soluble in water may enter the transpiration stream in plants and be transported to aboveground foliage where they may accumulate (Zapponi and Lupi, 1989).

The bioconcentration of heavy metals as ions from the soil varies with the type of metal, the soil conditions, and the plant species. Of those metals with human health significance, only cadmium is readily adsorbed within the roots and translocated to plant foliage (Alloway, 1990). Foliar absorption of metals from atmospheric fallout and soil dust are probably mostly responsible for foliage concentration observed.

Biodegradation (e.g., metabolism) of organic compounds by aquatic and terrestrial species can be a significant factor in their accumulation and bioconcentration. In vertebrates, the metabolic rate of organic chemicals follows these general relationships: small > large; omnivorous/herbivorous > carnivorous; mammals > birds > fish (Hoffman et al., 1990). Bioconcentration of azaarenes (PAHs associated with synthetic fuel production) in the fathead minnow (*Pimephales promelas*), a species used in standard aquatic toxicity tests, was less than that predicted on the basis of the K_{ow} of the compound due to metabolism by the fish (Southworth et al., 1980). Biotransformation rates for lipophilic organics in invertebrates tend to be slower or nonexistent, presumably due to their lack of appropriate oxidative enzyme systems, but may be higher for microorganisms (e.g., bacteria and fungi) with appropriate enzyme systems under proper environmental conditions. However, there is considerable variation in metabolism even within taxonomic groups, and there are biological (e.g., sex, age, reproductive status) and environmental factors (e.g., temperature) that affect internal processes. For example, laboratory animals have a higher capacity to metabolize fat-soluble xenobiotics (chemicals foreign to living organisms) than some aquatic vertebrates, and thus their use may fail to assess the potential for bioaccumulation in these species (Bridges et al., 1983). Physiological models have been developed that utilize knowledge of biotransformation processes to account for the internal fate of xenobiotics. They include pharmacokinetic models that predict the internal distribution, fate, and exposure to specific target tissues and can be used as predictive tools in estimating bioconcentration in aquatic and terrestrial wildlife. However, these models have not been widely applied in assessing the risk from environmental contaminants (Paustenbach, 1989a; Barron, 1990).

Microorganisms typically play the most significant role in mineralizing organic compounds in the nutrient cycling processes that occur in soil and water. Biodegradation by microorganisms is the most significant process contributing to the half-life for most organic chemicals in soils and sediments (Paustenbach, 1989b) and must be considered in the estimation of ecosystem exposure. Compilations of biodegradation studies for organic compounds in various media exist (Howard, 1989), and there are methods for predicting the vulnerability of compounds to microbial metabolism that utilize a chemical's structure, class, and fundamental properties (Lyman et al., 1982). However, care should be taken in the use of estimated biodegradation rates based on simple laboratory screening tests (e.g., conversion to CO_2) or field studies whose values cannot be extrapolated to actual site conditions (Liu et al., 1981). In addition, there may be a lag time of hours to weeks before biodegradation begins and while microorganisms are adapting to a new "food" source (Lyman et al., 1982). It is probably safe to assume, when dealing with the concentrations typical of soils at hazardous waste sites (HWS), that the degradation rate for a variety of organic contaminants in soil depends on the concentration of the contaminant (Roberts, 1981; Woltering and Bishop, 1989).

Studies have revealed microbial transformation of toxic chemicals to be a highly complex and variable process, yet one of great significance in determining the environmental fate of persistent contaminants such as some chlorinated organics and heavy metals. Microbial populations can resist the toxic effects of contaminants by developing the ability to convert them into less toxic metabolites. Bacterial inoculum, composed partially of Lake Ontario sediments, was shown to degrade a variety of organic compounds by cometabolism (when additional primary nutrients were also added) under aerobic and anaerobic conditions (Liu et al., 1981). A review of microbial degradation of PCBs by Abramowicz (1990) identified how complementary anaerobic and aerobic processes combined to convert even some highly chlorinated congeners to less-chlorinated and, therefore, more degradable and less toxic forms. Hankin and Sawhney (1984) incubated PCB-contaminated pond sediments containing Aroclors® 1248, 1254, and 1260 (commercial mixtures of PCB congeners) with eight agricultural and forest soils. They found that the lower chlorinated isomer (Aroclor® 1248) was more readily degraded (95%) over the 112-day test period than the higher chlorinated form (Aroclor® 1254) (40%) and that there was no degradation of the most chlorinated form (Aroclor® 1260). Although land disposal of PCBs will lead to varying degrees of bacterial degradation, it is possible that some will be taken up by growing plants and translocated to leaves thus entering the food chain. Vertical transport of contaminants out of the microbially rich topsoil down to bedrock does not necessarily indicate that biodegradation can be ignored. Swindoll et al. (1988) tested the degradation of xenobiotics by microbial communities from pristine aquifer solids in order to predict the persistence of organic compounds in contaminated aquifers. Results indicated that, at environmentally relevant levels of substrate, the microbes from this aquifer are able to degrade xenobiotic compounds over a wide range of concentrations.

ECOLOGICAL CONSIDERATIONS FOR BIOREMEDIATION

We have to become more knowledgeable about the rehabilitation of damaged ecosystems and of how to enlist the assistance of biological processes that naturally buffer and limit the effects of toxic chemical releases. Environmental health impacts must be critically evaluated where remediation utilizes release of materials into the environment (e.g., landfilling, incineration, injection wells, release of genetically engineered microbes, dispersants, or fertilizers). The fate and environmental impact of dispersants, nutrients, and microbial transformation products is an area of bioremediation that needs more consideration when performed *in situ* and may even be required on a site-specific basis (Sims et al., 1990). Ecological concerns fall under two major headings: (1) the potential for production and mobilization of chemical compounds with greater toxicity than those of original concern and (2) the disruption of microbial communities and processes leading to adverse delayed ecological impacts.

Transformation products from many organic constituents are not well known and may represent a threat to public health. Their environmental significance, fate, and behavior is uncertain, and synergistic effects might occur when microbial metabolism mobilizes soluble forms of adsorbed heavy metals such as mercury. The presence of co-contaminants (e.g., pesticides, heavy metals) might increase ecological risk if acted upon by microbes being stimulated in an *in situ* bioremediation process or may favor a microbial community resistant to their effects but which has subsequently lost much of their ability to degrade persistent organic compounds (Stigliani et al., 1991). Certain cyclodiene pesticides, PCBs, dioxins, and PAHs may be transformed by aquatic organisms to arene oxide intermediates which may form DNA and RNA adducts with the potential for genotoxic effects (Sims et al.,

1990). In addition, there are significant difficulties inherent in monitoring and analysis of microbial breakdown products and in measuring the amount of residual contamination in soils (Madsen, 1991).

Chemicals may be added to enhance *in situ* bioremediation, but they have the potential to cause harmful effects. Nutrients added to stimulate bacterial growth may lead to algal blooms or toxic levels of ammonia, especially if used in the watershed of small lakes. Some fertilizers could serve as a carbon source for microorganisms and, therefore, interfere with their use of the toxic chemicals as food, thereby delaying degradation. The high moisture requirement for efficient degradation of some toxic organics (e.g., PAHs) in soils may require a large infusion of water, causing leaching of nutrients (e.g., nitrate) or soluble contaminants out of the treatment area. Delivery of nitrogen (and oxygen) in the liquid phase to soils can allow downward migration of chemicals. Estimates of the toxicity of some detergents used as dispersants (for desorption) is based on laboratory studies and may not be accurately extrapolated to field conditions and at concentrations that might be applied in a bioremediation program. Also, the fate and environmental impact of electron acceptors other than oxygen (e.g., nitrate, peroxides) is an area that needs more consideration and may even be required on a site-specific basis. Proper monitoring of both air and leachate from soil remediation efforts is required to achieve an acceptable level of environmental health protection.

Major alterations in microbial ecology may occur by encouragement of toxic waste degrading microbes over the natural microbial fauna. Seeding with mixed cultures of nonnative, waste-acclimated microorganisms may be required to provide the variety of enzymes required for proper degradation of complex mixtures. Although genetically engineered bacteria generally compete poorly with other native microbes *in situ* (Golueke and Diaz, 1989), there is concern over their release and the unknown ecosystem impacts that could occur (such as disruptions of established nutrient cycles). In soils, the subsurface migration of microorganisms off-site may create potential impacts on adjacent areas if not properly controlled. Chapters 7 and 8 will present a more thorough analysis of the application of bioremediation techniques.

MODELING

Assessing the fate and behavior of chemical contaminants in environmental media frequently requires the appropriate selection and application of predictive models. When actual environmental data is unavailable, migration pathways for a variety of organic compounds may be identified and modeled using their fate and transport properties (Mackay et al., 1985; Kountzman and Tucker, 1989). Steady-state models are often used for initial screening, but the assumption that a chemical pollutant reaches an equilibrium among all environmental media can be misleading. Such an assumption is also unnecessary, given the increasing availability of reliable "typical" values for the estimation of transport parameters used by some models (Mackay et al., 1985). Time-variable models exist for predicting the longer-term effects of pollutant control and remediation programs. Thomann (1989) illustrated the application of long-term time variable chemical fate models for benzo[a]pyrene and cadmium in the Great Lakes that indicated the importance of the sediment as a reservoir of these chemicals. A bioconcentration model applied to PCBs in Lake Michigan (Breck and Bartell, 1989) also revealed sediments to be a continuing source of contamination through the food chain for lake trout despite reduced input into overlying waters. Selection of the best assumptions for model parameters should be based on site-specific conditions and the best available parameter estimates in the scientific literature.

Uncertainty about which model to use combined with uncertainty about parameter values can lead to uncertainties in model prediction spanning three orders of magnitude or more (Cox and Ricci, 1989). Sensitivity analysis allows identification of key inputs for which further data collection may be required in order to narrow the uncertainty range in the final exposure estimation (Maxim 1989). Thomann and Connolly (1984) tested the sensitivity of an age-dependent food chain model for PCBs in Lake Michigan lake trout and found that the choice of fish growth rate had a particularly significant effect on the model's output. Models that predict the partitioning behavior of chemicals can assist in identifying which ones persist and/or accumulate in biota and whose likely environmental concentration would be of concern. Chapter 9 provides an in-depth analysis of quantitative modeling in this field.

ECOTOXICOLOGY AND BIOTA

BIOLOGICAL EFFECTS ON INDIVIDUALS AND POPULATIONS

Organisms in any ecosystem usually exhibit a wide range of susceptibility to the effects of toxic chemicals, and such variability is particularly well demonstrated by examples for aquatic ecosystems. Susceptibility can span up to five orders of magnitude within major taxonomic groups of aquatic animals at all trophic levels in the food chain (Blanck, 1984). Toxicity may also vary depending on the organism's developmental stage when exposure occurs and a variety of other physical and chemical parameters. Factors such as water hardness, pH, suspended solids, total organic carbon, and temperature may play a key role (Cairns and Mount, 1990). Dissolved toxics may have a greater effect on bacteria, while microinvertebrates and fish may demonstrate greater responses to toxics adsorbed to particulates (Blum and Speece, 1990). Adsorption contaminants to sediments and solids in aquatic environments can influence removal mechanisms and become a potential source of increased exposure to organisms that live in and ingest sediments/solids. Such variability creates complex problems in hazard assessment and remediation strategy.

A variety of toxicity endpoints may be examined, including lethality, developmental and behavioral changes, and a variety of physiological "markers." Some of the best-documented effects of environmental toxins have been observed as reproductive failure in fish and wildlife species. A number of persistent pesticides and industrial chemicals have the potential to disrupt the endocrine system of animals and result in decreased fertility, birth defects, behavioral changes, and changes in sexual development (Colborn and Clement, 1992). In birds, exposure to organochlorine pesticides, such as DDT and its metabolite DDE, results in physiological changes, such as eggshell thinning, that are associated with lowered reproduction in a number of birds species (Ratcliff, 1970). However, in mammals the evidence for individual reproductive effects (e.g., as with DDT) is weak, and attributable population effects are even less likely (Peterle, 1991). Other physiological and behavioral processes associated with reaching sexual maturity, mating, and nesting and rearing activities may also be affected by environmental toxins and have contributed to reproductive failure as observed in fish-eating birds of the Great Lakes region. A critique of the evidence for reproductive failure in Great Lakes mink (*Mustela vison)* reveals little doubt as to their sensitivity to PCBs in fish consumed and is one of the few verified cases in mammals (other than humans) of population effects of low-level exposure to organic chemicals (Gilbertson, 1988). Their use as sentinels for potential human health effects from PCBs has been suggested (O'Brien et al., 1993). The relationship between observations of physiological damage in a wide variety of vertebrae species and population declines in these species has frequently not been determined. There is a need for more wildlife research on the role of environmental toxins in reproductive failure, its effect on vertebrate population dynamics, and the relevance of some wildlife species as sentinels for human health effects.

ANIMAL SENTINELS

When exposed to environmental levels of contamination, a "sentinel" animal exhibits measureable biological changes that can be used to evaluate the degree of contamination and provide an early warning of its potential threat to human health. The relevance of extrapolating from wildlife sentinels to potential human health effects is to identify potential long-term risks, such as cancer, neurodegenerative disease, and reproductive effects (also discussed in Chapter 6). Sentinels integrate the effects of chronic, real-life exposure to complex mixtures and may develop clinical signs more rapidly due to greater sensitivity and a shorter latency period between exposure and appearance of effects. In addition, it is relatively easy to sample livestock, companion animal, or wildlife populations, and their deliberate placement at a selected sites can be used for more controlled measurement of health hazard. The use of animal sentinels can reduce uncertainties in human risk assessment by providing data in a parallel "weight-of-evidence" approach that integrates a variety of health effects information in order to promote action and effect decisions on environmental management (e.g., toxic use reduction, control, and cleanup).

Selection of an appropriate animal sentinel is based on a number of considerations. It must share the contaminated environment and receive similar exposures to that of the human population being studied. Its position in the food chain can be similar to that of humans (top of chain) or to their food sources, and it must be nonmigratory with a restricted and identifiable home range. Knowledge of baseline parameters of the sentinel's physiology is important, as well as its pharmacokinetics of xenobiotics, and information on normal genotypes and levels of chromosome variation. The ability to propagate the sentinel species may be useful if more controlled studies are required. Selection of more than one species can add some taxonomic breadth to the weight-of-evidence provided to the study. Cross-species comparison of the effects of PCBs on the developing nervous system reveals similarities in neuronal and behavioral impairments between humans and lab animals (Tilson et al., 1990). However, estimated reference doses vary by three orders of magnitude, and the mechanism of effect is not understood. It is possible that behavioral studies of wildlife populations whose exposure is closer to that in humans could provide additional insight into this phenomenon and suggest possible follow-up studies in lab animal models. A fine review of the use of mammals as sentinels has recently been published (O'Brien et al., 1993).

INDICATORS OF ENVIRONMENTAL TOXICITY

Standard methods and endpoints for toxicity testing are readily available for a variety of aquatic organisms, some terrestrial animals, a few plants, and a few microorganisms, microcosms (ecosystems simulated in lab enclosure), and mesocosms (larger field enclosures) (Warren-Hicks et al., 1989). Evaluation of exposure hazard is typically performed using the quotient method, which examines the relationship between the concentration of a chemical in the environmental medium (e.g., water) and the LC_{50} (lethal concentration for 50% of an experimental population). The closer the quotient is to one, the greater the ecological risk involved. This method for assessing risk has numerous weaknesses, including its inability to adequately account for effects of incremental dosages or estimate indirect effects of toxicants (e.g., food chain interactions), its unknown reliability and uncertainty and its inadequate accounting for other ecosystem effects (e.g., predator-prey relationships, community metabolism, structural shifts, etc.) (Bascietto et al., 1990). Multiple bioassays that utilize a variety of species can be applied to gain a better understanding of toxicity at a variety of trophic levels and under field conditions, but they have had limited use in actual environmental toxicity assessments. Extrapolating adverse effects from what was observed in the lab to what might be observed in the field is one of the greatest uncertainties. Acclimation, adaptation, and compensation are typical responses of many aquatic and terrestrial populations to the effects of toxic substances that may not be considered when extrapolating from lab toxicity testing. Emphasis is shifting to include non-lab test species whose responses more accurately represent accumulated effects on the ecosystem being studied. For example, the lake trout *(Salvelinus namaycush)* was selected as the primary indicator for the Great Lakes as a terminal food chain species of significance to human health and because it integrates information over many levels of organization in the ecosystem (Robertson, 1988). In addition, the Great Lakes International Surveillance Plan, which was established for routine monitoring of organochlorine and heavy metal residues in fish, added the herring gull as a useful indicator of overall ecosystem pollution and the effects of complex mixtures of contaminants (NRC, 1991). The Great Lakes Water Quality Initiative requires that states consider the effects of environmental contamination on wildlife when setting water quality standards (IJC, 1992).

Criteria have been established by the EPA (USEPA, 1989c) for use of biomarkers in plants and animals for hazard ID to assess multimedia exposure and provide early warning of environmental health impacts. Potential endpoints for use as biomarkers include DNA damage at the molecular level, metabolites of xenobiotic chemicals at the biochemical level, and cellular damage to structure (e.g., chromosomes) or function (e.g., immune competence) (Shugart, 1993). Its advantages over traditional chemical monitoring of the environment include the ability to assess the hazard of complex mixtures of harmful chemicals, lower cost, and a more direct relationship to damaging effect on tissues than ambient monitoring. Microorganism, plant, or vertebrate embryo biomarker systems may be more sensitive than humans to the interaction of chemicals and ecological factors (Hoffman et al., 1990) and, therefore, reveal early signs of ecosystem damage that may ultimately threaten human health.

As an example of the application of biomarkers in animal sentinels, there is significant potential for the use of non-lab animals (NLAs) (e.g., domestic and companion animals and wildlife) as models for human genetic risk assessment. Such a research approach could provide real-life assessment of exposure to complex mixtures, better dose-response data than short-term *in vitro* tests, and insight into genetic susceptibility of populations or subpopulations. A variety of similar genetic endpoints, including somatic and germ cell mutations have been utilized in both NLA and human health effects studies. Research in the United States has focused on environmental biomonitoring of cytogenetic endpoints in bone marrow and sperm (e.g., using feral rodent species) (Tice et al., 1987; McBee, 1991), while research in Central and Eastern Europe has focused on chromosomal aberrations, SCEs, and morphological sperm abnormalities in various domestic livestock in highly polluted regions of countries, such as the Czech Republic (Rubeš et al., 1991, 1992).

Few studies, however, have been designed specifically to compare these effects in NLAs and humans with the goal of reducing the uncertainty in genetic risk assessment. It is clear that the ability to extrapolate from genetic damage in NLA models to humans will depend greatly on demonstrating that similar mechanisms of damage are operating; however, it appears that little effort has gone into examination of these mechanisms. The specific relevance of animal health for human health remains unclear despite the mounting evidence for the impact of environmental contamination on the health of individuals and populations of wild and domestic animals. A well-designed study examining genetic effects in cohabiting animals and humans would be valuable in clarifying the nature of this relationship. These biomarker techniques are discussed in more detail in Chapters 3 through 6.

ECOSYSTEM EFFECTS OF PERSISTENT TOXICANTS

More ecologically relevant assessment techniques are needed to reduce uncertainties inherent in extrapolating from pure chemical, single-species lab tests to predict significant ecological impacts in complex environmental matrices. Evaluation of a system of parameters that describes the normal operating range of an ecosystem can reveal when variations occur that are greater than those that occur due to normal ecosystem strain — the "stress" ecology of natural ecosystems. Toxic stress to ecosystems may be revealed in microcosm tests by values outside the domain of this normalized state (Kersting, 1984). However, Schindler et al., (1985) found that small-scale lab or microcosm studies would have been inad-

Table 1 Examples of Biological Testing Systems for Ecotoxcity

Aquatic organisms
Phytoplankton: photosynthetic activity; heavy metal uptake
Zooplankton: reproduction and productivity
Fish: sensitive early life stages — embryos; fry tumors and chromosomal abnormalities
Terrestrial animals (livestock, companion animals, wildlife) as sentinels
Molecular, biochemical, and cellular biomarkers
Behavioral studies
Plants
Seed germination
Photosynthesis rates
Mutation rates
Vegetation assessment through remote sensing (e.g., infrared)
Microbial toxicity
Metabolic inhibition of aerobic heterotrophs and methanogens
ATP assays
Enzyme activity inhibitors
Bioluminescence inhibition
Metabolic transformation rates
Multispecies toxicity test systems
Multiple bioassays
Predator-prey and competition tests
Model ecosystems
Microcosms (ecosystems simulated in the lab)
Mesocosms (large field enclosures usually in aquatic ecosystems)
Artificial streams
Field assessments
Structural endpoints
Relative abundance
Species richness and various other biotic indices
Invasion by exotics
Community organization (e.g., diversity indices)
Biomass
Functional endpoints
Cellular metabolism
Population growth rates
Community metabolism measurementsrates of nutrient transfer (e.g., primary productivity, nutrient cycling, and detritus processing)

equate to predict the magnitude of food-web disruption that actually occurred early in the experimental acidification of a small lake. While some stress indicators (e.g., changes in phytoplankton, cessation of fish reproduction, disappearance of benthic crustaceans, and appearance of filamentous algae) were observed, others (e.g., decreases in primary production, rates of decomposition, or changes in nutrient concentration) were not. It is obviously preferable that sensitive measurements of ecosystem-level processes be used to detect the early stages of ecosystem stress. There are crucial ecosystem processes, such as nutrient cycling, that are sensitive to pollutant stress and can be used to measure changes before impacts are observed in more highly visible and regularly monitored populations (Van Voris et al., 1980). However, obvious impacts on visible species may have no measurable impact on ecosystem function (Sheehan, 1984). Ecotoxicologists should attempt to quantify this "stress" by determination of the relationships between concentration and effect for *in situ* indicator organisms and many other species (Hopkin, 1993). Table 1 provides examples of some testing systems for ecotoxicity.

It is crucial to develop methods of measuring the loss or recovery of ecosystem integrity that results from contamination with toxic chemicals and to include social and economic costs in such an "ecosystem approach." The likelihood of impairment of beneficial human use or of an area's ability to support aquatic life have been used to designate Areas of Concern within the Great Lakes ecosystem and to generate remedial action plans under the various Great Lakes Water Quality Agreements (Hartig and Vallentyne, 1989). Criteria for maintaining the health of the Lake Superior ecosystem included a requirement that contaminant levels in lake trout not threaten their populations or their use as a human food source (Aquatic Ecosystem Objectives Committee, 1986). The effort to reduce PCB concentrations in fish to acceptable levels in Lake Michigan has so far failed because it has not taken an ecosystem approach. Such an approach should include control of nonpoint and atmospheric inputs, better understanding of internal cycling of PCBs, and more cooperation among researchers from various disciplines who have traditionally had little interaction (Schelske and Carpenter, 1992).

Traditional approaches to environmental remediation have not used ecosystem criteria in assessing the health effects of chemical contamination. While efforts must continue to improve the reliability of these criteria as indicators of ecosystem damage and the potential for human health effects, there needs to be a greater recognition of their value in determining strategies for and the success of remediation efforts. Harris et al., (1990) utilized stress ecology to assess the health of the Green Bay, Lake Michigan ecosystem. Their hypothesis was that the synergistic effects of nutrient, toxic (PCB), and sediment pollutants are the major stressors degrading this ecosystem. Impairments to the use of fishery resources and public swimming facilities, as well as to reproduction in some wildlife were the most significant effects of this degradation (Hartig and Vallentyne, 1989). Major problems existed at this advanced stage of contamination in tracking the complex movements of contaminated sediments and food chain effects in the search for a remediation strategy. A remedial action plan was developed using an ecosystem approach that involved extensive public participation in and community support of the process.

The Sixth Biennial Report on Great Lakes Water Quality (IJC, 1992) is a particularly relevant example of the implementation of an ecosystem approach to environmental pollution. In adopting a "weight-of-evidence" approach, the IJC has concluded that, despite limitations of scientific study design, the evidence indicates a causal relationship for direct injury and the likelihood of injury to both wildlife and human populations in the Great Lakes region from persistent toxic chemicals.

The repair of contaminated ecosystems and the restoration of their ability to sustain rather than damage human health and welfare requires an ecosystem approach that crosses scientific disciplines. In the case of extensive regions, such as the Great Lakes, an ecosystem approach is enhanced by international cooperation and the development of legal policies to control environmental contamination. Control and remediation of the severe environmental problems of large regions, such as Central and Eastern Europe, would appear to require such a method. (See the legal and international considerations addressed in Chapter 10.) In addition, an ecosystem approach to environmental remediation must include a major educational effort aimed at demonstrating to society the intimate connection between human and environmental health and, hopefully, influencing their underlying values.

REFERENCES

Abramowicz, D. A., Aerobic and anaerobic biodegradation of PCBs: A review, *Critical Rev. Biotech.,* 10(3), 241, 1990.

Allan, R. J., Factors affecting sources and fate of persistent toxic organic chemicals: Examples from the Laurential Great Lakes, in *Aquatic Ecotoxicology: Fundamental Concepts and Methodologies,* Vol. 1, Boudou, A. and Ribeyre, F., Eds., CRC Press, Boca Raton, FL, 219, 1989.

Alloway, B. J., *Heavy Metals in Soils,* Halsted Press, New York, 1990.

Aquatic Ecosystem Objectives Committee. *1985 Annual Report,* Great Lakes Science Advisory Board, International Joint Commission, Windsor, Ontario, 1986.

Barrett, G. W. and Rosenberg, R., *Stress Effects on Natural Ecosystems,* John Wiley & Sons, New York, 1981.

Barron, M. G., Bioconcentration, *Environ. Sci. Technol.,* 24(11), 1612, 1990.

Bascietto, J., Hinckley, D., Plafkin, J., and Slimak, M., Ecotoxicity and ecological risk assessment, *Environ. Sci. Technol.,* 24(1), 10, 1990.

Blanck, H., Species dependent variation among aquatic organism in their sensitivity to chemicals, *Ecol. Bull.,* 36, 107, 1984.

Blum, D. J. W. and Speece, R. E., Determining chemical toxicity to aquatic species, *Environ. Sci. Technol.,* 24(3), 284, 1990.

Bonazountas, M., Mathematical modeling of petroleum products in soil systems, in *Soils Contaminated by Petroleum,* Calabrese, E. J. and Kostecki, P. T., Eds., Lewis Publishers, Chelsea, MI, 1989, 31.

Boudou, A. and Ribeyre, F., Fundamental concepts in aquatic ecotoxicology, in *Aquatic Ecotoxicology: Fundamental Concepts and Methodologies, Vol. 1,* Boudou, A. and Ribeyre, F., Eds., CRC Press, Boca Raton, FL, 1989, 35.

Breck, J. E. and Bartell, S. M., Approaches to modeling the fate and effects of toxicants in pelagic systems, in *Toxic Contaminants and Ecosystem Health: A Great Lakes Focus,* Evans, M. S., Ed., John Wiley & Sons, New York, 1988, 427.

Brett, S. M., Schlesinger, J. S., Turnbull, D., and Machado, R. J., Assessment of the public health risks associated with the proposed excavation of a hazardous waste site, in *The Risk Assessment of Environmental Hazards,* Paustenbach, D. J., Ed., John Wiley & Sons, New York, 1989, 427.

Bridges, J. W., Chasseaud, L. F., Cohen, G. M., and Walker, C. H., Application of pharmacokinetics, in *Animals and Alternatives in Toxicity Testing,* Balls, M., Riddell, R. J., and Worden, A. N., Eds., Academic Press, New York, 1983, 31.

Cairns, J. and Mount, D., Aquatic toxicology, *Environ. Sci. Technol.,* 24(2), 154, 1990.

Colborn, T. and Clement, C., Eds., *Chemically Induced Alterations in Sexual and Functional Development: The Wildlife/Human Connection,* Vol. 21, Princeton Scientific Publishing Co., Inc., 1992.

Connell, D. W., Ed., *Bioaccumulation of Xenobiotic Compounds,* CRC Press, Boca Raton, FL, 1990.

Connell, D. W. and Miller, G. J., *Chemistry and Ecotoxicology of Pollution,* John Wiley & Sons, New York, 1984.

Conway, R. A., Ed. *Environmental Risk Analysis for Chemicals,* Van Nostrand Reinhold Co., New York, 1982.

Cox, L. A. and Ricci, P. F., Risk, uncertainty and causation: Quantifying human health risks, in *The Risk Assessment of Environmental Hazards,* Paustenbach, D. J., Ed., John Wiley & Sons, New York, 1989, 125.

Dawe, C. J., Implications of aquatic animal health for human health, *Environ. Health Perspect.,* 86, 257, 1990.
DiToro, D. M. and Horzempa, L. M., Reversible and resistant component model of hexachlorobiphenyl adsorption-desorption resuspension and dilution, in *Physical Behavior of PCBs in the Great Lakes,* Mackay, D., Paterson, S., Eisenreich, S. J., and Simmons, M. S., Eds., Ann Arbor Science, Ann Arbor, MI, 1983, 89.
Draggan, S., Cohrssen, J. J., and Morrison, R. E., Eds., *Preserving Ecological Systems,* Praeger, New York, 1987.
Eisenreich, S. J., Hollod, G. J., and Johnson, T. C., Accumulation of polychlorinated biphenyls (PCBs) in surficial Lake Superior sediments. Atmospheric deposition, *Environ. Sci. Technol.,* 13(5), 569, 1979.
Evans, M. S. and McNaught, D. C., The effects of toxic substances on zooplankton populations: A Great Lakes perspective, in *Toxic Contaminants and Ecosystem Health: A Great Lakes Focus,* Evans, M. S., Ed., John Wiley & Sons, New York, 53, 1988.
Fries, G. F., Assessment of potential residues in foods derived from animals exposed to TCDD-contaminated soil, *Chemosphere,* 16, 2123, 1987.
Gilbertson, M., Epidemics in birds and mammals caused by chemicals in the Great Lakes, in *Toxic Contaminants and Ecosystem Health: A Great Lakes Focus,* Evans, M. S., Ed., John Wiley & Sons, New York, 1988, 133.
Golueke, C. G. and Diaz, L. F., Biological treatment for hazardous wastes, *BioCycle,* December, 1989.
Hallet, D. J., Ecosystem management of persistent toxic chemicals — Back to basics, in *Toxic Contaminants and Ecosystem Health: A Great Lakes Focus,* Evans, M. S., Ed., John Wiley & Sons, New York, 1988, 491.
Hankin, L. and Sawhney, B. L., Microbial degradation of polychlorinated biphenyls in soil, *Soil Sci.,* 137(6), 401, 1984.
Harris, H. J., Sager, P. E., Regier, H. A., and Francis, G. R., Ecotoxicology and ecosystem integrity: The Great Lakes examined, *Environ. Sci. Technol.,* 24(5), 598, 1990.
Hartig, J. H. and Vallentyne, J. R., Use of an ecosystem approach to restore degraded areas of the Great Lakes, *Ambio,* 18(8), 423, 1989.
Hoffman, D. J., Barnett, A. R., and Russell, J. H., Wildlife toxicology, *Environ. Sci. Technol.,* 24(3), 276, 1990.
Hopkin, S. P., *In situ* biological monitoring of pollution in terrestrial and aquatic ecosystems, in *Handbook of Ecotoxicology,* Calow, P., Ed., Blackwell Scientific, Oxford, 1993.
Howard, P. H., *Handbook of Environmental Fate and Exposure Data for Organic Chemicals,* Lewis Publishers, Chelsea, MI, 1989.
Humphrey, H. E. B., Chemical contamination in the Great Lakes: The human health aspect, in *Toxic Contaminants and Ecosystem Health: A Great Lakes Focus,* Evans, M. S., Ed., John Wiley & Sons, New York, 1988, 153.
International Joint Commission, *Sixth Biennial Report on Great Lakes Water Quality,* IJC, Windsor, Ontario, 1992.
Kenaga, E. E., Correlation of bioconcentration factors of chemicals in aquatic and terrestrial organisms with their physical and chemical properties, *Environ. Sci. Technol.,* 14(5), 553, 1980.
Kersting, K., Normalized ecosystem strain: A system parameter for the analysis of toxic stress in (micro-)ecosystems, *Ecol. Bull.,* 36, 150, 1984.

Konasewich, D., The most difficult problem: Toxins and the lakes, in *The Enduring Great Lakes,* Rousmaniere, J., Ed., W.W. Norton, New York, 1979, 63.

Kountzman, J. A. and Tucker, W. A., Multimedia transport and exposure analysis of contamination from an NPL site, in *Proc. 6th Natl. Conf. Hazardous Wastes and Hazardous Materials,* Hazardous Materials Control Research Institute, Washington, D.C., 1989, 171.

Liu, D., Strachan, W. M. J., Thomson, K., and Kwasniewska, K., Determination of the biodegradability of organic compounds, *Environ. Sci. Technol.,* 15(7), 788, 1981.

Lyman, W. J., Ruhl, W. F., and Rosenblatt, D. H., *Handbook of Chemical Property Estimation Methods,* McGraw-Hill, New York, 1982.

Mackay, D., Paterson, S., Cheung, B., and Neely, W. B., Evaluating the environmental behavior of chemicals with a fugacity level III model, *Chemosphere,* 15, 335, 1985.

Mackay, D., Paterson, S., Eisenreich, S. J., and Simmons, M. S., Eds., *Physical Behavior of PCBs in the Great Lakes,* Ann Arbor Science, Ann Arbor, MI, 1983.

Madsen, E. L., Determining in situ biodegradation, *Environ. Sci. Technol.,* 25(10), 1663, 1991.

Maxim, L. D., Problems associated with use of conservative assumptions in exposure and risk analysis, in *The Risk Assessment of Environmental Hazards,* Paustenbach, D. J., Ed., John Wiley & Sons, New York, 1989, 526.

McBee, K., Chromosomal aberrations in native small mammals (*Peromyscus lucopus)* at a petrochemical waste disposal site. II. Cryptic and inherited aberrations detected by G-band analysis, *Environ. Toxicol. Chem.,* 10, 1321, 1991.

McCally, M. and Cassel, C. K., Medical responsibility and global environmental change, *Ann. Intern. Med.,* 113, 467, 1990.

McCarthy, J. F. and Zachara, J. M., Subsurface transport of contaminants, *Environ. Sci. Technol.,* 23(5), 496, 1989.

Moilanen, K. W., Crosby, D. G., Soderquist, C. J., and Wong, A. S., Dynamic aspects of pesticide decomposition, in *Environmental Dynamics of Pesticides.* Haque, R. and Freed, V. H., Eds., Plenum Press, New York, 1975, 45.

Murchelano, R. A., Fish health and environmental health, *Environ. Health Perspect.,* 86, 257, 1990.

Nalepa, R. F. and Landrum, P. F., Benthic invertebrates and contaminant levels in the Great Lakes: Effects, fates, and role in cycling, in *Toxic Contaminants and Ecosystem Health: A Great Lakes Focus,* Evans, M. S., Ed., John Wiley & Sons, New York, 77, 1988.

National Research Council, *Animals as Sentinels of Environmental Health Hazards,* National Academic Press, Washington, D.C., 1991.

O'Brien, D. J., Kaneene, J. B., and Poppenga, R. H., The use of mammals as sentinels for human exposure to toxic contaminants in the environment, *Environ. Health Perspect.,* 99, 351, 1993.

Oliver, G. and Charlton, M. N., Chlorinated organic contaminants on settling particulates in the Niagara River vicinity of Lake Ontario, *Environ. Sci. Technol.,* 18, 903, 1984.

Paustenbach, D. J., Important recent advances in the practice of health risk assessment: Implications for the 1990's, *Reg. Toxicol. Pharm.* 10, 204, 1989a.

Paustenbach, D. J., *The Risk Assessment of Environmental and Human Health Hazards: A Textbook of Case Studies,* John Wiley & Sons, New York, 1989b.

Peterle, T. J., *Wildlife Toxicology,* Van Nostrand Press, New York, 1991.

Pitelka, L. F. and Raynal, D. J., Forest decline and acidic deposition, *Ecology,* 70(1), 2, 1989.

Potter, L., Kidd, D., and Standiford, D., Mercury levels in Lake Powell: Bioamplification of mercury in man-made desert reservoir, Environ. Sci. Technol., 9(1), 41, 1975.

Potter, J., Analysis of petroleum contaminated soil and water: An overview, in *Soils Contaminated by Petroleum,* Calabrese, E. J. and Kostecki, P. T., Eds., Lewis Publishers, Chelsea, MI, 1989, 97.

Rasmussen, L., Ecotoxicology, *Ecol. Bull.,* 36, 1984.

Ratcliffe, D. A., Changes attributable to pesticides in egg breakage frequency and eggshell thickness in some British birds, *J. Appl. Ecol.,* 71, 67, 1970.

Roberts, J. R., McGarrity, J. T., and Marshall, W. K., *A Screening Level Analysis of Lipophilic Organic Compounds in Aquatic Ecosystems,* NRCC No. 18570, Part II, National Research Council of Canada, 1981.

Robertson, A., Development of ecosystem objectives for the Great Lakes, in *Toxic Contaminants and Ecosystem Health: A Great Lakes Focus,* Evans, M. S., Ed., John Wiley & Sons, New York, 1988, 477.

Rubeš, J., Borkovec, L., Hořínová, Z., Urbanová, J., Proroková, I., and Kulíková, L., Cytogenetic monitoring of farm animals under conditions of environmental pollution, *Mutat. Res.,* 283, 199, 1992.

Rubeš, J., Hořtínová, Z., Gustavsson, I., Borkovec, L., and Urbanová, J., Somatic chromosome mutations and morphological abnormalities in sperms of boars, *Hereditas,* 115, 139, 1991.

Schelske, C. L. and Carpenter, S. R., Lake Michigan, in *Restoration of Aquatic Ecosystems,* National Research Council, Washington, D.C., 380, 1992.

Schindler, D. W., Mills, K. H., Malley, D. F., Findlay, D. L., Shearer, J. A., Davies, I. J., Turner, M. A., Linsey, G. A., and Cruikshank, D. R., Long-term ecosystem stress: The effects of years of experimental acidification on a small lake, *Science,* 228, 1395, 1985.

Sheehan, P. J., Effects on individuals and populations, in *Effects of Pollutants at the Ecosystem Level,* Sheehan, P. J., Miller, D. R., Butler, G. C., and Bourdeau, P., Eds., John Wiley & Sons, New York, 1984, 23.

Shugart, L. R., State of the art — Ecological biomarkers, in *Use of Biomarkers in Assessing Health and Environmental Impacts of Chemical Pollutants,* Travis, C. C., Ed., Plenum Press, New York, 1993, 237.

Sims, J. L., Sims, R. C., and Matthews, J. E., Approach to bioremediation of contaminated soil, *Hazard. Wastes Hazard. Mater.,* 7(2), 117, 1990.

Southworth, G. R., Keffer, C. C., and Beauchamp, J. J., Potential and realized bioconcentration: A comparison of observed and predicted bioconcentration of azaarenes in the fathead minnow (*Pimephales promelas*), *Environ. Sci. Technol.,* 14(12), 1529, 1980.

Stevens, J. B. and Swackhamer, D. L., Environmental pollution: A multimedia approach to modeling human exposure. *Environ. Sci. Technol.,* 23(10), 1180, 1989.

Stigliani, W. M., Doelman, P., Salomons, W., Schulin, R., Smidt, G. R. B., and Van der Zee, S., Chemical time bombs, *Environment* 33(4), 4, 1991.

Swindoll, C. M., Aelion, C. M., Dobbins, O. J., Long, C., and Pfaender, F. K., Aerobic biodegradation of natural and xenobiotic organic compounds by subsurface microbial communities, *Env. Toxicol. Chem.,* 7, 291, 1988.

Thomann, R. V., Deterministic and statistical models of chemical fate in aquatic systems, in *Ecotoxicology: Problems and Approaches,* Levin, S. A., Harwell, M. A., Kelly, J. R., and Kimball, K. D., Eds., Springer-Verlag, New York, 1989, 245.

Thomann, R. V. and Connolly, J. P., Model of PCB in the Lake Michigan lake trout food chain, *Environ. Sci. Technol.,* 18(2), 65, 1984.

Tice, R. R., Ormiston, B. G., Boucher, R., Luke, C. A., and Paquette, D. E., Environmental biomonitoring with feral rodent species, in *Short-term Assays in the Analysis of Complex Mixtures,* Sandu, S. S., DeMarini, D. M., Mass, M. J., Moore, M. M., and Mumford, J. L., Eds., Plenum Press, New York, 1987, 175.

Tilson, H. A., Jacobson, J. L., and Rogan, W. J., Polychlorinated biphenyls and the developing nervous system: Cross-species comparisons, *Neurotoxicol. Teratol.,* 12, 239, 1990.

Travis, C. C. and Arms, A. D., Bioconcentration of organics in beef, milk, and vegetation, *Environ. Sci. Technol.,* 22(3), 271, 1988.

U.S. Environmental Protection Agency, *Superfund Exposure Assessment Manual,* EPA/540/1-88/001, Office of Remedial Response, Washington, D.C., 1988.

U.S. Environmental Protection Agency, *Exposure Factors Handbook,* EPA/600/8-89/043, Exposure Assessment Group, Office of Health and Environmental Assessment, Washington, D.C., 1989a.

U.S. Environmental Protection Agency, *Assessing Human Health Risks from Chemically Contaminated Fish and Shellfish: A Guidance Manual,* EPA/503/8-89-002, Office of Marine and Estuarine Protection, Washington, D.C., 1989b.

U.S. Environmental Protection Agency, *Risk Assessment Guidance for Superfund, Vol. 1, Human Health Evaluation Manual,* EPA/540/1-89/002, Office of Emergency and Remedial Response, Washington, D.C., 1989c.

Van Voris, P., O'Neill, R. V., Emanuel, W. R., and Shugart, H. H., Jr., Functional complexity and ecosystem stability, *Ecology,* 61(6), 1352, 1988.

Warren-Hicks, W., Parkhurst, B. R., and Baker, S. S., Jr., Eds., *Ecological Assessment of Hazardous Waste Sites: A Field and Laboratory Reference,* EPA/600/3-89/013, U.S. Environmental Protection Agency, Environmental Research Laboratory, Corvallis, OR, 1989.

Woltering, D. M. and Bishop, W. E., Evaluating the environmental safety of detergent chemicals: A case history of cationic surfactants, in *The Risk Assessment of Environmental Hazards,* Paustenbach, D. J., Ed., John Wiley & Sons, New York, 1989, 345.

Woodwell, G. M., On toxins and toxic effects: Guarding life in a small world, in *Preserving Ecological Systems,* Draggan, S., Cohrssen, J. J., and Morrison, R. E., Eds., Praeger, New York, 1987, 41.

Zamuda, C. J., Superfund risk assessments: The process and past experience at uncontrolled hazardous waste sites, in *The Risk Assessment of Environmental Hazards,* Paustenbach, D. J., Ed., John Wiley & Sons, New York, 1989, 266.

Zapponi, L. and Lupi, C., Environmental and health impact assessment of soil pollutants, in *Risk Management of Chemicals in the Environment,* Vol. 12, Seip, P. and Heiberg, S., Eds., NATO, Brussels, 1989, 89.

CHAPTER 3

Integration of Complex Mixture Toxicity and Microbiological Analyses for Environmental Remediation Research

Larry D. Claxton, Virginia S. Houk, and S. Elizabeth George

CONTENTS

1-56670-141-4/96/$0.00+$.50

UNDERSTANDING THE PROBLEMS

Accidental spills and intentional disposal of toxic waste have contaminated thousands of sites throughout the United States (see Chapter 10). Because the hazardous substances escaping from these sites into the air, soil, and water can pose serious human health risk to exposed populations, the public has called upon government and industry to regulate, properly administer, and clean up the nation's worst hazardous waste sites (see Chapters 1 and 10). Many of these sites present as mixtures of multiple toxicants with other pollutants (Waters et al., 1983). Synergism,* therefore, becomes a major concern. The complexity of these environmental situations also complicates the monitoring, evaluation, and remediation of hazardous waste sites (HWS) (see Chapters 5, 7, and 8). Most types of remediation (including incineration, physical treatment, and bioremediation) produce additional components that add to the complexity of evaluating the potential health effects of a contaminated site before, during, and after remediation efforts. Many of the substances emitted into the environment and produced within the environment from anthropogenic materials are unidentified and/or have unknown toxicity. These nonconventional pollutants (NCPs) pose special problems to those who must evaluate human health risks. According to the Committee on Environmental Epidemiology of the National Research Council (1991), "There is evidence that NCPs are a potentially important source of hazardous exposure. Some preliminary toxicologic studies suggest that NCPs have important biologic properties, environmental persistence, and mobility. ... In the broadest sense, these unidentified substances represent risk of unknown magnitude."

This chapter will discuss the challenges that face public health researchers as they consider the use of bioremediation as an alternative approach for waste treatment and cleanup efforts. Although we will mention the present methods used

*Synergism is the combined effect of two or more agents being greater than the sum of the effects of the individual agents.

for health evaluations, we will stress recent research efforts aimed at improving health effects evaluations. The first two sections will provide an understanding of the complexity of HWS and how these complexities are sometimes multiplied by remediation efforts. The next three sections will illustrate the research strategies that can be employed in order to better understand health hazards before, during, and after remediation efforts. The chapter will conclude by suggesting how these research efforts could have an impact upon our understanding of risks associated with hazardous wastes and their remediation. In order to reduce the complexity of the discussion, the review will be confined to the area of genetic toxicology (carcinogenesis and mutagenesis). Complexity also will be reduced by limiting the dialogue to HWS efforts. The reader, however, can apply the principles to related topics such as health assessments for pollution prevention efforts and industrial effluents.

EFFECT OF THE DYNAMIC COMPLEXITY OF HAZARDOUS WASTE SITES ON UNDERSTANDING POTENTIAL TOXIC EXPOSURES

ENVIRONMENTAL TRANSFORMATION OF POLLUTANTS

Even a single pollutant placed in an environmental situation can create a complex situation for the public health researcher who must evaluate potential health effects. Figure 1 illustrates this statement using a common environmental pollutant, pentachlorophenol (PCP). A general-use biocide, PCP is used as a wood preservative, a fungicide, and an insecticide. Depending upon the organisms involved and the conditions of degradation, PCP can be metabolized to one of several products. Ideally, PCP degradation yields CO_2 and methane. Mikesell and Boyd (1986) showed that under certain conditions, the anaerobic microorganisms found in municipal sewage sludge converted 66% of the introduced PCP to CO_2 and methane. They later showed that this consortium of bacteria mixed in soil converted PCP to TeCP,* TCP,** and DCP*** (Mikesell and Boyd, 1988). In sterile soil, the main product was TCP. However, when the sludge microorganisms were added to soil containing PCP and other microorganisms, the main product was TeCP. In contrast, one of the main PCP degradation products of white rot fungus is TCHD† (Mileski et al., 1988). Steiert and Crawford (1986) found that another bacterium produced chlorinated hydroquinones upon the metabolism of PCP. From these and other studies (Schenk et al., 1990; Lamar and Dietrich, 1990), it is quite apparent that complex mixtures of products are produced during the biodegradation of even single compounds. The products are dependent upon both the organisms present and the environmental circumstances.

Because these intermediate products of PCP produced by microorganisms might accumulate in the environment, the public health researcher must be concerned about the intermediates and by-products as well as the originally identified

---

* TeCP is 2,3,4,5-tetrachlorophenol.
** TCP is 3,4,5-trichlorophenol.
*** DCP is 3,5-dichlorophenol
† TCHD is 2,3,5,6-tetrachloro-2,5-cyclohexanediene-1,4-dione.

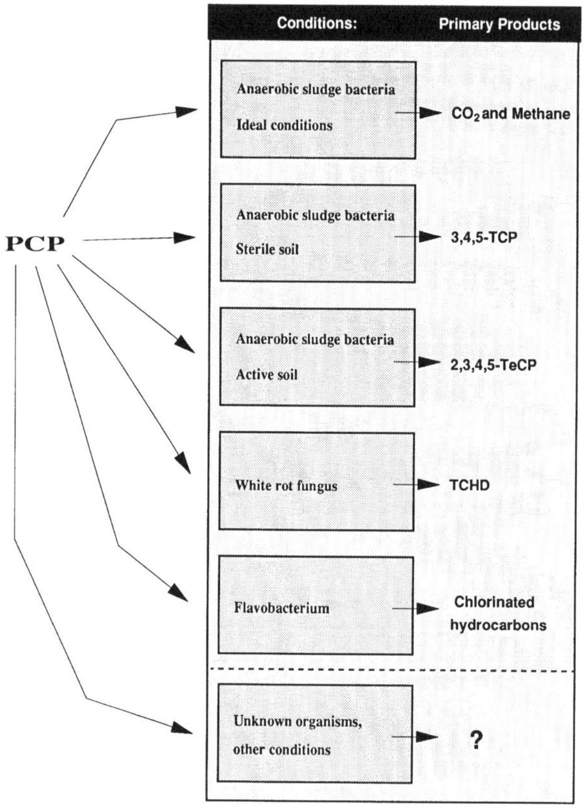

Figure 1 The complexity arising from microbial metabolism of a single pollutant (PCP).

toxic chemical. Table 1 illustrates what is known about the genetic toxicology of PCP and its microbial metabolites. Outside of the commercial product PCP, little is known about the other compounds. Judging from this table, one cannot be sure if microbial degradation is a reasonable approach. Many questions remain. Relative to PCP, how toxic is each of these metabolites? Are they more or less likely than PCP to contaminate products to which humans are apt to be exposed? Are there other toxic metabolites that are likely to accumulate in the environment because of other types of transformations?

When complex substances and/or mixtures of substances are present, the situation is even more complicated. Figure 2 illustrates that multiple pollutant exposures provide very dynamic situations. Because compounds differ in their volatility, water solubility, ability to diffuse within soils, and ability to be metabolized by indigenous organisms, the concentrations of individual compounds varies by both site and time. In this illustration, two of the pollutants, creosote and crude oil, actually consist of hundreds of organic chemicals (Nylund et al., 1992; Blackburn et al., 1986). The lower molecular weight compounds in crude oil and creosote will volatilize into the air. The compounds within crude oil with the highest molecular weights will attach to particles and will sink to the bottom of water masses. The other organic compounds in the illustration will disperse through soil and ground

Table 1 Pentachlorophenol and Selected Metabolites

Compound	Genetic toxicology
PCP	Liver carcinomas/adenomas in mice[5]
	Negative for sex-linked recessive lethals in *Drosophila*[6]
	Non-mutagenic in *hprt* V79 cells[4]
	Chromosomal aberrations in CHO cells[1]
	Non-mutagenic in *Salmonella*[3,7]
	Prophage induction in *Escherichia coli*[1]
2,3,4,5-TeCP	Prophage induction in *E. coli*[1]
	Non-mutagenic in *Salmonella*[3,7]
3,4,5-TCP	No chromosomal aberrations in CHO cells[2]
	Non-mutagenic in *Salmonella*[3,7]
	Prophage induction in *E. coli*[1]
3,5-DCP	Questionable mutagenicity in *Salmonella*[3]
	Weak positive for prophage induction in *E. coli* [1]
TCHD	?

[1]DeMarini et al., 1990; [2]Galloway et al., 1987; [3]Haworth et al., 1983; [4]Jansson and Jansson, 1986; [5]National Toxicology Program, 1989; [6]Vogel and Chandler, 1974; [7]Zeiger et al., 1988.

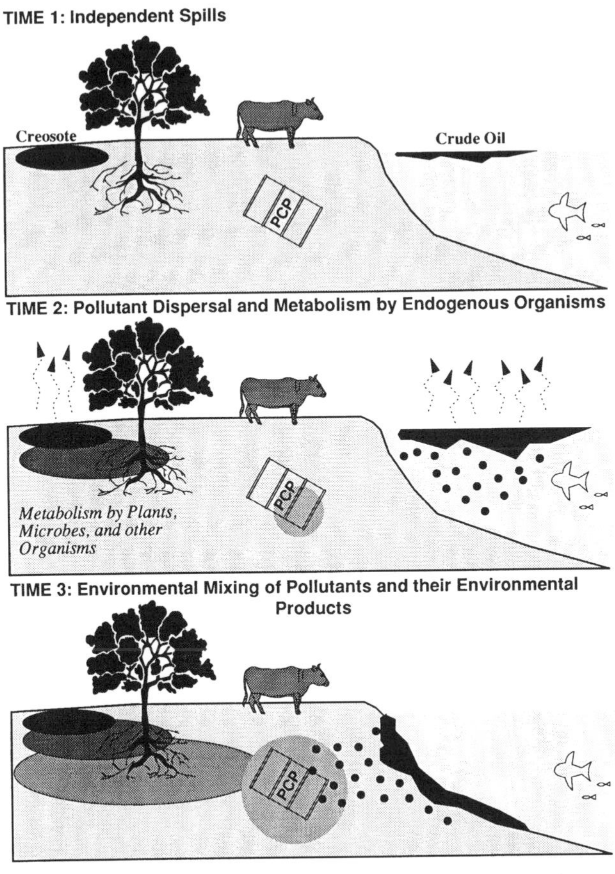

Figure 2 The dynamic complexity of a environmental hazardous contamination site.

Table 2 Genotoxic Components Found in Coastal Sediments near Barcelona, Spain

Chemical class	Example compounds	Possible sources	Contribution to mutagenicity
Organohalogenated compounds (27 identified)	Pentachlorobenzene *Dioxins* Chlordane DDTs Dieldrin	Pesticides Thermal degradation of dieldrin	Minor
Polycyclic aromatic hydrocarbons and sulfur heterocycles (45 identified)	Phenanthrene Benz[a]anthracene Chrysene Perylene	Fossil fuels and oils	Considerable
Oxygenated polycyclic aromatic compounds (23 identified)	Aromatic ketones and carboxyaldehydes Benzophenone Dibenzofluorenone Coumarines	Oxidation during combustion of fossil fuels and wood Photooxidation of soot particle organics Degradation products of PAHs in sunlight	Limited
Nitroarenes (11 identified)	1-Nitropyrene 6-Nitrochrysene Nitrobenzopyrenes	Atmospheric transport of vehicular emissions Street runoff Urban particulate matter	Moderate
Other nitrogen compounds (13 identified)	Azobenzene Acridine Benzacridine Carbazole	Industrial synthetic intermediates Degradation products of dyestuff components	Moderate
Other compounds (19 identified)	Aromatic carboxylic acids Anhydrides Esters Phenols	Dichloromethyl benzoate Nitrophenol Benzenecarboxylic acid	Minor

Extracted summary of paper by Fernández et al., 1992. With permission.

water systems (see Chapters 2, 7, 8, and 9 for other discussions relating to environmental dispersion and interactions). Because the compounds disperse at differing rates, the relative mixture of compounds differ from location to location. In addition, environmental transformation* of the compounds is occurring due to the interaction of compounds and to the metabolism of some compounds by microbial, plant, and animal systems. It is this highly complex mixture of substances, therefore, that contaminates water and food supplies and ambient air. As with the previous PCP illustration, these processes are transforming the substances present in the environment and are therefore altering the overall toxicity of the contaminated areas. These processes may either increase or decrease toxicity, and the effect may range from dramatic to undetectable.

Fernández et al. (1992) examined the components within coastal sediments near Barcelona, Spain, for genotoxic components. Table 2 summarizes their findings. They found the sediment samples to be genotoxic. The genotoxicity was due

*Transformation is the physical or biological alteration of a compound to another compound or compounds.

to a wide variety of compounds. Interestingly, they found that the sediments collected in the immediate vicinity of the "main pollutant sources" exhibited lower mutagenic activity than the coastal sediments. Based on the chemical composition of the components responsible for the mutagenicity, they attribute the genotoxicity to pyrolytic sources whose products are transported to the sediments by either street runoff or atmospheric deposition. In addition, one should consider other types of environmental transformation, for example, degradation of pesticides, marine fuels and oils, and photooxidation of other organic sources.

Normally, all types of remediation efforts will influence these environmental transformation processes. Landfilling usually will slow these processes because the pollutants are contained, organism exposure is minimized, and nutrients for microorganisms are not supplied. Chemical treatments are designed to alter specific pollutants but also may alter other pollutants and naturally occurring substances. Chemical treatments also may affect indigenous organisms, especially microorganisms. Incineration generally emits particles, volatile organic materials, and various gases into the atmosphere. Bioremediation enhances the rate at which the pollutant is metabolized. All types of remediation, therefore, can alter the rate and type of pollutant dispersal, the toxicity of the environmental contaminants, and thus the hazards to which people are exposed.

The purpose of all remediation efforts, including bioremediation, is to both decrease toxicity and to limit exposure to remaining toxic components. Because so many compounds exist in most situations, there is always some uncertainty concerning the amount and toxicity of the compounds to which human populations are potentially exposed. Analytical chemistry methods are used to detect and quantitate only a limited number of known toxic chemicals. In most situations, it is not feasible, due to cost and available technology, to identify and quantitate all polluting substances. Even if one were able to identify all compounds present in an environmental situation, toxicological information would not be available for most. For example, it is not yet possible to identify all of the organic compounds that exist within crude oils. For those compounds that have been identified, most have either no associated toxicological information or information for only a few toxicological endpoints (carcinogenicity, hepatotoxicity, neurotoxicity, immunotoxicity, etc.). Public health researchers, therefore, are developing methods and strategies for evaluating these complex environmental situations. These newer methods must be more exhaustive in the numbers and types of toxic substances detected, they must be technically feasible, and they must be affordable.

COMPLEXITIES INTRODUCED BY REMEDIATION MICROORGANISMS

Most bioremediation efforts encompass the use of microorganisms. This use of microorganisms introduces another level of complexity into the evaluation of potential health risks. As mentioned previously, microbes can metabolize toxicants and other pollutants thus changing the toxicity profile of an environmental mixture. Not only do microorganisms have the capacity to persist within the environment, but they may actually increase in number and be spread by a variety of mechanisms. Although pathogens are very unlikely to be used in bioremediation efforts, microorganisms can have adverse effects for humans by other mechanisms. Therefore, the complexities introduced by the use of microorganisms are being addressed by public health scientists.

Environmental Transformation of Pollutants and Natural Products

Bacteria and fungi are especially capable of producing a wide variety of enzymes that can degrade organic compounds and mineralize other substances. Saprophytic microorganisms are those that are capable of feeding on "dead organic matter." This would include organic pollutants. Molholt (1989) points out that pseudomonades* have been used in the biodegradation of a number of synthetic organic compounds that are environmental toxicants. Some of the compounds Molholt mentions are 1-octene, benzene, toluene, trichlorobenzene, pentachlorophenol, PCBs, diazinon, 2,4,5-T, and dieldrin. Microorganisms also have been used to degrade complex mixtures such as creosote (Mueller et al., 1991) and crude oils (Leahy and Colwell, 1990). As discussed in the previous section with PCP, intermediate metabolites from these substances also may be toxic. If the degradation process does not take the contaminants to intermediary metabolites or immobilize the substance, the toxicological evaluation becomes even more complex.

HWS are identified because of the presence of one or more toxic substances. However, other nontoxic pollutants are generally present at these same sites. A question that has not yet been addressed discernably is the extent to which remediation efforts, including microbial methods, might convert some of these nontoxic substances into toxic components. A second consideration is that microbes designed for *in situ* remediation are chosen because they are expected to persist within the environment. Unlike chemicals within the environment, some microorganisms may actually survive and thrive. Other bacteria can be maintained in a small, nondetectable niche until the environment is more suitable for growth. From this it can be concluded that (1) the environmental bacterial concentrations are in a constant state of flux, (2) the concentration of bacteria to which individuals would be exposed is therefore variable, (3) changing numbers of bacteria influence the rate by which all pollutants are metabolized, and (4) because of the rate of bacterial growth and metabolism of pollutants, the type and levels of toxicants are in a dynamic state.

Health Concerns for the Remediating Microorganisms

The greatest concern of the public is not the ability of microbes to metabolize environmental compounds. The concern, instead, is whether or not the organisms themselves can cause health effects. Relatively few bacterial species, however, are recognized pathogens. Providing precise numbers of existing species and pathogenic species is difficult. According to Brenner (1991), the 1966 edition of *Index Bergeyana* included some 29,000 bacterial names. Starting on January 1, 1980, however, approved names of bacterial species must be published within the Ap-

*Pseudomonades are a grouping of bacteria characterized as Gram-negative aerobic motile rods with polar flagella that contain an oxidase enzyme. The free-living forms are found in soil, freshwater, and marine environments, and they are important decomposers of organic matter; as a human pathogen, they infect the urinary tract, wounds, and burns.

proved Lists of Bacterial Names in the *International Journal of Systematic Bacteriology*. By 1989, only approximately 2700 species were accepted into this approved list. This number has grown at a rate of about 5% per year. The most comprehensive work for identifying both pathogens and nonpathogens is the four-volume *Bergey's Manual of Systematic Bacteriology* (Holt and Krieg, 1984; Sneath, 1986; Staley, 1986; Williams, 1989). Creager et al., (1990) state that 16 of the 33 sections of *Bergey's Manual* are medically important sections. Most of the species within these 16 sections are not recognized pathogens. In addition, a large number of microbes are found in a symbiotic* association with humans. Creager et al. (1990) list 31 bacterial groups of organisms associated with skin, upper respiratory tract, mouth, intestine, and urogenital tract that are considered normal indigenous flora for humans. Some pathogens (e.g., *Staphylococcus aureus*) are frequently among indigenous flora, but normally they do not obtain the competitive advantage needed to cause disease. Other organisms are called "opportunistic pathogens" because they must take advantage of very special circumstances before they can cause disease. Still other bacterial genera (e.g., *Lactobacillus*) are most always found in relatively large numbers and rarely, if ever, cause disease.

In spite of these considerations and in spite of the fact that pathogens are avoided for use in bioremediation, there are some human health reasons to be cautious about the environmental use of microorganisms. Isenberg and D'Amato (1991) state, "No hard and fast rule divides the human microbiota into clear-cut categories of harmless, commensal organisms and pathogenic species." In other words, a microorganism not presently recognized as a pathogen or opportunistic pathogen may behave as a pathogen. For example, *Legionella pneumophila* is a natural inhabitant of both terrestrial and aquatic habitats. This soil and water organism will grow in damp locations (such as air conditioner systems) and become aerosolized in high numbers. Upon exposure, a person can develop fever, headache, diarrhea, and chest pains, and may die from respiratory or kidney failure (Creager et al., 1990). *L. pneumophila* was not recognized as a pathogen until after a lethal outbreak of atypical pneumonia occurred in Philadelphia in 1976. It was difficult to identify because it is typically a soil microbe with complex nutritional requirements. Occurrences such as these have warned researchers to be cautious when people are exposed to high concentrations of any microorganism or any material that might contain high concentrations of microorganisms.

Direct pathogenicity is not the only mechanism by which microorganisms inhabiting a niche within the body can have an adverse effect. If a nonpathogenic microbe or other factors disturb the balance established among normal body flora, pathogens may have an opportunity to invade, colonize, multiply, and have an infectious effect upon the host (George et al., 1990; Deitch and Berg, 1987). Moreover, when the normal flora of an individual are altered, the metabolism of pollutants within that individual may be altered. Studies using the gastrointestinal tract microflora of rodents provide important examples of these two phenomena.

Microorganisms designed for environmental use can colonize the gastrointestinal tract and alter the distribution of normal intestinal microflora. In several related studies, George et al. (1989, 1991) challenged the intestinal flora of male mice with PCB-degrading bacteria (mutagen-induced strains of *Pseudomonas*

*Symbiotic refers to the state of two dissimilar organisms living together in a mutually beneficial manner.

aeruginosa). In a 14-day study in which each mouse was challenged with a relatively high number of bacteria (1×10^9), the pseudomonad could be recovered after 14 days. Although the balance of the normal flora was disturbed in some of these studies, these studies did not show any overt effects to the mice themselves. Presently, studies are under way to determine whether or not these types of alterations will allow pathogens to more easily colonize the intestinal tract.

The potentially harmful chemicals to which people are exposed gain entrance to the gastrointestinal tract via ingested food, water, and inspired air. Perhaps as many as 90% of the genotoxicants require some type of metabolic transformation before interacting with their ultimate target (e.g., DNA) and initiating a disease process (e.g., cancer) (Chadwick et al., 1992). The review by Chadwick et al. demonstrates that intestinal microflora can play a highly significant role in the bioactivation of genotoxicants. It stands to reason that alteration of this microflora population will affect the metabolism of xenobiotics. Once again, however, how the interaction of large numbers of normal microflora, challenging microbes (e.g., bioremediation treatment organisms), and xenobiotics will affect health is not easily predicted.

RESEARCH STRATEGIES FOR ORGANIC POLLUTANTS

Because of the complexity described above, comprehensive and reliable environmental health assessments are extremely difficult to accomplish. When one considers cancer and heritable effects, the dimension of time also plays a major role. The time between exposure to the carcinogen that initiates the development of a cancerous cell and the diagnosis of the actual cancer may be as great as 20 years (Macieira-Coelho, 1986). The time between exposure to a mutagen and the recognition of adverse heritable effects is measured in generations. Although the health researcher is faced with these challenges, a variety of research tools exist for developing an understanding of the hazards and risks emanating from a hazardous compound, source, or site. This section describes some of the tools and approaches that health researchers can take to identify and evaluate environmental hazards. The reader should also refer to Chapters 4, 5, and 6 for related discussions.

RELEVANCE OF EPIDEMIOLOGICAL, CHEMICAL, AND BIOASSAY APPROACHES

One of the challenges confronting the environmental toxicologist is how to estimate the human health effects that could occur in the future as a result of past, present, and continued disposal of hazardous chemicals. This is especially evident in the field of genetic toxicology, where the time between exposure and outcome can be great. In very broad terms, today's health researcher must depend upon the methods associated with epidemiology, analytical chemistry, toxicology testing, and biological analyses.

The most convincing evidence for hazard assessments comes from humans. The study of human populations to document the incidence and to determine the factors that control the presence or absence of disease is epidemiology. The Committee on Environmental Epidemiology of the National Research Council defines *environmental epidemiology* as "the study of the effect on human health of physical, biologic, and chemical factors in the external environment, broadly conceived. By examining specific populations or communities exposed to different ambient environments, it seeks to clarify the relationship between physical, biologic, or chemical factors and human health." In other words, the epidemiologist compares the frequency of disease among exposed populations with that among nonexposed or less exposed populations in order to examine the cause (not mechanism) of disease. The greatest challenge to the environmental epidemiologist is the identification, understanding, and avoidance of confounding factors. A person diagnosed with a kidney tumor may have been exposed in the past to personal (e.g., cigarette smoke), occupational (e.g., coke oven fumes), and environmental (e.g., automotive emissions) pollutants that have genotoxic activity; therefore, one cannot distinguish which source initiated the observed disease. Medical examinations (e.g., histology) cannot distinguish between lung carcinoma caused by cigarette smoking and lung carcinoma caused by coke oven fumes. Moreover, evidence for an environmental exposure (e.g., to vinyl chloride) may not exist 5 to 10 years later when the evidence for a cancer becomes apparent. Perhaps the greatest drawback to cancer epidemiology is the fact that people must have been exposed to a potentially toxic agent and enough time must have elapsed for an adverse effect to have occurred. Epidemiology, therefore, is not a suitable approach for a newly manufactured product or pollutant. See Table 3 for a summary of some of the advantages and disadvantages of environmental cancer epidemiology efforts. In spite of these limitations, epidemiologists assimilated the evidence to identify asbestos, vinyl chloride, and cigarette smoking as cancer-causing agents (Sagan, 1984).

For many environmental pollutants, multiple exposures, long-term exposures to low concentrations of pollutants, long time intervals, and lack of unique disease endpoints for pollutants make epidemiology an unsuitable approach. This is especially true for newly recognized or previously unstudied pollutants. Risk assessments, therefore, often employ measurements of pollutant levels to which humans are exposed merged with toxicological information coming from nonhuman test systems. Traditionally, toxicity testing to predict effects on humans examined one chemical at a time. Within the biological test system, the relationship between exposure and outcome can usually be established in precise terms. However, the predictive power of toxicity-test data is limited by our understanding of interspecies differences, extrapolations from the high test doses to the low environmental exposure doses, and other considerations listed in Table 3.

Another approach is to use biological systems to test the toxicity of actual environmental samples. By examining a complex mixture with a biological assay, one diminishes (1) the concerns over the interactions of multiple compounds, (2) the time between exposure and effect, and (3) the need to identify and test the toxicity of all the pollutants within the environmental sample. As pointed out in Table 3, however, there can be some critical disadvantages. Because of the cost of collecting and preparing environmental samples for bioassay, *in vitro* assays are more commonly used than whole animal bioassays. This, in turn, makes extrapolation to the human situation less certain. Because the actual toxicant(s) causing the observed effect in the bioassay may not be identified, the toxicity of some samples may be missed because the appropriate bioassay is not used. By not identifying actual toxicants, it may be more difficult to understand the results of bioassays when they are used over time to track toxic pollutants. Although the biological evaluation of complex mixtures (compared to single compounds) usually varies

Table 3 Some Example Advantages and Disadvantages for Environmental Health Assessment Monitoring for Carcinogenic Effects and Genotoxicants

Approach	Advantages	Disadvantages
Epidemiology	Direct relevance. Actual exposure levels involved. Occupational to environmental extrapolations possible. No interspecies extrapolations. Chemical and biological interactions automatically incorporated.	Applicable only after human effects have occurred. Confounding sources of exposures to environmental substances. Long time periods between exposure and effect. Most exposures at very low levels; therefore, large populations needed. Sensitive populations (e.g., children) may not be considered. Unmonitored substances may contribute to the effects seen.
Toxicity testing of environmental compounds plus analytical chemistry monitoring	It is possible to have: Accurate determination of environmental toxicant levels; Small, highly controlled populations for toxicity testing; and Studies concerning the mechanisms of disease. Can be predictive of effects (prior to exposure).	Interspecies extrapolation needed. High dose to low dose extrapolations usually needed. May not know what substances to monitor. Interactions (multiple exposures) often inadequately studied. Extrapolation to environmental situation usually needed. More relevant tests are generally the more costly and longer in duration. Must have quantitative, analytical chemistry methods for pollutants.
Biological monitoring and assessment of environmental samples	Can be predictive of effects (prior to exposure). Can monitor mixtures without knowing all components. Can be used to identify unknown hazardous pollutants. May be less expensive than first two approaches. May be used to monitor the the toxicity of dynamic and complex situations (changing environmental situations).	Interspecies extrapolations needed. High dose to low dose extrapolations usually needed. *In vitro* assays usually required or heavily relied upon; thus, toxicity endpoints may be limited. Actual toxicant may not be identified.
Integrated approaches	Can expand on advantages of the other approaches. Can monitor for both known and unknown toxicants. Can be used to identify new important pollutants. Can assess interactions. Can be used to identify sources of pollutants.	Increased cost. Increased levels of expertise and scientific capabilities needed. Increased complexity in data analysis.

only in extent and not in kind of testing, the experimental design, sample collection and preparation, and interpretation of results are usually more complex. Biological assays have, however, played an important role in identifying complex mixtures that are potential human toxicants (Claxton and Huisingh, 1979; Claxton and Kohan, 1981; Claxton, 1981; Graham et al., 1984; Houk and Claxton, 1986; Driver et al., 1990; Houk, 1992, 1994).

If the reader is starting to despair over the ability to evaluate the relationship between environmental agents and human disease, this review has too strongly emphasized the difficulties encountered when using the available techniques and approaches. A wise public health scientist will examine *all* of the evidence available concerning the environmental substances being investigated. This wise investigator will understand the limitations of any single study and will attempt to use multiple approaches to theoretically and/or practically fill any data gaps. The research scientist, in other words, is not limited to the three described approaches. The most useful information is generally gained from combinations of approaches, techniques, information gathering, and data analysis.

For example, analytical chemistry methods can enhance the understanding of exposure in epidemiology studies. As one illustration, the level of polycyclic aromatic hydrocarbons (PAHs) that are carcinogens (such as benzo(a)pyrene, BaP) can be measured not only in the environment in which people reside, but also at critical target sites within human tissues. In prospective epidemiological studies, measuring the dose of a toxin delivered to a target tissue or cellular component permits for more accurate correlations between exposure and effect. It is now possible to use a new method to determine the exposure of human DNA to PAHs. This method, called ^{32}P-postlabeling of aromatic DNA adducts, allows hundreds of bulky aromatic compounds that interact with DNA to be detected simultaneously at extremely low detection limits without structural knowledge of the specific adducts being detected. This method has been used for monitoring occupational exposures (Phillips et al., 1988), cigarette smoking exposures (Phillips et al., 1990), and environmental exposures in an industrial area (Hemminki et al., 1990). For a more thorough description of such biomarkers and their application see Chapter 4.

Environmental and public health researchers also can combine analytical chemistry methods with bioassays in order to enhance monitoring methods, to identify hazardous pollutant sources or sites, and to identify the actual environmental chemicals that are responsible for specific toxic effects. For example, Lewis et al. (1988) combined the analytical information about air particles with Salmonella mutagenicity information concerning the organic portion of the same air particles. The analysis of the combined information provided estimates of the contribution from different sources (e.g., automobile emissions and woodsmoke) of airborne genotoxicants for Albuquerque, NM. This type of "source apportionment" can be applied to any media for which the appropriate information exists.

In her review of the genotoxicity of industrial wastes and effluents, Houk (1992) demonstrates the utility of both comparative bioassay results and determining the genotoxicity of individual constituents in a hazardous waste. The use of toxicological assay systems to direct the chemical fractionation and identification of compounds within complex mixtures makes the identification of environmental toxicants more efficient and more relevant (Claxton, 1982, 1983). The integration of environmental, analytical chemistry, and bioassay analysis will be addressed more fully in the next section.

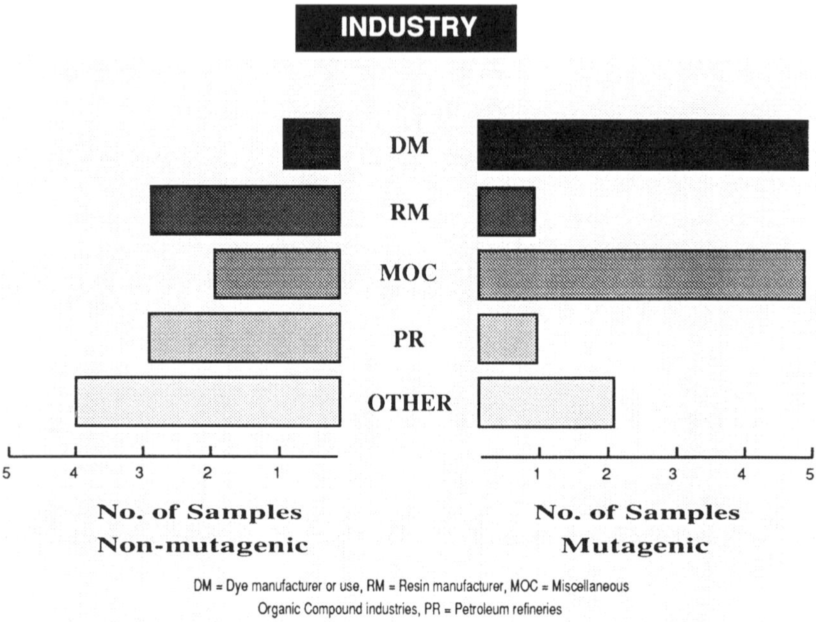

Figure 3 Number of industrial emission samples by mutagenicity response and industrial category (summary of information from McGeorge et al., 1985).

IDENTIFYING HAZARDOUS COMPLEX MIXTURES, THEIR SOURCES, AND THEIR GENOTOXIC COMPONENTS

Although epidemiological efforts (see Chapter 10) and the study of indigenous plants and animals (see Chapters 5 and 6) can be used to identify and assess HWS and effluents, this section will address only the use of integrated analytical chemistry and bioassay approaches for the assessment of hazardous environmental contamination. The reader is referred to the three chapters mentioned for the discussion of other approaches. In addition, instead of trying to present an exhaustive review, a single example for each of the four discussion areas will illustrate basic approaches.

Identifying Potentially Hazardous Complex Mixtures Including Environmental Sites

Although more than 200 genetic toxicology assays are available, very few have been used to identify hazardous environmental sites or wastes. In a review of industrial wastes and effluents, Houk (1992) noted that approximately 60% of the samples described in the open literature were bioassayed with a *Salmonella* mutagenicity assay. In contrast, only about 3% were assayed using *in vivo* assays. McGeorge et al. (1985) present an example of how bioassays can be used to monitor environmental samples. They used the *Salmonella* assay to evaluate effluent and intake waters from 27 industrial facilities. Figure 3 summarizes the qualitative results of their study. Approximately 40% of the samples were mutagenic. Another 40% were nonmutagenic, and about 20% were either equivocal or toxic to the test system. The authors conclude, "The Ames [*Salmonella*] assay provides one mechanism for monitoring, and possibly controlling, the introduction of mutagens into surface waters." Studies such as these demonstrate that bioassays are an efficient and reliable means for determining whether or not environmental sites contain potentially hazardous materials. This also illustrates that biological assays can be an important adjunct when needing to establish priorities between sites or sources that require remediation.

Identifying Individual Toxicants Within Complex Mixtures

During biodegradation, the toxicity of the contaminating material will likely fluctuate (increase and decrease) when constituents of the mixture are weathered and metabolized. The most comprehensive studies integrate chemical and toxicological data. Chemical characterizations are performed to identify as many chemicals as possible in the mixture and to reveal the predominant constituents. Evaluations include, but are not limited to, gas chromatography (GC), mass spectrometry (MS), and/or high pressure liquid chromatography (HPLC). Information can then be paired with preexisting toxicity databases to predict the overall toxicity of the mixture. The obvious advantage of this approach is that especially hazardous chemicals can be identified, whereupon control measures can be implemented. Sometimes, however, chemistry data are incomplete. Even a relatively detailed chemical analysis often fails to detect or identify all toxic compounds present in a sample due to the insensitivity of the analytical technique, the relatively small amount of an individual chemical present in the mixture (even though the chemical may be highly toxic), or the sheer number of components to be resolved. Moreover, critical toxicity information may be unavailable. An estimated 400 organic compounds were detected in the wastes at Love Canal, including carcinogens, mutagens, neurotoxins, hepatotoxins, and renal, pulmonary, and fetal embryo toxins (USEPA, 1980a). For over half the identified chemicals, no toxicological data were available to assess potential hazards (Lowrance, 1981). For these reasons, it is prudent to include bioassays in addition to chemical characterizations to analyze toxicity. One method, bioassay-directed fractionation, identifies the biologically active components of a mixture by incorporating fractionation, biological testing, and chemical identification. The environmental sample is sequentially separated into fractions according to chemical properties. The fractions are then tested individually for biological effect. Further fractionation and chemical characterization of biologically active fractions is performed until the key toxic constituents are isolated and identified. An example of this approach is provided by Donnelly et al. (1987a), who partitioned a wood-preserving waste into acid, base, and neutral fractions. Fractions were tested for genotoxicity, and analysis by GC/MS identified some of the sources of the toxicity to be fluoranthene and pentachlorophenol. With such complex environmental samples, this method is more practical and resourceful than attempting to use analytical chemistry to detect and quantify all toxicants.

Estimating the Contribution of Individual Toxicants to Total Estimated Toxicity

To estimate the toxicity of a complex environmental mixture, the sample can be treated as a single entity or as a composite of its individual parts. In the latter case, either the toxicity of each component must be determined and appropriate mathematical models applied to predict overall toxicity, or complex combinations of components that simulate the whole mixture must be tested. The selection of approach depends upon the complexity of the mixture, available toxicity data, and the goals of the assessment (NRC, 1988). The most common strategy for predicting toxic response to a mixture is the additivity model, which assumes that the toxicity of the complex mixture is a direct summation of the toxicity of its individual components. This is a relatively reasonable assumption if the exposures to the toxic components are small. At high doses, however, deviations from additivity are observed, due in part to interactions between the chemicals that make up the mixture (NRC, 1988). An example of some of the problems encountered in the estimation of toxicity is illustrated in the work of Schaeffer and Kerster (1985). These authors performed mutagenicity testing on a raw, unfiltered effluent from a coke plant and an organic extract of this wastewater. Both samples were mutagenic. Over 130 compounds were identified in the effluents (Schaeffer et al., 1980), but only 10 of these were reported to be mutagens, and their weak activities were insufficient to account for the total activity of the effluent. Either nonadditive (synergistic) mechanisms were responsible for the response, or unidentified mutagens were present in the wastewaters.

EXAMINING DYNAMIC SITUATIONS

Utility of Different Types of Bioassays, Bioassay Selection, and Endpoint Selection

Bioassays can assist remediation efforts by identifying sources of toxicity, explaining the processes leading to the formation of the toxic constituents, and monitoring the effectiveness of the treatment method. It must be remembered, however, that different bioassays convey different information. To illustrate, more than 200 short-term tests have been developed over the last 20 to 25 years to aid in the detection of toxic compounds that pose a genetic hazard to humans (Waters et al., 1988). These genetic toxicity tests can be grouped on the basis of the biological system employed and the toxic endpoint detected. Prokaryotic (bacterial) bioassays detect agents that induce gene mutations and primary DNA damage. Included in this category is the popular *Salmonella* mutagenicity assay. Eukaryotic assays (fungal tests, plant assays, and mammalian cells in culture) detect a broader range of genetic damage, including gene mutations, DNA damage and repair, chromosomal damage, and aneuploidy. Selection of an appropriate assay depends on the physical and chemical composition of the sample, the validity of the test system and its database, the utility of the assay for complex environmental samples, the purpose for which data are intended, and relative cost and simplicity. For example, eukaryotic assays, which are more expensive to perform and require greater technical expertise, are used less frequently than bacterial assays. Because of the differences in detection capabilities among assays, toxicologists often include several test systems in a single study. By using a battery of tests, various endpoints can be studied to ascertain the specific type(s) of damage induced by a sample. Judicious selection of a test matrix is required to screen samples in a cost-effective manner without an unacceptable loss of detection capability. For example, Houk and DeMarini (1988) used the *Salmonella* assay and a prophage-induction assay in *E. coli* to evaluate the types of genetic damage induced by 14 hazardous industrial wastes. The prophage induction assay, which measures DNA damage and repair, was included because it detects carcinogenic metals, solvents, and chlorinated compounds that are not mutagenic to *Salmonella,* but which nonetheless were common contaminants of the industrial wastes tested in the study. Five waste samples that were not mutagenic to *Salmonella* induced prophage, illustrating the usefulness of selecting bioassays that reflect some consideration of the chemical composition of the waste. Sentinel plants and animals represent an additional alternative for the detection of potential hazards associated with contaminated sites (see Chapters 2, 5, and 6).

Critical Factors: Sampling, Strategy and Design, Purpose of Monitoring, Resources

In order to examine or evaluate such dynamic and complex situations, it is important to understand the available research alternatives and goals. To optimize research and monitoring efforts, the goals of an evaluation are carefully defined in advance. The objectives of the study dictate the kind of testing prescribed. For example, a waste site that poses the likelihood of repeated or extended exposure for a large number of individuals may warrant a full-scale investigation of exposed and control populations as part of an epidemiological study, including biomarker analysis. Alternatively, multiple target-organ toxicities could be evaluated in test animals. Analytical chemistry coupled with bioassay results may be sufficient if hazard identification is all that is required. Samples may be collected from wastewaters, soils, leachates, sludges, sediments, air, and groundwater. Each situation requires different sampling techniques and equipment. Representative sampling methods have been described by the U.S. EPA (1980b). For toxicity tests, an adequate amount of sample must be collected to provide sufficient mass or volume following sample preparation procedures (such as concentration). Preferable containment, preservation, and storage techniques are sample-specific, but in general, samples should be refrigerated and protected from light. Obtaining a homogenous sample that is representative of a particular waste mixture can present a challenge. Continually changing conditions will be encountered at a waste site as compounds volatilize, solubilize, disperse, interact, and metabolize at different rates. These changes can be caused by weathering, the action of microbial populations, soil characteristics, and remediation efforts, to name a few. Consequently, the chemical composition of a sample will vary according to sampling site and time of collection. The dynamics of the situation invariably alter the toxicity of the contaminated area, and monitoring activities should reflect this consideration. It is important to follow randomized sampling techniques, to determine reasonable time intervals for collection of treated samples, and to collect appropriate control samples. Chemical analyses and measurements of toxicity before, during, and after remediation are necessary to accurately interpret the mechanics of degradation and by-product generation.

The Alaskan Oil Spill Remediation Effort Example

On March 24, 1989, approximately 11 million gallons of crude oil spilled into the waters of Prince William Sound, Alaska, contaminating one of the most pristine and sensitive environments in the United States. As part of an effort to clean up the spill, the U.S. EPA conducted a bioremediation study using indigenous microorganisms to degrade the oil. Two fertilizers (a water-soluble mixture that slowly releases nutrients as it dissolves and an oleophilic formulation designed to adhere to oil) were applied to selected beaches to enhance microbial growth and to accelerate the degradation of the oil. The shoreline consisted of two types of beaches: a cobblestone beach and a sand-and-gravel beach. Oil penetration varied from approximately 2 to 30 cm in depth. Both beach types were treated with both fertilizer formulations, and soil samples were collected from all four sites. In addition, a reference, or untreated, site at each beach was included as a control. Samples were collected over a 3-month period: at the initiation of the application of the nutrient fertilizers and at two 6-week intervals subsequent to treatment. Samples were collected from randomly selected grids within each treatment area and placed in glass jars with Teflon®-lined caps. The crude oil, its weathered counterpart collected from oil-impacted water, and the two fertilizer formulations were also sampled. One of the methods used to assess potential health effects was the examination of mutagenic activity associated with the spilled oil and its remediation by-products (Claxton et al., 1991a, b). The *Salmonella* mutagenicity assay was selected to provide this information. The objective of the study was to evaluate the effect of bioremediation on the toxicity of oil spill organics by monitoring fluctuations in mutagenicity over time. It was important to determine whether toxicity increased or decreased as constituents of the oil were degraded or generated by microbial metabolism. Results showed that both the crude oil and its weathered counterpart were weakly mutagenic. Neither of the fertilizers was mutagenic. The mutagenicity of the samples collected from the contaminated beaches declined over time, indicating that mutagenic compounds present in the oil were diminishing. In conjunction with the data obtained from chemical analyses, the decrease in toxicity was shown to result from both the naturally occurring and bioremediation processes.

FOR COMPARATIVE ASSESSMENTS

Between Sites and Over Time

Comparative studies are based on the premise that chemical or toxicological assessments of one sample will provide relevant information regarding the chemistry/toxicity of a related sample. Even with complex environmental mixtures, this assumption can be valid. Wastes from similar industries do exhibit similar chemical and toxic properties, and patterns can emerge. For example, in her review of the genotoxicity of industrial wastes, Houk (1992, 1994) found that generalities can be made for petroleum refining wastes. Independent investigations of genotoxicity consistently showed that the highest level of mutagenic activity resided in specific chemical fractions, that the mutagenicity profiles (type of genotoxic damage induced) were similar, and that mammalian metabolism processes enhanced the toxic response. Chemicals commonly associated with these wastes were the alkylated PAHs and heavy metals. However, for waste sites, this kind of evaluation is problematic due to the dissimilarities between sites, not to mention between locations within the same site. Comparative assessments of remediated waste sites can be performed only within a limited context. Chemical and biological analyses of waste sites, however, can provide information on similarities/dissimilarities in terms of classes of compounds, chemical concentrations, and toxicities. Establishing chemical profiles over time can reveal something about changes that are taking place. Toxicological assessments can provide information on relative hazards, and monitoring toxicity over time can address the success of a remediation project. The Alaskan oil spill remediation effort described above provides an example of a concerted effort in which chemical and toxicological data were combined to evaluate the effectiveness of bioremediation and to determine the processes involved in changes observed at the site.

Between Potential Technologies

Selection of an appropriate waste treatment technology should be based on the reduction of exposure and toxicity and, thus, the minimization of risk. Comparisons of different treatment options are therefore critical to prudent waste management. Chemical characterizations alone are insufficient for such evaluations because treatment reactions may inadvertently yield by-products that enhance the toxicity of the parent waste. For example, when incineration is used for remediation, a destruction and removal efficiency (DRE) of 99.9999% is a target narrowly focused on a specific, known toxicant of the waste. The removal of this toxicant, however, cannot guarantee that other toxic compounds will not remain or be formed. Comprehensive determinations of relative risk should incorporate both chemical and toxicological data. A prime example is provided by a compilation of studies performed on the effluent from the pulp and paper mill industry and the numerous treatment technologies applied to reduce its toxicity. This body of literature illustrates the advantages of assessments that integrate chemical and toxicological methods. Approximately 300 chemical compounds have been identified in pulp and paper mill effluents, including 3 known carcinogens, 16 "suspected" carcinogens, and numerous other compounds known to induce genetic damage. With the combined aid of fractionation, bioassay, and chemical analysis, two compounds (MX* and 2-CP**) have been shown to account for most of the mutagenic activity detected in the effluent. Numerous treatment methods have been devised to reduce or eliminate the genotoxicity (and, simultaneously, the aquatic toxicity) of the effluent. Options include biological treatment (aerated lagoons and activated sludge), chemical treatment (increasing the pH of the effluent to alkaline levels, substitution of chlorine dioxide or sulfur dioxide for chlorine in the bleaching process, treatment with nucleophiles, ozonation), and physical treatment (adsorption onto an ion exchange column). A few of these options will be discussed here; for a more detailed discussion of this research, including references, the reader is referred to Houk (1994).

* MX is 3-chloro-4-dichloromethyl-5-hydroxy-2(5H)-furanone.

** 2-CP is 2-chloropropenal.

Biological treatment significantly reduces the mutagenicity of pulp and paper industry effluent, and several mechanisms contribute to this loss, including microbial degradation, physical adsorption onto the biomass, and chemical instability of the compounds. Biotreatment is only partially effective, however, for process streams that contain large amounts of chlorinated organics (such as chlorination-stage effluents), which are more resistant to biological degradation. The most effective way to reduce the mutagenicity of the chlorination-stage effluent is by increasing its pH with the addition of alkali. Several studies explain the chemical reactions that occur during alkaline treatment. In general, the disappearance of mutagenicity results from the pH lability of chloroacetones, including the destruction of the highly potent 2-CP. These examples underscore the importance of combined chemical and biological testing to elucidate mechanisms of action and to evaluate the efficacy of a particular treatment technology. Measurements of multiple endpoints are advantageous to the decision-making process. For example, the replacement of chlorine with chlorine dioxide in the bleaching stage not only reduced the mutagenicity of the effluent, but the BOD,* COD,** TOC,*** color, and acute toxicity as well.

For Setting Priorities

The prioritization of waste sites targeted for mitigation or remediation relies upon the availability of information regarding the magnitude and diversity of potential toxic effects. Obviously, epidemiological evidence is the most compelling piece of information when human health effects are of concern. Rarely, however, are these data available. Consequently, chemical and toxicological information often drive decisions. A chemical analysis will identify many known toxicants of great concern in terms of exposure (i.e., those at highest concentrations, those leaching into groundwater, etc.), and toxicity information will identify those of greatest concern in terms of effect. Risk analysis modeling using this exposure and toxicity information attempts to estimate the total lifetime risk to the exposed population.

Another way to establish the magnitude of the threat and thereby set priorities is to place the risks posed by a particular waste site in context with other waste sites and with other environmental sources of contamination. This can be done by ranking the toxic potency measured at a site. Such rankings draw attention to those sites that pose the greatest toxic burden, identify sites that need more testing, and allow priority setting for regulatory action. For example, McGeorge et al. (1985) tested effluents from 27 industrial facilities in New Jersey for mutagenic activity as part of a broad assessment of effluents conducted by the New Jersey Department of Environmental Protection. The mutagenicity assays (included as monitoring tools to identify and rank effluents) identified those industries that contributed mutagenic activity to surface waters. Comparisons to similar sources of contamination can provide additional information regarding relative hazard. Driver et al. (1990) examined combustion emissions released by a medical/pathological waste incinerator for mutagenic activity. In a comparison of results to other combustion

* BOD is biological oxygen demand.

** COD is chemical oxygen demand.

*** TOC is total oxygen demand.

sources, it was found that the mutagenic emission rates for the incinerator (under routine operating conditions) were less than estimates published for wood stoves, automobile engines, and residential furnaces.

RESEARCH STRATEGIES FOR MICROORGANISMS

CHOOSING THE MICROORGANISMS TO TEST

There are three general classes of microbes used for environmental bioremediation: (1) naturally occurring, (2) "customized," and (3) genetically engineered microorganisms. Because *in situ* bioremediation generally occurs through enrichment of indigenous microorganisms at the site of environmental contamination, it probably is not reasonable to isolate, identify, and test individual isolates. *In situ* bioremediation generally is not accomplished by a single species, but by a consortium of microbial strains. For example, following the Valdez oil spill in Alaska, previous research had indicated that multiple soil organisms were present that could degrade oil (Pritchard and Costa, 1991). In the bioremediation effort, a decision was made to enrich the naturally occurring population instead of applying nonindigenous microbes. Following enrichment, it would be difficult to ascertain if all of the microorganisms involved in the process had been isolated and identified. On the other hand, federal regulation (Anon., 1986) dictates that all genetically engineered microorganisms (constructed by cloning of "foreign" DNA into the chromosome or plasmid DNA of the recipient) must meet federal approval prior to environmental release. Therefore, it is necessary to obtain as much information as possible on all microorganisms designated for environmental release that fall into this category. One of the first approved microbes for environmental release, the *Pseudomonas syringae* Ice-strain, went through lengthy review, and finally was released in California in 1987 (Seidler and Hern, 1988). Extreme safety precautions were taken including site enclosure, monitoring, and protective apparel worn by personnel applying the bacteria to the plants. It should be noted that "customized" microorganisms (e.g., those with enhanced degradative capabilities due to mutagenesis) do not fall under any federal jurisdiction and therefore do not have to be scrutinized prior to environmental release. Because these organisms are released into the environment in unnaturally high numbers, it is in the user's best interest to have some understanding of what health effects may be associated with these "customized" strains.

HEALTH OUTCOMES TO CONSIDER

When determining which health endpoints are important, one considers the primary route of exposure. Human exposure to biotechnology agents has the highest probability of occurring during manufacturing or product application. The three most likely routes of exposure are pulmonary, dermal, and ingestion (Levy, 1986). The skin is the first line of defense against disease. In most cases, an allergic response would be the worse-case scenario following dermal exposure. Ingestion of the released microbes ultimately might lead to colonization of the gastrointestinal tract and alteration in the resident flora. Once the normal balance of the flora is upset, either the biotechnology agent or harbored pathogens may have an opportunity to colonize and cause disease. If this condition persists, bacteria can translocate from the GI tract into the spleen, liver, and mesenteric lymph nodes, culminating in a systemic infection (Steffen and Berg, 1983). If the biotechnology agents are inhaled, the agent might colonize the lung, causing pneumonia or an allergic response. In addition, the agent may cause a subclinical infection in the nasal cavity, thereby persisting over an extended period. Pulmonary exposure may lead to inoculation of the intestinal tract causing the adverse health effects previously mentioned for ingestion. These theoretical worse-case scenarios determine the type of health research needed for environmentally released microorganisms (ERMs).

SELECTING A TESTING STRATEGY

Although determining the primary route(s) of exposure is an important "first step," exposure to microbes is insignificant if no adverse outcomes are associated with the microbes. Methods are needed, therefore, to identify potentially adverse actions of microorganisms (George et al., 1987). Even though an organism may not normally behave as a pathogen, the right set of circumstances may cause the organism to colonize, produce pathogenicity factors, and cause disease. These responses need to be anticipated by the health researcher. A normal progression for determining potential adverse effects is to (1) survey known information, (2) use screening tests to identify other problematic areas, and (3) use animal models and human studies to determine the magnitude of any potential outcomes.

Survey of Known Information on Pathogenicity

When examining microorganisms, it is important to identify the organism in question as precisely as possible using biochemical characteristics. By using references such as *Bergey's Manual,* and computerized databases such as MICRO-IS (National Institute of Dental Research, Bethesda, MD), Microbial Identification System (MIDI, Newark, NJ), and Biolog (Biolog, Inc., Hayward, CA), modern identification methods are very accurate. Once the biotechnology agent has been unquestionably identified, there are several references that are invaluable in determining pathogenicity. These include: *Manual of Clinical Microbiology,* 5th edition (Balows et al., 1991) and Medline (National Library of Medicine, Silver Spring, MD).

Screening Tests and the Comparative Parallelogram Approach

One must recognize the fact that not all bacterial strains that play a role in human disease have been associated with human disease. The example of *Legionella pneumophila* (responsible for Legionaire's Disease) is a good reminder of this concept. Many soil organisms fall into the category of "opportunistic pathogens." *Pseudomas fluorescens* and *P. putida* are common soil and water isolates and occasionally are isolated from clinical specimens. Both organisms cause disease in cancer patients and are involved in infection following transfusion with contaminated blood. *P. fluorescens* has been associated with pelvic inflammatory disease and

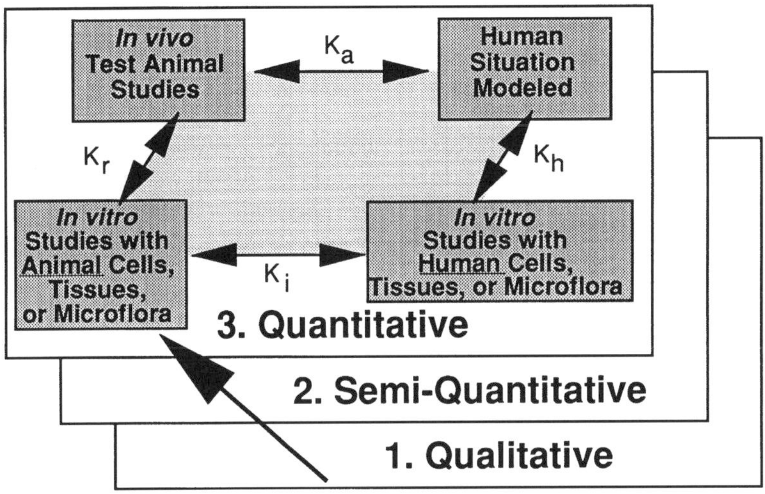

Figure 4 The parallelogram approach for evaluating potential human health effects of environmentally applied microorganisms. (Process can be qualitative, semiquantitative, or quantitative. Assumptions are K_r approximates K_h and K_a approximates K_i.)

urinary tract and postoperative infections (Balows et al., 1991). Therefore, given the right set of circumstances, these organisms can colonize, replicate, and cause disease. It is important, therefore, to screen ERMs.

Because microorganisms generally must survive and persist in order to cause adverse effects, one first examines persistence (survival), then the ability to behave as pathogens, and then response in animals and man. If the intestinal tract is the primary target, perseverance of the biotechnology agent and the balance of the microbiota can be monitored from the feces. Persistence of the strain in the lungs and nasal cavity can be detected in bronchial lavages or nasal swabs if pulmonary exposure is suspected. Some research from our laboratory suggests that certain pathogenicity factors are predictive of adverse health effects in animal models (George et al., 1991). *P. aeruginosa* strains that produce exoenzyme S and proteases are lethal to mice; those that are negative show poor colonization. An inflammatory index (sum percent of lymphocytes and polymorphoneutrophils) can be calculated to investigate the severity of the bacterial challenge. In research animals, lung weights are indicative of a pulmonary insult (lung weight increases with influx of fluid and white blood cells). Clinical infection in mice is manifested by weight loss, listlessness, ruffled fur, and conjunctivitis.

Rarely can one use human subjects in direct experimentation. Therefore, one must rely on animal models. Although simplistic, effects upon humans can be estimated with the comparative parallelogram approach (Figure 4). For example, if one is interested in colonization and competition of a biotechnology agent in the intestinal tract, one can treat animals and look for recovery of the administered strain in the GI flora. Alternatively, a microcosm that simulates the intestinal tract *in vitro* can be implemented, mimicking the interactions of intestinal flora. The biotechnology agent can be added to the system, and survival and competition determined. The relationship of survival *in vivo* and *in vitro* can be designated by a constant (K_r). The human intestinal flora can also be modeled *in vitro* using human fecal inoculum instead of animal inoculum in the microcosm and again, survival of the biotechnology agent and competition with the microbiota determined. The relationship of survival in the human microcosm and the animal microcosm can be established by another constant (K_i). Because the human *in vitro* values are known through experimentation, the relationship of animal *in vivo:in vitro* can be used in concert to mathematically determine a range of theoretical *in vivo* human effects, such as survival time (i.e., solve for K_a and K_h).

This type of work is currently under way in our laboratory. Animals have been exposed to biotechnology agents; and *in vivo* and *in vitro* survival and competition studies are ongoing. Once a mathematical relationship is calculated, *in vitro* survival and competition studies can be done in the same type of microcosm using microbiota from a human source. Initial studies in our laboratory indicate that there is little difference between the human and animal scenario in the survival time or competition with the flora (Nelson et al., 1992).

INTEGRATED POLLUTANT/MICROORGANISM STUDIES

For Environmental Samples

As previously described, the bioremediation of hazardous complex chemical mixtures is a multifaceted process. Ideally, one would like to fully understand the biochemistry involved with each step of the process *and* experimentally perform a bench scale study at each site. However, this is not always feasible because unraveling the metabolic steps in the biodegradation may take years. Therefore, testing each metabolite in a suitable bioassay is not a realistic expectation. On the other hand, bench scale degradation studies are relatively easy. Culture samples can be taken at regular intervals, concentrated, and prepared for bioassay. Several culture options are available and include: (1) utilization of target chemical(s) as sole carbon and energy source, (2) utilization of target chemical(s) in the presence of added nutrients, (3) addition of sterile soil to mineral salts culture containing target chemical(s), (4) addition of nonsterile soil to mineral salts culture containing target compound, (5) utilization of sterilized soil from an HWS as carbon and energy source, (6) utilization of unsterilized soil from an HWS as carbon and energy source, and (7) addition of chemical(s) to unsterilized soil for carbon and energy source. Suitable control cultures must be run concurrently. For example, if option 4 was selected, the researcher would compare results derived from the following conditions: (1) controls with no additions, (2) nonsterile soil only, (3) cells only, (4) chemical(s) only, (5) nonsterile soil plus chemicals, (6) nonsterile soil plus cells, (7) cells and chemicals, and (8) soil plus chemicals and cells. Once the results were tabulated, one would be able to unravel what influences the contributing factors (cells, chemicals, and nonsterile soil) have on the biodegradation process.

For illustration purposes, two studies will be discussed that explore the genotoxicity of hazardous chemical degradation. The first study (George et al., 1992) examines pure culture degradation of 2,4,5-trichlorophenoxyacetic acid (2,4,5-T), one of the active ingredients in Agent Orange, and the second study (Donnelly et al., 1987b) looks at the genotoxicity decrease in wood-preserving waste applied to two nonsterile soils. In the 2,4,5-T experiments, mineral salts medium was supplemented with 2,4,5-T and/or bacterial cells and the degradation process monitored for 5 days. One of the metabolites of 2,4,5-T is 2,4,5-trichlorophenol (TCP) which is approximately 100-fold more genotoxic in the *E. coli* prophage induction bioassay than the parental compound 2,4,5-T. Culture samples were withdrawn at time intervals and filtered to remove cell or other debris. No concentration of the samples was done because 2,4,5-T was added to the culture medium at a concentration within the detection limits of the bioassay. The results indicated that 2,4,5-T concentration decreased as cell numbers increased over time. Genotoxicity, as evidenced by the *E. coli* prophage induction bioassay, decreased also. All results were negative in *Salmonella* histidine reversion bioassay because this assay is not sensitive to the halogenated hydrocarbons. In this situation, if the only test used was the *Salmonella* reversion bioassay, no information concerning the actual genotoxic hazard could be determined due to insensitivity. Also, because 2,4,5-T was added to the culture medium as sole carbon and energy source, no 2,4,5-TCP accumulated causing an increase in genotoxicity.

In the second study by Donnelly et al., (1987b), wood-preserving waste was added to trays containing two different nonsterile soils. Immediately following the application of the hazardous waste, soil samples were taken. Over a period of 1200 days, the soil was sampled and the chemicals extracted, concentrated, and tested in the *Salmonella* histidine reversion bioassay. Mutagenicity decreased over time; however, one of the major chemical constituents was PCP, which is negative in the *Salmonella* assay. If an assay was utilized that could detect the genotoxicity of PCP and its probable chlorinated metabolites, such as the *E. coli* prophage induction bioassay, more information about potential hazard could be determined.

The final results rely on the sensitivity of the bioassay selected and the efficiency of the chemical purification steps. Many compounds and environmental complex mixtures have been tested using the *Salmonella* reversion bioassay and have been found to be mutagenic (Waters et al., 1985). However, this straightforward test is fairly insensitive to halogenated hydrocarbons, which can constitute a high percentage of hazardous wastes (Houk, 1992). If halogenated hydrocarbons are present, the *E. coli* prophage induction assay can detect genotoxicity linked with this chemical class (Houk and DeMarini, 1987).

For Mammalian/Human Exposure and Metabolism Studies

The use of rodent models in toxicology to investigate associated carcinogenicity, teratogenicity, and neurotoxicity is well documented. Therefore, the testing of rodents with chemical mixtures targeted for bioremediation is a logical step. The difficulty comes in deciding which chemical or chemical mixtures to use in the animal testing. When one treats animals with mixtures of chemicals, interactions (potentiation or suppression) may occur. Recent research shows that treatment of rats individually with three different chlorinated aromatic hydrocarbons (Aroclor® 1254, PCP, and 2,4,5-T) impacts the bioactivation of the promutagen/procarcinogen, 2,6-dinitrotoluene (2,6-DNT) (Chadwick et al., 1991a, b; George et al., 1992). Results indicated that Aroclor® 1254 and PCP potentiated 2,6-DNT-induced urine mutagenicity and hepatic DNA adduct formation. Both of these results would indicate increased potential for carcinogenicity outcomes. Conversely, treatment with 2,4,5-T reduced 2,6-DNT urine mutagenicity. Presently, this apparent complexity forces researchers to examine whole mixtures, individual components of mixtures, and combinations of the individual components to assess potential toxicity.

The toxicity of the target chemical's metabolites formed by microbial degradation also needs to be investigated. If metabolites are known, animals can be exposed to them singularly or in a mixture. Another approach is to use bench scale bioremediation as discussed in the previous section, and then expose the animal to the metabolized material. Because exposure to the biotechnology agent most likely will occur simultaneously with chemical exposure, animal models can be used to better understand the overall effect. For example, animals can be treated with the hazardous compound(s) together with the biotechnology agent. Modulations in the intestinal flora can be examined. The implications of this were discussed in a previous section. Intestinal enzymes important in procarcinogen or promutagen activation, such as nitroreductase, azo reductase, or β-glucuronidase, can be assayed. Enzyme changes could be indicative of enhanced carcinogen formation from other mixture components. Urine can be collected, concentrated, and bioassayed for genotoxicity in the *Salmonella* reversion and/or *E. coli* prophage induction bioassays. All of these endpoints are a measure of potential adverse health effects.

Even though humans are not intentionally administered toxicants or microbial bioremediation products, unintentional exposures to the microorganisms, the target chemicals, and their by-products may occur. At these times, human fluids and cells can be monitored for genotoxicants similarly to that done in rodents. Urine and fecal specimens can be collected and analyzed chemically prior to bioassay for genotoxicity. Biotechnology agents can be enumerated from stool samples. Blood can be collected and examined for DNA and protein adduct formation. Findings from human samples can then be compared to those from rodents. Extrapolations from animal to human results can be performed to better understand the potential genotoxicity or other adverse health effects that may occur following exposure to biotechnology agents and hazardous chemicals.

IMPACT OF INTEGRATED BIOLOGICAL ANALYSES ON EXPOSURE AND RISK ASSESSMENT METHODS

Under the regulation (NCP*) that implements Superfund (CERCLA**), risk information generated by the human health evaluation process is to be used for the evaluation of Superfund sites (USEPA, 1989a). The Superfund *Human Health Evaluation Manual* (USEPA, 1989a, 1990, 1991) states, "Although risk information is fundamental to the RI/FS [remedial investigation/feasibility study] and to the remedial response program in general, Superfund site experience has led EPA to balance the need for information with the need to take action at sites quickly and

* NCP (National Oil and Hazardous Substances Pollution Contingency Plan) is the regulation that implements CERCLA.

** The act that created "Superfund" is CERCLA (Comprehensive Environmental Response, Compensation, and Liability Act of 1980).

to streamline the remedial process." In addition, an earlier manual (USEPA, 1986) indicates that there is an understanding that HWS are extremely complex making public health evaluations very difficult. The authors state:

> Public health evaluation cannot be reduced to simple, "cookbook" procedures. ... state-of-the-art public health evaluation techniques have not been fully accepted by all scientists, and important chemical data are frequently unavailable. For instance, toxicity testing has not yet kept pace with the need for information on many chemicals, and procedures used in exposure assessment often require many assumptions. The universe of uncontrolled hazardous waste sites is both variable and complex, with each site posing a unique set of circumstances. It would be unrealistic to expect that all data necessary to determine precisely the health risks associated with every site will be available. (USEPA, 1986, p. 1)

Health researchers, therefore, need to continually improve health assessment information in order to make it more relevant and more responsive to the needs of decision makers. As demonstrated within this chapter, bioremediation efforts many times will also require that decision makers examine the potential effects of microorganisms.

How can one make informed decisions about such complex situations? Following the NCP guidelines under CERCLA, the U.S. EPA presently uses a risk assessment process to decide whether or not a site needs remediation efforts, to choose between cleanup options, and to evaluate sites after cleanup attempts. The U.S. EPA guidelines for human health risk assessments are outlined in a three-volume manual (USEPA, 1989a, 1990, 1991) jointly referred to as *Risk Assessment Guidance for Superfund: Volume 1 - Human Health Evaluation Manual.* A similar manual is available for environmental assessments (USEPA, 1989b). While some of these volumes await final approval, they are presently issued as interim reports. Figure 5A provides an overview of the baseline risk assessment process used for Superfund sites. Figure 5B outlines the steps for the assessment of carcinogens. In most situations, this present assessment process depends on the chemical identification of individual carcinogens, the knowledge concerning these individual carcinogens, dose extrapolation models, and the assumption of additivity when more than one carcinogen is present. In practical terms, therefore, a risk assessment effort first would use chemical analysis to determine the environmental levels of *identified* toxicants. Using available physiochemical and toxicological information about individual chemicals, the risk analysis method then calculates exposure potentials and whether or not toxicants' exposures exceed established guidelines. When multiple toxicants are involved, the effects are generally assumed to be additive. Because monitoring methods cannot monitor for all potential toxicants and because there are gaps of knowledge concerning the toxicity of many environmental chemicals (which includes environmental transformation products), the uncertainties associated with this type of risk assessment are quite evident.

As demonstrated by this chapter, state-of-the-art risk research is beginning to provide alternative approaches. These approaches have the potential to provide less expensive, more responsive methods that examine for both the identified and unidentified carcinogens that may exist at a HWS. Figure 5C should remind the

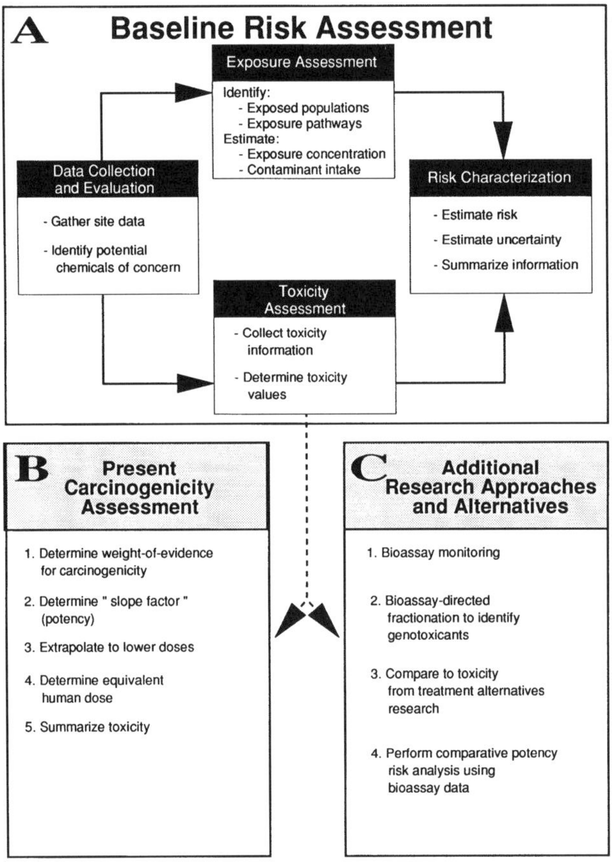

Figure 5 Baseline risk assessment process for Superfund sites (USEPA, 1989a). A, Overview of the basic process; B, summary of steps presently taken for carcinogenicity assessment; C, potential approaches that can be used by researchers not presently a part of the process.

reader of some of the key new approaches surveyed within this chapter. Relative potencies within biological assays (which examines the effects of a series of substances in one or a few bioassays) can be used to estimate risk (NRC, 1988). The NRC states that this comparative-potency approach is particularly useful for the evaluation of complex mixtures (NRC, 1988). Figure 6 diagrammatically illustrates how this comparative-potency method is performed. First, one determines within the bioassay to be used, the potency within that bioassay of known human carcinogens (or other toxic endpoint). The mathematical relationship between human potency and bioassay potency is then established (the regression line labeled "1" determined from values A, B, C, and D). Now, new similar environmental samples are tested using this bioassay and the potency within the bioassay is established (value 2). The assessor then determines where on the relationship curve the bioassay potency value finds an equivalent point (label 3). The estimated human potency (value 4) of this substance can then be calculated using the established

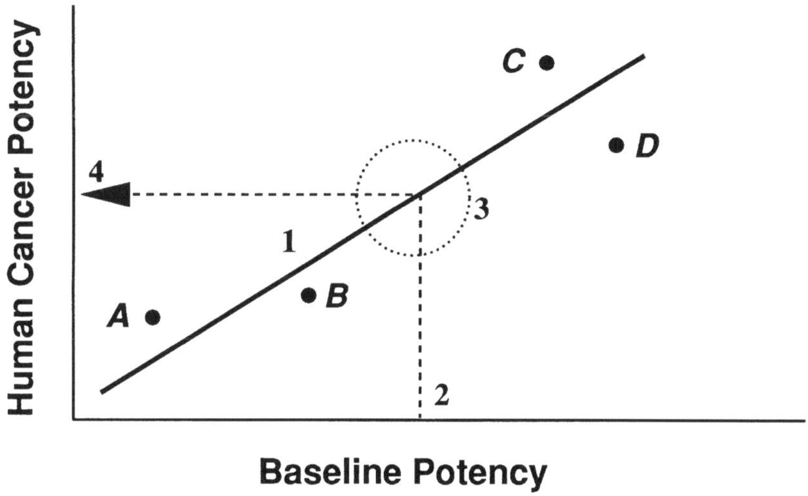

Figure 6 Illustration of the comparative-potency risk calculation approach. Points A, B, C, and D represent toxicants, e.g., complex mixtures like cigarette smoke, for which bioassay and human data are available. A, B, C, and D are used to establish the mathematical relationship between the bioassay and the human results, i.e., the curve. Point 2 represents the bioassay result for a toxicant for which no human data exist. Point 3 represents the area of the curve to which the results extrapolate and point 4 represents the extrapolated human risk estimation. See National Research Council (1988).

mathematical relationship. Using this method, it is not necessary to identify all the toxicants within a complex mixture. The primary restraint for this method is that the substances used to establish the mathematical relationship should be very similar to the substance of concern and that the molecular mechanisms of the bioassay must be related to the human toxicological endpoint being addressed. In order to establish the utility of comparative-potency and other innovative risk assessment methods, extensive environmental studies must be conducted with these goals in mind so that the proper samples, information, and bioassay data needed to validate such approaches be established.

Risk assessment will continue to evolve in a manner similar to all sciences. Due to the complex and dynamic nature of waste sites, potential and actual remediation situations will continue to provide some of the most difficult challenges to risk assessment scientists. By integrating biological test systems with the other sciences used to evaluate waste sites, however, these challenges can be met.

SUMMARY AND CONCLUSIONS

Environmental situations requiring remediation typically present as mixtures of multiple toxicants and other pollutants. Most types of remediation efforts, including biological remediation endeavors, produce additional compounds that add to the complexity of evaluating the potential health effects of a contaminated site during and after remediation efforts. Likewise, when environmental applications of microorganisms are used for remediation purposes, the organisms and their products must be considered for safety evaluation.

Because the products of remediation activities differ in physical properties from the original pollutants, the type and distribution of compounds within the different environmental media constitute an evolving dynamic situation. Due to the presence of multiple pollutants and the potential production of other pollutants by remediation processes, most of the actual toxicants within a typical remediation site are not identified. The Committee on Environmental Epidemiology of the National Research Council (1991) states "There is evidence that NCPs [nonconventional pollutants] are a potentially important source of hazardous exposure. Some preliminary toxicologic studies suggest that NCPs have important biologic properties, environmental persistence, and mobility. ... In the broadest sense, these unidentified substances represent risk of unknown magnitude." These NCPs, therefore, limit the ability to do risk characterizations of remediation sites when only analytical chemistry is used for exposure assessment studies. By incorporating biological tests into assessment studies, it is possible to improve the estimations of potential human toxicant exposure before, during, and after remediation efforts. When appropriately coupled with analytical chemistry, bioassays also can be used to identify the major toxic pollutants. In addition, during the development of remediation methods, bioassays can be used for comparative assessments between differing technological approaches.

Naturally occurring, mutant, and genetically engineered microorganisms have great potential for use in environmental remediation. Environmental releases, however, may result in direct human exposure to these organisms. In the 1970s, using primarily *Escherichia coli,* researchers allayed public and scientific concerns that recombinant strains could survive and transfer genes for toxins to normal human microflora to a degree that would cause major health concerns. However, because some microorganisms when present in large numbers can cause adverse health effects by other mechanisms, it is important to explore the potential health effects of environmentally released organisms (EROs). Recent efforts examine competition and survival of EROs with normal human flora, their effect upon xenobiotic metabolism (environmentally and *in vivo),* and how they affect the competition and survival of pathogens.

This chapter discussed the use of bioassays for comparative assessments of complex mixtures, for identifying individual toxicants within complex mixtures, and for enhancing exposure assessment. The discussion also introduced developing methods for the health assessment of microorganisms that are targeted for environmental release. Clearly, much more research is needed in order to aid risk assessors, environmental managers, scientists, and the general public in their attempts to make the most appropriate decisions. The following recommendations are given to kindle discussion as to the areas where further research is most urgently needed. It should be clearly understood that these recommendations do not reflect the views of any government agency; they represent only the views of this chapter's authors attempting to share their experiences. These also are not recommendations for policy change but recommendations for areas of research that need to be expanded. The specific recommendations are as follows.

More research is needed to (1) determine whether or not nontoxic pollutants at HWS can be converted into toxic substances by the remediation efforts, (2) examine the extent to which bioremediation workers are exposed to toxicants and microbes and a determination of the survivability of those microbes, (3) validate and compare existing exposure models for estimating exposure to both toxicants and environmentally applied microbes, (4) compare presently applied risk evaluations method for actual HWS with newer alternative methods for risk assessment, (5) reevaluate remediated ("cleaned") sites using biological monitoring methods that would detect nonconventional pollutants, (6) test the additivity hypothesis for complex mixtures that commonly occur at HWS (e.g., crude oil, creosote, etc.), (7) test the comparative-potency approach to present risk calculation approaches for substances commonly found at HWS, (8) compare multiple cleanup options at single sites in order to evaluate directly items such as cost, efficiency, production of other toxicants, exposure to workers, etc., (9) establish an epidemiological health registry for bioremediation workers, and (10) test the parallelogram comparative potency approach using both known pathogens and microorganisms targeted for environmental release. Although this list of research recommendations is not exhaustive, the authors hope that the recommendations will stimulate the imagination of other researchers and public health scientists and will provoke further discussions about how to advance the scientific assessment of complex environmental situations. People are seldom exposed to single compounds. Most substances, both natural and artificial, to which humans are exposed are mixtures of chemicals. As pointed out by the National Research Council (1988) "The impetus, therefore, for studying complex mixtures is the need to generate reliable predictions on which to base estimates of potential health risks. ..." Hopefully, this chapter will stimulate others to consider new and innovative approaches for evaluating this most complex situation.

ACKNOWLEDGMENTS

We would like to thank Drs. Arthur Bloom and Frederick de Serres for the opportunity to participate in this endeavor. We also want to thank Drs. Lawrence Reiter, Robert Dyer, Elaine Grose, Mike Waters, and Robert Menzer, who provided the resources and incentive within the U.S. EPA for conducting research that led to this publication. Thanks also are extended to other committee members and especially to Drs. Joseph Bonaventura and Shabeg Sandhu for review and helpful comments concerning this manuscript.

Although this document has been reviewed in accordance with U.S. EPA policy and approved for publication, it does not necessarily reflect the views or policies of the Agency and no official endorsements should be inferred.

REFERENCES

Anon., Coordinated framework for regulation of biotechnology; announcement of policy and notice for public comment, *Fed. Regist.,* 15, 23302, 1986.

Balows, A., Hausler, W. J., Jr., Herrmann, K. L., Isenberg, H. D., and Shadomy, H. J., Eds., *Manual of Clinical Microbiology,* 5th ed., American Society for Microbiology, Washington, D.C., 1991.

Blackburn, G. R., Deitch, R. A., Schreiner, C. A., Mehlman, M. A., and Mackerer, C. R., Predicting carcinogenicity of petroleum distillation fractions using a modified *Salmonella* mutagenicity assay, *Cell Biol. Toxicol.,* 2, 63, 1986.

Brenner, D. J., Taxonomy, classification, and nomenclature of bacteria, in *Manual of Clinical Microbiology,* 5th ed., Balows, A., Hausler, W. J., Jr., Herrmann, K. L., Isenberg, H. D., and Shadomy, H. J., Eds., American Society for Microbiology, Washington, D.C., 1991, chap. 2.

Chadwick, R. W., George, S. E., Chang, J., Kohan, M. J., Allison, J., Hayes, Y., and Crownover, E., Aroclor® 1254 alters intestinal enzyme activity and biotransformation of 2,6-dinitrotoluene in rats, *The Toxicologist,* 11, 255, 1991a.

Chadwick, R. W., George, S. E., Chang, J., Kohan, M. J., and Dekker, J. P., Potentiation of the genotoxicity of 2,6-dinitrotoluene by pretreatment of Fischer 344 rats with pentachlorophenol, *Pest. Biochem. Physiol.,* 39, 68, 1991b.

Chadwick, R. W., George, S. E., and Claxton, L. D., Role of gastrointestinal mucosa and microflora in the bioactivation of dietary and environmental mutagens or carcinogens, *Drug Metab. Rev.,* 24, 425, 1992.

Claxton, L. D., Mutagenic and carcinogenic potency of diesel and related environmental emissions: *Salmonella* bioassay, *Environ. Int.,* 5, 389, 1981.

Claxton, L. D., The utility of bacterial mutagenesis testing in the characterization of mobile source emissions: A review, in *Toxicological Effects of Emissions from Diesel Engines,* Lewtas, J., Ed., Elsevier Science, New York, 1982, 69.

Claxton, L. D., The integration of bioassay and physio-chemical information for complex mixtures, in *Short-Term Bioassays in the Analysis of Complex Environmental Mixtures III,* Waters, M. D., Sandhu, S. S., Lewtas, J., Claxton, L. D., Chernoff, N., and Nesnow, S., Eds., Plenum Press, New York, 1983, 153.

Claxton, L. D. and Huisingh, J. L., Comparative mutagenic activity of organics from combustion sources, in *Pulmonary Toxicology of Respirable Particles,* U.S. Department of Energy Symposium Series (CONF-7911002), Washington, D.C., 53, 453, 1979.

Claxton, L. and Kohan, M., Bacterial mutagenesis and the evaluation of mobile-source emissions, in *Short-Term Bioassays in the Analysis of Complex Environmental Mixtures II,* Waters, M. D., Sandhu, S. S., Huisingh, J. L., Claxton, L. D., and Nesnow, S., Eds., Plenum Press, New York, 1981, 299.

Claxton, L. D., Houk, V. S., Williams, R., and Kremer, F., Effect of bioremediation on the mutagenicity of oil spilled in Prince William Sound, Alaska, *Chemosphere,* 23, 643, 1991a.

Claxton, L. D., Houk, V. S., Williams, R. and Kremer, F., Oil spill cleanup, *Nature,* 353, 24, 1991b.

Creager, J. G., Black, J. G., and Davison, V. E., *Microbiology: Principles and Applications,* Prentice-Hall, Englewood Cliffs, NJ, 1990, 248.

DeMarini, D. M., Brooks, H. G. and Parkes, D. G., Jr., Induction of prophage lambda by chlorophenols, *Environ. Mol. Mutagen.,* 15, 1, 1990.

Deitch, E. A. and Berg, R., Bacterial translocation from the gut: A mechanism of infection, *J. Burn Care Rehab.,* 8, 475, 1987.

Donnelly, K. C., Brown, K. W., and Kampbell, D., Chemical and biological characterization of hazardous industrial waste. I. Prokaryotic bioassays and chemical analysis of a wood-preserving bottom-sediment waste, *Mutat. Res.,* 180, 31, 1987a.

Donnelly, K. C., Davol, P., Brown, K. W., Estiri, M., and Thomas, J. C., Mutagenic activity of two soils amended with a wood-preserving waste, *Environ. Sci. Technol.,* 21, 57, 1987b.

Driver, J. H., Rogers, H. W., and Claxton, L. D., Mutagenicity of combustion emissions from a biomedical waste incinerator, *Waste Manage.,* 10, 177, 1990.

Fernández, P., Grifoll, M., Solanas, A. M., Bayona, J. M., and Albalgés, J., Bioassay-directed chemical analysis of genotoxic components in coastal sediments, *Environ. Sci. Technol.,* 26, 817, 1992.

Galloway, S. M., Armstrong, M. J., Reuben, C., Colman, S., Brown, B., Cannon, C., Bloom, A. D., Nakamura, F., Ahmed, M., Duk, S., Rimpo, J., Margolin, B. H., Resnick, M. A., Anderson, B., and Zeiger, E., Chromosome aberrations and sister chromatid exchanges in Chinese hamster ovary cells: Evaluations of 108 chemicals, *Environ. Mol. Mutagen.,* 10 (Suppl. 10), 1, 1987.

George, S. E., Chadwick, R. W., Chang, J. J., Kohan, M. H., Dekker, J. P., and Hayes, Y., 2,4,5-Trichlorophenoxyacetic acid influence on 2,6-dinitrotoluene-induced urine genotoxicity in the Fischer 344 rat: Effect on GI microflora and enzyme activity, *Fundam. Appl. Toxicol.,* 18, 240, 1992.

George, S. E., Kohan, M. J., Walsh, D., and Claxton, L. D., Methodology for evaluating potential human health effects of microorganisms that degrade hazardous wastes, in *Proc. 3rd Annu. Symp. Solid Waste Testing and Quality Assurance,* Vol. 1, 1987, 175.

George, S. E., Kohan, M. J., Walsh, D. B., Stead, A. G., and Claxton, L., Polychlorinated biphenyl degrading pseudomonades: Survival in the mouse intestines and competition with the normal microbial flora, *J. Toxicol. Environ. Health,* 26, 19, 1989.

George, S. E., Kohan, M. J., Whitehouse, D. A., Creason, J. P., and Claxton, L. D., Influence of antibiotics on intestinal tract survival and translocation of environmental Pseudomonas species, *Appl. Environ. Microbiol.,* 56(6), 1559, 1990.

George, S. E., Kohan, M. J., Whitehouse, D. A., Creason, J. P., Kawanishi, C. Y., Sherwood, R. L., and Claxton, L. D., Distribution, clearance and mortality of environmental pseudomonades in mice upon intranasal exposure, *Appl. Environ. Microbiol.,* 57, 2420, 1991.

Graham, J. A., Claxton, L. D., O'Neil, J. J., Otto, D. A., and Goldstein, B. D., Automobile emissions: Primary health effects concerns, *SAE Tech. Pap. Ser.,* 840908, 1, 1984.

Haworth, S., Lawlor, T., Mortelmans, K., Speck, W., and Zeiger, E., *Salmonella* mutagenicity test results for 250 chemicals, *Environ. Mutagen.,* 5 (Suppl. 1), 3, 1983.

Hemminki, K. E., Grzybowska, E., Chorazy, M., Twardowska-Saucha, K., Srocynski, J. W., Putman, K. L., Randerath, K., Phillips, D. H., Hewer, A., Santella, R. M., Young, T. L., and Perera, F. P., DNA adducts in humans environmentally exposed to aromatic compounds in an industrial area of Poland, *Carcinogenesis,* 11, 1229, 1990.

Holt, J. G. and Krieg, N. R., *Bergey's Manual of Systematic Bacteriology,* Vol. 1, Williams & Wilkins, Baltimore, 1984, 1092.

Houk, V. S., The genotoxicity of industrial wastes and effluents. A review, *Mutat. Res.,* 277, 91, 1992.

Houk, V. S., The effect of waste treatment and remediation alternatives on the genotoxicity of industrial wastes, effluents, and waste sites, 1994, manuscript in preparation.

Houk, V. S. and Claxton, L. D., Screening complex hazardous wastes for mutagenic activity using a modified version of the TLC/Salmonella assay, *Mutat. Res.,* 169, 81, 1986.

Houk, V. S. and DeMarini, D. M., Induction of prophage lambda by chlorinated pesticides, *Mutat. Res.,* 182, 193, 1987.

Houk, V. S. and DeMarini, D. M., Use of the microscreen phage-induction assay to assess the genotoxicity of 14 hazardous industrial wastes, *Environ. Mol. Mutagen.,* 11, 13, 1988.

Isenberg, H. D. and D'Amato, R. F., Indigenous and pathogenic microorganisms of humans, in *Manual of Clinical Microbiology,* 5th ed., Balows, A., Hausler, W. J., Jr., Herrmann, K. L., Isenberg, H. D., and Shadomy, H. J., Eds., American Society for Microbiology, Washington, D.C., 1991, chap. 2.

Jansson, K. and Jansson, V., Inability of chlorophenols to induce 6-thioguanine-resistant mutants in V79 Chinese Hamster cells, *Mutat. Res.,* 171, 165, 1986.

Lamar, R. T. and Dietrich, D. M., In situ depletion of pentachlorophenol from contaminated soil by *Phanerochaete* spp., *Appl. Environ. Microbiol.,* 56 (10), 3093, 1990.

Leahy, J. G. and Colwell, R. R., Microbial degradation of hydrocarbons in the environment, *Microbiol. Rev.,* 54(3), 305, 1990.

Levy, S. B., Human exposure and effects analysis for genetically modified bacteria, in *Biotechnology Risk Assessment — Issues and Methods for Environmental Introductions,* Fiksel, J. and Covello, V., Eds., Pergamon Press, New York, 1986, 56.

Lewis, C. W., Baumgardner, R. E., Stevens, R. K., Claxton, L. D., and Lewtas, J., Contribution of wood smoke and motor vehicle emissions to ambient aerosol mutagenicity, *Environ. Sci. Technol.,* 22, 968, 1988.

Lowrance, W. W., *Assessment of Health Effects at Chemical Disposal Sites,* Proc. Symp., June 1–2, Rockefeller University, New York, 1981.

Macieira-Coelho, A., Cancer and aging, *Exp. Gerontol.*, 21, 483, 1986.

McGeorge, L. J., Louis, J. B., Atherholt, T. B., and McGarrity, G. J., Mutagenicity analyses of industrial effluents: Results and considerations for integration into water pollution control programs, in *Short-Term Bioassays in the Analysis of Complex Environmental Mixtures, IV,* Waters, M. D., Sandhu, S. S., Lewtas, J., Claxton, L., Strauss, G., and Nesnow, S., Eds., Plenum Press, New York, 1985, 247.

Mikesell, M. D. and Boyd, S. A., Complete reductive dechlorination and mineralization of pentachlorophenol by anaerobic microorganisms, *Appl. Environ. Microbiol.*, 52, 861, 1986.

Mikesell, M. D. and Boyd, S. A., Enhancement of pentachlorophenol degradation in soil through induced anaerobiosis and bioaugmentation with anaerobic sewage sludge, *Environ. Sci. Technol.*, 22, 1411, 1988.

Mileski, G. J., Bumpus, J. A., Jurek, M. A., and Aust, S. D., Biodegradation of pentachlorophenol by the white rot fungus *Phanerochaete chrysosporium, Appl. Environ. Microbiol.*, 54(12), 2885, 1988.

Molholt, B., Innovative biotechnologies for detection and disposal of hazardous materials and wastes, in *Management of Hazardous Materials and Wastes: Treatment, Minimization and Environmental Impacts*, Majumdar, S. K., Miller, E. W., and Schmalz, R. F., Eds., Pennsylvania Academic Press, San Diego, 1989, 58.

Mueller, J. G., Middaugh, D. P., Lantz, S. E., and Chapman, P. J., Biodegradation of creosote and pentachlorophenol in contaminated ground water: Chemical and biological assessment, *Appl. Environ. Microbiol.,* 57(5), 1277, 1991.

National Research Council (NRC), Committee on Methods for the In Vitro Toxicity Testing of Complex Mixtures, Board on Environmental Studies and Toxicology, Commission on Life Sciences, *Complex Mixtures: Methods for In Vivo Toxicity Testing,* National Academy Press, Washington, D.C., 1988, 227.

National Research Council (NRC), Committee on Environmental Epidemiology, *Environmental Epidemiology: Public Health and Hazardous Waste,* Vol. 1, National Academy Press, Washington, D.C., 1991, 282.

National Toxicology Program (NTP), Toxicology and Carcinogenesis Studies of Pentachlorophenol (CAS No. 87–86–5) in $B6C3F_1$ Mice (Feed Studies), NIEHS, Research Triangle Park, NC, NTP Technical Report No. 349, NIH Publ. No. 88–2804, 1989.

Nelson, G. M., Claxton, L. D., Creason, J. P., and George, S. E., Competition potentials of environmentally applied bacteria with human fecal microbiota, *Environ. Toxicol. Chem.*, 11, 1627, 1992.

Nylund, L., Heikkila, P., Hameila, M., Pyy, L., Linnainmaa, K., and Sorsa, M., Genotoxic effects and chemical compositions of four creosotes, *Mutat. Res.*, 265, 223, 1992.

Phillips, D. H., Hemminki, K., Alhonen, A., Hewer, A., and Grover, P. L., Monitoring occupational exposure to carcinogens: Detection by ^{32}P-postlabeling of aromatic DNA adducts in white blood cells from iron foundry workers, *Mutat. Res.,* 204, 531, 1988.

Phillips, D. H., Schoket, B., Hewer, A., Bailey, E., Kostic, S., and Vincze, I., Influence of cigarette smoking on the levels of DNA adducts of human bronchial epithelium and white blood cells, *Int. J. Cancer,* 46, 569, 1990.

Pritchard, P. H. and Costa, C. F., Bioremediation of the Alaska oil spill, *Environ. Sci. Technol.,* 25, 372, 1991.

Sagan, L. A., Problems in health measurements for the risk assessor, in *Technology Risk Assessment,* Ricci, P. F., Sagan, L. A., and Whipple, C. G., Eds., The NATO Advanced Science Series (Series E), Martinus Nijhoff, The Hague, Netherlands, 1984, 1.

Schaeffer, D. J. and Kerster, H. W., Estimating the mass of mutagens in indeterminate mixtures, *Ecotoxicol. Environ. Saf.,* 10, 190, 1985.

Schaeffer, D. J., Tigwell, D. C., Somani, S. M., Glave, W. R., and Janardan, K. G., Efficiencies of liquid-liquid extraction and XAD-4 and XAD-7 resins in collecting organic compounds from a coke plant's effluent, *Bull. Environ. Contam. Toxicol.,* 25, 569, 1980.

Schenk, T., Muller, R., and Lingens, F., Mechanism of enzymatic dehalogenation of pentachlorophenol by *Arthrobacter* sp. strain ATCC 33790, *J. Bacteriol.,* 172(12), 7272, 1990.

Seidler, R. and Hern, S., Release of ice minus recombinant bacteria, Special Report, U.S. Environmental Protection Agency, Environmental Research Laboratory, Corvallis, OR, 1988.

Sneath, P., *Bergey's Manual of Systematic Bacteriology,* Vol. 2, Williams & Wilkins, Baltimore, 1986, 552.

Staley, J. T., *Bergey's Manual of Systematic Bacteriology,* Vol. 2, Williams & Wilkins, Baltimore, 1986, 744.

Steffen, E. K. and Berg, R. D., Relationship between cecal populations levels of indigenous bacteria and translocation to the mesenteric lymph nodes, *Infect. Immun.,* 39, 1252, 1983.

Steiert, J. G. and Crawford, R. L., Catabolism of pentachlorophenol by a *Flavobacterium* sp. *Biochem. Biophys. Res. Commun.,* 141(2), 825, 1986.

U.S. Environmental Protection Agency, Chemicals found in waste dumps (Love Canal). Results of toxicity testing, Washington, D.C., 1980a.

U.S. Environmental Protection Agency, Samplers and sampling procedures for hazardous waste streams, EPA600/2-80-018, January 1980, Washington, D.C., 1980b.

U.S. Environmental Protection Agency, Office of Emergency and Remedial Response, Superfund Public Health Evaluation Manual, EPA/540/1-86/060, October 1986, Washington, D.C., 1986.

U.S. Environmental Protection Agency, Office of Emergency and Remedial Response, Risk Assessment Guidance for Superfund: Vol. I, Human Health Evaluation Manual (Part A): Interim Final, EPA/540/1-89/002, December 1989, Washington, D.C., 1989a.

U.S. Environmental Protection Agency, Office of Emergency and Remedial Response, Risk Assessment Guidance for Superfund: Vol. II, Environmental Evaluation Manual, Interim Final, EPA/540/1-89/001A (OSWER Directive 9285.7–01), Washington, D.C., 1989b.

U.S. Environmental Protection Agency, Office of Emergency and Remedial Response, Risk Assessment Guidance for Superfund: Vol. I, Human Health Evaluation Manual (Part B, Development of Preliminary Remediation Goals): Interim, December 1989, Washington, D.C., 1990.

U.S. Environmental Protection Agency, Office of Emergency and Remedial Response, Risk Assessment Guidance for Superfund: Vol. I, Human Health Evaluation Manual (Part C, Risk Evaluation of Remedial Alternatives): Interim, December 1989, Washington, D.C., 1991.

Vogel, E. and Chandler, J. L. R., Mutagenicity testing of cyclamate and some pesticides in *Drosophila melanogaster*, *Experientia*, 30, 621, 1974.

Waters, M. D., Sandhu, S. S., Lewtas, J., Claxton, L., Chernoff, N., and Nesnow, S. *Short-Term Bioassays in the Analysis of Complex Environmental Mixtures, III,* Plenum Press, New York, 1983.

Waters, M. D., Sandhu, S. S., Lewtas, J., Claxton, L., Strauss, G., and Nesnow, S. *Short-Term Bioassays in the Analysis of Complex Environmental Mixtures, IV,* Plenum Press, New York, 1985.

Waters, M. D., Stack, H. F., Brady, A. L., Lohman, P. H. M., Haroun, L., and Vainio, H., Use of computerized data listings and activity profiles of genetic and related effects in the review of 195 compounds, *Mutat. Res.,* 205, 295, 1988.

Williams, S. T., *Bergey's Manual of Systematic Bacteriology,* Vol. 2, Williams & Wilkins, Baltimore, 1989, 440.

Zeiger, E., Anderson, B., Haworth, B., Lawlor, T., and Mortelmans, K., *Salmonella* mutagenicity tests. IV. Results from the testing of 300 chemicals, *Environ. Mol. Mutagen.*, 11 (Suppl. 12), 1, 1988.

CHAPTER 4

Biomarkers of DNA Damage

Lee R. Shugart

CONTENTS

BIOMARKERS: AN OVERVIEW

At the present time there is a strong drive to clean up the environment; however, since costs increase rapidly with the degree of cleanup, there is a pressing need to have a practical, defensible strategy that provides information for establishing

1-56670-141-4/96/$0.00+$.50

both priorities for environmental restoration and endpoints for regulatory compliance.

Unfortunately, a large gap exists in our capabilities either to quantify exposure to chemical and physical agents in the environment or to assess the biological significance of such exposure. Exposure cannot readily be quantified by measuring body burdens of contaminants because many toxic agents do not accumulate, but are rapidly metabolized. Furthermore, the relationship between body burden and toxic response is complex and not fully understood. Assessing the significance of exposure to complex mixtures of chemicals is even more problematic because possible synergistic or antagonistic interactions within biota can invalidate predictions based on the toxicity of individual chemicals.

An informative approach to quantifying exposure and its potential impact is to monitor biological endpoints (biomarkers) in living organisms, including humans, as indicators of exposure and effects of environmental contaminants (Shugart et al., 1992a; Shugart, 1993a). The biomarkers are any of a series of biochemical or molecular responses to compounds that have entered an organism, reached sites of toxic action, and are exerting an effect on the organism. In this context, the organism functions as integrator of exposure, accounting for abiotic and physiological factors that modulate the dose of toxicant taken up from the environment. These biological markers can be used to quantify exposure to harmful agents and to detect change in the organism to environmental insults (Shugart et al., 1989; McCarthy and Shugart, 1990; Shugart et al., 1992a; Huggett et al., 1992; Peakall, 1992; Shugart, 1993a).

The concept of biomarkers has been described (McCarthy et al., 1989; Shugart et al., 1989; NRC/CEE, 1991). Current usage relies heavily on analytical methodologies with greater inherent sensitivities than previously used in occupational and public health research and practice. Investigators are now able to describe a variety of biological events along the continuum between exposure and clinical disease at the individual, tissue, or subcellular levels and to evaluate the physiological condition of an organism in response to contaminants in their environment (McCarthy and Shugart, 1990; Huggett et al., 1992; Peakall, 1992). Implicit in the use of biomarkers is the *in situ* approach where organisms in their natural environment are monitored and evaluated for responses that indicate exposure to and harm from hazardous substances. Chapter 5 discusses the potential usefulness of monitoring animal and plant species as sentinels for genetic hazards. This theme is carried over in Chapter 6, which describes biomarkers of general toxicity, immunotoxicity, and genotoxicity that are applicable to many wild animals.

In order to predict the health risks associated with environmental exposure, an understanding of the critical cellular events that may occur between exposure and effects must be known. Therefore, it is often necessary to regress down the conceptual sequence of organismal responses to toxicant exposure and select appropriate biological responses that are causally related to, or predictive of, longer-term effects. Biochemical and cellular events tend to be more sensitive to exposure and more responsive than other biomarkers at higher levels of biological organization. This is because these biological responses are closely linked in time and cellular target to the insulting agents.

Strategies for the utilization of biomarkers often recommend a suite of diagnostic assays applied in a multi-tiered approach (Jenkins and Sanders, 1992). Initial screening would address broad questions pertaining to the health of the organisms at the locations being evaluated. Detection of adverse effects at this level would dictate that subsequent testing identify classes of contaminants and define the spatial distribution of the impact. The purpose of the biomarker approach during remedial action is to provide a degree of environmental surveillance that avoids unacceptable and often irreversible effects at higher levels of biological organization (e.g., disease in humans, mass mortality, and loss of commercially or ecologically important species).

Biomarkers are often divided into those that indicate exposure to chemical(s), those that demonstrate adverse effects resulting from that exposure, and those that indicate the sensitivity of individuals or subpopulations to environmental pollutants.

A biomarker of exposure is defined as the identification of an exogenous substance within a biological system, the interactive product between a xenobiotic compound and endogenous components, or other event in the system related to exposure. These may include measures of internal dose or of the biologically effective dose (i.e., the amount of material interacting with critical subcellular, cellular, and tissue targets or with an established surrogate) (CBMNRC, 1987).

A biomarker of effect is defined as an indicator of an endogenous component of the biological system, a measure of the functional capacity of the system, or an altered state of the system that is recognized as impairment or disease. This includes an actual health impairment or recognized disease, an early precursor of a disease process that indicates a potential for impairment of health, or an event peripheral to any disease process, but correlated with it and thus predictive of development of impaired health (CBMNRC, 1987).

A biomarker of susceptibility is an indicator of an inherent or acquired limitation of an organism's ability to respond to the challenge of exposure to a specific xenobiotic substance (CBMNRC, 1987)

Biomarkers of exposure, effect, or susceptibility can be useful in biological monitoring. Because of the commonality of biochemical and cellular structure and function among organisms, biomarker approaches are potentially applicable over a broad range of species and across most ecosystem types. Additional information on biomarkers and their potential for application in evaluating environmental contamination is available in proceedings of recent symposia and workshops (McCarthy and Shugart, 1990; Shugart et al., 1992a; McCarthy et al., 1992; Huggett et al., 1992; Peakall and Shugart, 1993; Shugart, 1993a; Shugart and Halbrook, 1993).

DNA DAMAGE AS A BIOMARKER

Perhaps one of the more serious societal concerns associated with the accelerated release of pollution in the environment is the potential for exposure to substances that are genotoxic. This concern arises because some of the pollutants known to be present have been shown through corresponding laboratory studies to be chemical carcinogens and mutagens with the capacity to affect both the structural and biological integrity of DNA (Wogan and Gorelick, 1985). In fact, changes to the integrity of DNA have been proposed as useful endpoints for assessing the hazardous nature of environmental pollutants on human health (Kohn, 1983; CBMNRC, 1987) or biota therein (Shugart, 1990a). Early attempts at assessing the effects of genotoxins on species present in the environment most often involved direct observation such as the visual occurrence of neoplasms and chromosomal aberrations in various plants, wild terrestrial mammals, and aquatic vertebrates (Sandhu and Lower, 1989). Public awareness of increasing contamination of streams, rivers, and oceans and the potential for associated health risks has sparked renewed efforts, and reports are appearing that suggest a correlation between environmental factors, including pollution, and cancerlike lesions in fish (Harshbarger and Clark, 1990).

Recently, a range of assays has been developed at either quantifying the extent of exposure to known genotoxins or monitoring exposed individuals for evidence that a genotoxic agent has interacted with and/or modified the DNA of somatic tissue (Ashby, 1988; Marnett, 1993). Thus, it has become possible to extrapolate backward from the type of DNA-damage event observed to identify the potential agent(s) responsible for the insult, or conversely, to extrapolate forward and predict the consequence(s).

BASIC MECHANISMS OF DNA DAMAGE

Cellular metabolism of genotoxic chemicals, once they become bioavailable, can be a relatively complex phenomenon and the lack of complete detoxication sometimes lead to the formation of highly reactive electrophilic metabolites (Philips and Sims, 1979; Harvey, 1982; Phillips, 1983; Groopman and Kensler, 1993). These intermediates can undergo attack by nucleophilic centers in macromolecules such as lipids, proteins, DNA, and RNA that often results in cellular toxicity. Interaction with DNA is manifested primarily by structural alterations to the DNA molecule and can take the form of adducts (where the chemical or its metabolite becomes covalently attached to the DNA), of strand breakage, or of chemically altered bases (Shugart et al., 1992b).

Conceptually, the preceding discussion on structural alterations by genotoxic agents can be depicted (Figure 1) as a model that reflects events and cellular processes related to DNA integrity (Shugart, 1990a; Shugart et al., 1992b). DNA is present in cells as a functionally stable, double-stranded entity without discontinuity (strand breaks) or abnormal structural modifications (adducts or chemically altered bases). As such, it is considered to have high integrity. The rigid maintenance of this integrity is important for survival and is reflected in the low mutational rate observed in living organisms, which has been estimated to be on the order of one mutation per average gene per 200,000 years (Alberts et al., 1989). DNA damage can occur as the result of: (1) wear and tear by normal cellular events such as metabolism and random thermal collisions (pathway 1); (2) interaction with physical agents such as ultraviolet light and ionizing radiation (pathway 2); and, (3) interaction with chemical agents (pathway 3). These various processes give rise to structural alterations, which are usually rapidly repaired, but in the process produce a transient population of DNA with strand breaks and low integrity. Some chemicals work through free-radical mechanisms and cause strand breaks directly, whereas other genotoxic agents can interfere with the fidelity of normal repair or DNA modifications (pathway 6). The loss of bases from the DNA molecule (abasic sites) are frequent lesions that occur as a result of random thermal collisions or from breakdown of chemically unstable adducts. Even the normal cellular process of replication (pathway 5) produces DNA with strand breaks. Therefore, at any one

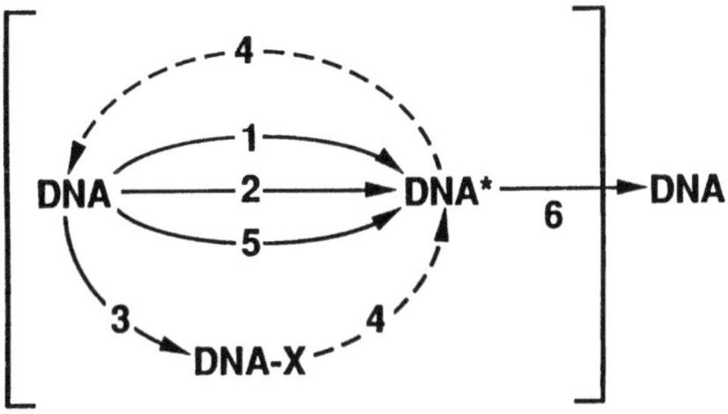

INSULT

1. Normal wear and tear
2. UV and ionizing radiation (γ and X irradiation)
3. Chemical

REPAIR

4. Incision excision resynthesis ligation

SYNTHESIS

5. Replication
6. Postreplication modification

DNA: Normal double-stranded DNA with no strand breaks.
DNA-X: Chemically modified DNA.
DNA*: DNA with strand breaks.

Figure 1 A schematic representation of the status of DNA in relation to insults that disrupt DNA integrity and cellular processes that maintain DNA integrity. (From Shugart, L. R., *Biological Markers of Environmental Contaminants,* McCarthy, J. F., and Shugart, L. R., Eds., Lewis Publishers, Boca Raton, FL 1990a, 205. With permission.)

time, a background level of DNA with low integrity (DNA with various types of structural alterations) may exist in the cell. Fortunately, most cells have DNA repair mechanisms (Sancar and Sancar, 1988) that under normal circumstances efficiently eliminate DNA of low integrity (pathway 4).

The consequence of the structural perturbations discussed above can be innocuous due to repair of the damage or death of the cells containing the damaged DNA. However, those lesions not properly repaired may result in alterations that become fixed and are eventually transmitted into daughter cells. It has been postulated that the reaction of chemicals with DNA and ensuing changes that result may cause deleterious, pathological conditions (Harvey, 1982). Affected cells often exhibit altered cellular functions indicative of subclinical manifestation of genotoxic disease. For example, structural alterations to DNA caused by genotoxic chemicals may potentiate irreversible changes to the DNA molecule and result in the expression of cellular responses such as chromosomal aberrations and oncogene activation. Such responses have been associated with the latent events of chemical carcinogenesis (Weinstein, 1978).

From the preceding discussion it is evident that upon exposure to genotoxic chemicals, a cascade of events may be induced that can change the integrity of an organism's DNA and result in various types of damage. If the damage is not repaired or if it persists, then an opportunity to test for genotoxicity is provided (Shugart, 1990a). Thus, the detection and quantitation of various events in this

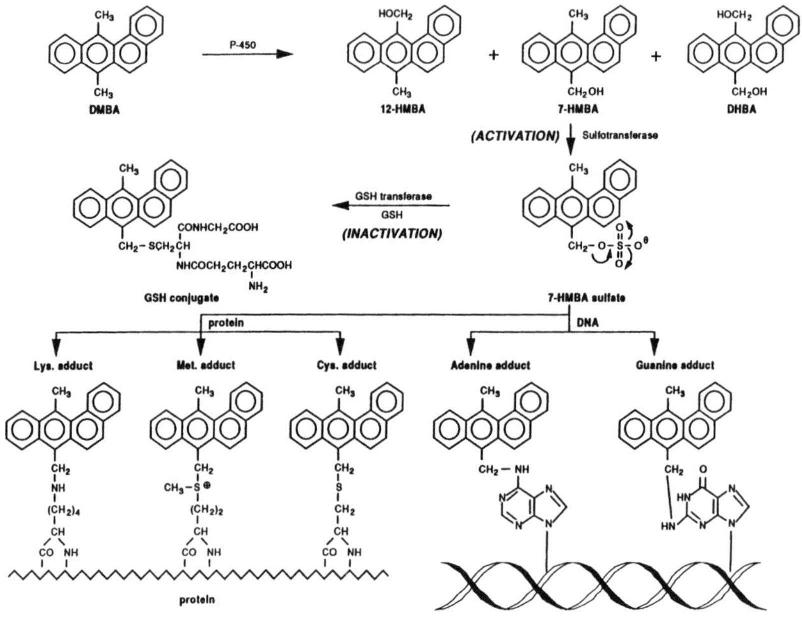

Figure 2 Formation of protein and DNA adducts from the metabolism of 7,12-dimethylbenz[a]anthracene (DMBA).

sequence may be employed as biomarkers of exposure and effects in organisms exposed to genotoxic agents in their environment (Marnett, 1993).

CHANGES IN DNA INTEGRITY ACCESSIBLE FOR ANALYSIS

Structural Damage

Structural modifications of DNA are probably the best understood genotoxic event for which analytical techniques with the appropriate selectivity and sensitivity are available (Wogan and Gorelick, 1985; Bartsch et al., 1988; Santella, 1988; Shugart, 1988). For this reason, many investigations concerned with genotoxic agents in the environment have as their main focus the analysis of the DNA molecule for structural alteration (Stein et al., 1994).

Soon after an organism is exposed to genotoxic chemicals, the presence of an exogenous substance or its interactive product may be detected in the form of a covalently bound DNA adduct. Metabolic activation of the carcinogenic chemical 7,12-dimethylbenz[a]anthracene (DMBA) to several hydroxymethylated intermediates (12-HMBA, 7-HMBA, and DHBA) is depicted in Figure 2. Subsequent enzymatic activation may proceed to a sulfate derivative (7-HMBA sulfate) that can become covalently attached to bases on the DNA molecule. Note also that other cellular macromolecules such as proteins may also become adducted with this chemical. This latter situation and its diagnostic importance to DNA adduct formation will be discussed subsequently. Methods of varying sensitivity have been devised to measure DNA adducts, including physiochemical (Rahn et al., 1982), immunological (Santella et al., 1987), and ^{32}P-postlabeling (Randerath et al., 1981). The methods most commonly employed are ^{32}P-postlabeling and immunoassays using adduct-specific antibodies.

Exposure to genotoxic agents may cause, in addition to or concomitant with adduct formation, other types of damage to the DNA molecule. These structural aberrations can manifest themselves in various ways including strand breaks in the DNA polymer or chemically modified bases (other than adducts).

On the basis of current scientific knowledge, it would be surprising if changes in DNA integrity as a result of genotoxic insult lead to a stoichiometric occurrence of some subsequent deleterious event (mutation, disease, etc.). Cell proliferation and DNA repair are confounding cellular events that preclude this assumption (Thilly and Call, 1986).

Irreversible Events

Structural damage to DNA by genotoxins, if allowed to persist or not properly repaired, can lead to various other anomalies in genetic material such as chromosomal aberrations, sister chromatid exchanges, gene mutation, and cellular effects (e.g., appearance of tumors). In relation to the initiating damage to the DNA molecule, these other events are irreversible and latent; however, some anomalies may occur shortly (relatively speaking) after the structural damage to DNA has become established due to faulty repair and cellular replication (Thilly and Call, 1986).

SURROGATES FOR DNA DAMAGE

Since evidence of DNA alterations in most tissues, including that of humans, is sometimes difficult to obtain, damage to ancillary molecules may serve as a surrogate (e.g., protein adducts) for estimating exposure to genotoxic agents. Blood proteins such as hemoglobin and serum albumin have proven useful in this context (Ehrenberg et al., 1983; Shugart and Kao, 1985; Shugart, 1993b; Pohl, 1993). This concept is illustrated in Figure 2 and shows that in addition to DNA bases, amino acids on proteins may also be modified. In the example provided it should be noted that it is the same metabolite of 7,12-dimethylbenz[a]anthracene that becomes adducted to the two different cellular macromolecules (protein and DNA). For this reason, protein adducts may serve as surrogates for DNA adducts. Furthermore, since protein adducts are not subject to repair, they persist for the lifetime of the protein and accumulate in a dose-related manner.

Because it meets a number of essential requirements, hemoglobin has been proposed as a surrogate for DNA for estimating the *in vivo* dose of chemicals subsequent to exposure (Shugart and Kao, 1985). First, it has reactive nucleophilic sites and the reaction products with electrophilic agents are stable. More than 60 compounds to date have been shown to yield covalent reaction products with hemoglobin in animal experiments. These compounds include representatives of most of the important classes of genotoxic chemicals currently known. No mutagenic or cancer-initiating compound has failed to produce covalent reaction products with hemoglobin. Second, hemoglobin has a well-established lifespan, is readily available in large quantities in humans and animals, and its concentration is not subject to large variation. Third, modification of hemoglobin has been shown to give an indirect measure of the dose to the DNA in cells that are potential targets for genotoxic agents.

The detection and measurement of adducts to hemoglobin has been seriously studied in humans exposed to hazardous occupational chemicals for some time, and a considerable scientific literature exists on methodologies and approaches in this area (Wogan and Gorelick, 1985; Bartsch et al., 1988). It is only recently that protein adducts (particularly hemoglobin) have received attention as biomarkers to assess environmental contamination (Shugart et al., 1989; Shugart, 1993b).

STRATEGIES FOR BIOMARKER USE

From the standpoint of environmental health, biomarkers of DNA damage can be used to identify susceptibility to toxic effects of environmental pollutants, exposure to such toxicants, and biologic effects of exposure. Such responses may range from transient and reversible changes with virtually no health effects to permanent changes that are associated with increased morbidity and mortality. The characterization of relationships among biomarker test results, toxic exposures, and health effects requires special emphasis upon developing reference ranges and validating their sensitivity and specificity for biologic health endpoints. These factors are central to applying any screening test and should not be overlooked when examining populations that may have been exposed to low levels of substances associated with hazardous waste sites (HWS) (CDC/ATSDR Report, 1990). Therefore, the collection of data from an appropriate reference/control site or population cannot be overemphasized. Once adequate baseline information concerning the suite of biomarkers is established for a species or set of species within a defined area, one would be able to sample that species or species subset from selected sites and compare the results with established baselines. Such comparisons will help define the influence of, or contribution of, intrinsic factors that define variability (Peakall and Shugart, 1993).

SELECTION OF BIOMARKERS

A suite of biomarkers of genotoxicity that could be used in a remedial project designed to test for the influence of genotoxins is given in Table 1. This list is derived from the previous discussion that described the types of DNA damage that genotoxic agents cause as well as some of the consequences of the damage. Each marker represents a unique type of change in the integrity of the DNA molecule corresponding, to some extent, to the progressive nature of carcinogenesis. Furthermore, existing analytical capabilities have been described (Huggett et al., 1992) that have the sensitivity necessary to detect these genotoxic responses.

It is obvious, from both logistical and economical points of consideration, that a remedial action project cannot include a large number of biomarkers. Established guidelines will dictate to a great extent the type and number considered suitable, but even at this point in the selection process, the list of potential biomarkers might be too large. If several general attributes and characteristics of biomarkers are

Table 1 Biomarkers of Genotoxicity/Carcinogenicity

Biomarker	Biological response	Level of biological organization	Temporal occurrence[a]	Restriction[b] Detection level	Limitations
DNA damage	Adducts	Molecular	Early	Low	Repair/ analysis
DNA damage	Strand breaks	Molecular	Early	Moderate	Repair
DNA damage	Repair	Molecular	Early	Low	Analysis
Mixed function oxidase	Enzyme induction	Molecular	Early	Moderate	Species variability
DNA damage	Minor nucleoside	Biochemical	Early/middle	High	Analysis
Gene mutation	Oncogene activation	Gene	Middle	High	Analysis
Chromosomal aberrations	Abnormal DNA	Subcellular	Middle/late	High	Analysis
Aneuploidy	Abnormal DNA	Chromosome	Middle/late	High	Analysis

[a] Temporal occurrence subsequent to exposure: early—hours to days; middle—days to weeks/months; late—weeks/months to years.

[b] Constraints on biomarker application: (1) detection level—anticipated probability of detecting biomarker and (2) limitation—factor(s) contributing to detection or affecting use of biomarker.

recognized and considered, then selection can be fine-tuned without diminishing effectiveness. Factors to be considered are:

1. Response — It should not be anticipated *a priori* that exposure of an organism to contaminants in its environment will elicit a biological response, or that responses will be observed through different levels of biological organization. Where possible, biomarker selection should be based on the known toxicological mechanism of action of the contaminants present, thus maximizing the probability of observing a biological response(s).
2. Temporal Occurrence — Some biomarkers are measurable early (hours) after exposure and are observable at the biochemical/molecular level of biological organization. These early responses may exhibit a temporal existence. Others appear much later (years) and are seen at higher levels of organization. Implicit in the concept of biomarker use is the potential for correlating responses among various levels of biological complexity.
3. Variability — Sources of variability that may influence the measurement of a biological response generally fall into two categories: those that are inherent to the laboratory method, procedure, or assay selected, and those intrinsic to the species being sampled. Sample collection, preparation, and storage as well as reagent purity and instrument selection and calibration are examples of the first category. They are more easily controlled through adherence to quality assurance and quality control policies. Age, sex, and disease state of the organism being sampled, or environmental stresses such as climate or food availability are examples of factors that contribute variability in the second category. The effects of these latter factors on the biomarker assay is difficult to predict; however, they can be documented and accounted for by

establishing baseline data from appropriate noncontaminated or reference sites.

4. Limitations — Specific limitations and restrictions apply to the use and interpretation of many of the biomarkers, and these factors should be verified by consulting the current scientific literature.

PRIORITIZATION OF BIOMARKERS

The prioritization process involves the selection of biomarkers that are appropriate to the objectives of a given biological monitoring program. The suitability of a particular biomarker should be judged on the anticipated probability of obtaining information that will document either exposure, status of cellular compensatory mechanisms, or potential for harm. An in-depth survey of biomarkers was conducted recently (Huggett et al., 1992). From such a survey it is possible to identify a small subset of biomarkers with which to examine and evaluate both specific or general types of contamination problems. The identification of this subset and its supplementation represents the first step in the prioritization process. The list of biomarkers described in Table 1 would constitute a reasonable core for evaluating genotoxicity. The suite of biomarkers chosen addresses the identified concern by including biomarkers that could provide information on exposure (whether it has occurred), on the status of compensatory mechanisms (whether the organism's capacity to maintain homeostasis has been compromised), and indications of harm (whether irreversible events have occurred). As with any suite of biomarkers, the interpretation of the observed response must be tempered by the current state of our scientific knowledge (i.e., specific limitations and restrictions). Some responses will be definitive indicators of exposure and even predictive of long-term adverse effects, whereas other responses will only be a signal of a potential problem (Shugart et al., 1989).

The importance of integrating results from biomarker studies in the laboratory with those from the field (and vice versa) cannot be overemphasized (McCarthy et al., 1992). The elucidation of relationships between known contaminants and biological response under controlled laboratory conditions will help in the selection of biomarkers for field studies. Conversely, biomarker data from field studies can direct the choice of biomarkers for more detailed laboratory studies.

RISK ASSESSMENT

As a result of the many varied practices by our society today, numerous substances and materials are constantly entering the environment. Many of these pose a long-term threat to public health and indigenous wildlife because they are nonbiodegradable. Some of the more persistent and widespread environmental chemicals can affect the structural and biological integrity of DNA, thus inducing genetic perturbations and potentiating subsequent deleterious effects. Traditional physicochemical methods work well for evaluating concentrations of exogenous chemical constituents in tissues or in well-defined subset environmental compartments, but they do not detect biological effects. Biomarkers of DNA damage can offer a sensitive, practical, and cost-effective approach for monitoring the occurrence and effects of genotoxins in the environment and would greatly aid in assessing ecosystem health and delineating the effects of anthropogenic pollutants (Lyne et al., 1992).

Biomarkers of DNA damage can improve the risk assessment process by providing the data needed to determine how genotoxic chemicals enter the body, reach target sites, and elicit biological responses. Such information can provide irrefutable evidence of exposure without relying on traditional assessment techniques, including assumptions based on data from laboratory test species. Furthermore, these measurements may indicate changes and trends in environmental levels, information that is important to environmental policy and regulation (USEPA/ORD Document, 1991).

Although it is difficult at this time to employ biomarkers to predict effects at higher levels of environmental organization (i.e., population-, community-, and ecosystem-level responses), it is important to note that appropriate use of biomarkers may provide very sensitive, early warning signals of incipient ecological damage as well as effects on human health (NRC/CEE, 1991).

FUTURE DIRECTIONS

As mentioned previously, it is difficult to demonstrate the direct effect of environmental pollution at the ecosystem level. Therefore, biomarker strategy has been to measure responses at various levels of biological organization within the individual and then to attempt to demonstrate relatedness among these levels; that is, exposure to deleterious substances may induce a cascade of biological events that eventually leads to some dysfunction. Once impairment at the individual level is observed, it becomes less difficult to extrapolate the potential for harm to the population and community levels.

Another way to approach this problem is to ask mechanistic questions about how an organism relates to other organisms and in turn to its physical environment. Ecosystems result from the dynamic interactions of living and inert matter where the living material acclimates and adapts to environmental change. These processes are physiological and have a genetic basis, so that understanding change at the genetic level should help define the more complex changes at the ecosystem level. Recent advances in the discipline of molecular biology provide the experimental tools with which to precisely investigate key biological mechanisms that regulate and limit the response of an individual organism to physical factors in their environment (e.g., chemicals). This is a fruitful area for new biomarker research, as it offers an opportunity to rapidly advance our knowledge and understanding of the effect of pollution on ecosystems (Shugart and Theodorakis, 1994).

STRATEGIES FOR BIOMARKER EVALUATION

Many biomarkers are measurements of cellular or physiological processes that are normal components of an organism's attempt to maintain a constant internal balance or to deal with normal metabolic processes. The measured values of these processes will fluctuate within some normal range within an organism. If the organism is exposed to a xenobiotic contaminant, it may respond to this exposure by compensatory increases or decreases in one or several of these cellular or physiological processes. A measure of this abnormal increase or decrease in the cellular or physiological process is the biomarker.

The dose-response model that is the generally accepted paradigm for evaluating the usefulness of a biomarker has some very beneficial uses but may not be the only paradigm for this purpose. Dose-response is a concept formulated for controlled laboratory studies in which an organism is exposed to a single chemical and a response (biomarker) is measured over a range of doses. Information obtained from these laboratory responses are then extrapolated to organisms living under natural environmental conditions. The accuracy of this extrapolation has been the subject of much concern. Natural populations are not usually exposed to a single chemical but rather to mixtures of chemicals and are subjected to many stresses not normally encountered in laboratory studies (e.g., heat and cold stress, nutritional and energy deficiency, and predator/prey stress). These variables greatly confound the extrapolation of laboratory studies to field situations. An organism under natural field conditions may respond quite differently to exposure than an organism in a laboratory situation. In fact, individuals within the same population may respond differently to the same exposure scenario, depending on individual idiosyncracies, genetic variability, varying states of nutrition, and varying microenvironments. Also, it is clear that interacting stresses and multiple exposures as well as natural physiological processes will affect biomarker dose-response.

An organism stressed by multiple contaminants or ecological events would be expected to express progressively greater biomarker responses to increased doses of a single toxicant; however, that dose-response relationship may be quite different from the dose-response relationship exhibited by an individual of the same species that is free of additional stresses (or afflicted with different stresses). In the natural environment, it is unlikely that a single unique dose-response relationship exists that can relate a biomarker response to a dose of a chemical. It is probably more reasonable to think in terms of a family of dose-response relationships, each corresponding to different combinations and permutations of interacting chemical and environmental stresses (McCarthy et al., 1992; Peakall and Shugart, 1993).

The premise of this concept is an expansion of ideas recently published by Depledge (1989) to include aspects of biomarker studies (Peakall and Shugart, 1993). The basic scenario is that an organism in its natural environment maintains homeostasis (Figure 3). As that organism is exposed to xenobiotic contaminants, physiological compensatory mechanisms become active, an increase/decrease in one or more physiological processes or functions occurs, and/or structural changes occur. All of these responses are biomarkers to the contaminant and, as such, indicate that exposure has occurred. If the exposure continues or increases, the organism's compensatory mechanisms become overwhelmed, damage occurs, and physiological repair mechanisms become active. At this point and with increased exposure or dose, additional biomarkers (e.g., reduced body weight, increased liver, spleen, and adrenal weights) are more prevalent and the susceptibility to disease and/or decimating factors (predation, freezing, starvation, and stress) increases. Under natural environmental conditions, as an organism progresses through these phases (homeostasis, compensatory response, and repair response), the energy required for normal maintenance is reduced as energy is needed for compensatory response and repair. The organism becomes progressively "weaker" and will most likely be quickly eliminated from the population. Therefore, *in situ* surveys

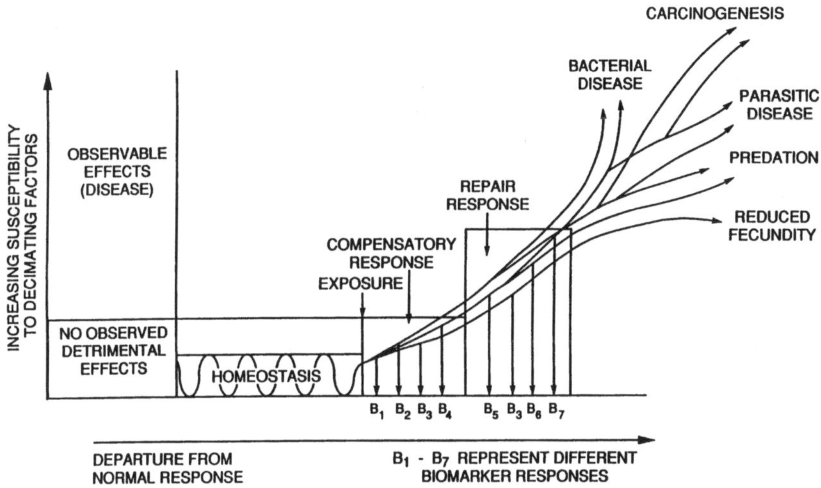

Figure 3 Multiple-response paradigm (modified from Depledge, 1989). (From McCarthy, J. F., Halbrook, R. S., and Shugart, L. R., *Conceptual Strategy for Design, Implementation, and Validation of a Biomarker-Based Biomonitoring capability,* ORNL/TM-11783, Oak Ridge National Laboratory, Oak Ridge, TN, 1992. With permission.)

of animal populations may not normally detect abundant organisms in a diseased condition even though exposure and effects have occurred.

This paradigm has a number of useful features for interpreting biomarker responses to evaluate environmental health. It recognizes and accounts for the multiple chemical and ecological stresses to which organisms are exposed and expresses the net cumulative exposure in terms of a functional assessment of the overall health of an organism (and the population as a whole). In essence, we are directly probing the biological system to determine the extent of a "health impact" of the contaminant exposure and expressing that information in terms that are relevant to environmental protection.

In this multiple-response paradigm (illustrated in Figure 3), the goal of biomarker measurements is not to quantitatively define the dose of different toxicants to which an organism is exposed but rather to determine where a population of organisms is located on the continuum between normal homeostasis and disease. Responses of a suite or battery of biomarkers serve as indicators of whether the animal is challenged but readily coping with toxicant stress (compensatory phase) or is being damaged and is having to expend resources to repair that damage. The goal of biomarker research is to identify which biomarker responses correspond with different levels of departure from normal homeostasis.

IN SITU STUDIES

It is generally recognized that contamination of aquatic environments by anthropogenic chemicals has led to increased occurrences of neoplasia in aquatic species (Harshbarger and Clark, 1990). Genotoxic agents in environmental samples (e.g, sediment, food, or water) in all probability will occur as complex mixtures. Even if the chemical composition of these mixtures were known, it would be impossible to predict the biological consequences of exposure. This is true for several reasons. First, toxicological properties of chemicals in mixtures do not necessarily follow the properties of individual components, and second, the compensatory and repair processes of organisms vary considerably. Therefore, meaningful information concerning the potential for harm from environmental exposure to genotoxic agents or attempts to remediate sites contaminated with such agents can be obtained by examining the integrity of DNA of these organisms. The two case studies that follow are examples of such an approach.

CASE STUDY: EAST FORK POPLAR CREEK, TENNESSEE

East Fork Poplar Creek (EFPC) is the receiving stream for industrial effluent from the Y-12 plant, a facility in Oak Ridge, Tennessee, managed by Martin Marietta Energy Systems for the U.S. Department of Energy. It also receives some urban and agriculture runoff. Water and sediments in EFPC downstream of the Y-12 plant contain metals, organic chemicals, and radionucleides discharged over many years of operation. Sediment samples contained 10 priority pollutant organics (including 7 polycyclic aromatic hydrocarbons, polychlorinated biphenyls, and total phenols) and 7 metals (As, Cd, Pb, Hg, Ni, Ag, and Zr) at concentrations above background. Monitoring of the discharge into EFPC identified potential toxicity problems due to copper, ammonia, residual chlorine, perchloroethylene, total nitrogen, and oil and grease (Loar, 1988). An extensive remedial action program has been implemented to reduce discharge of industrial pollution into the stream, and a Biological Monitoring and Abatement Program (BMAP), as stipulated under an interim National Pollutant Discharge Elimination System permit issued to the Oak Ridge Y-12 plant, was established. This comprehensive monitoring program includes tasks that are evaluating the toxicity of the ambient stream water, the bioaccumulation of contaminants in fish and mollusc, and the structure and function of the aquatic community, in addition to the measurement of biomarkers. Detection of strand breaks in the DNA of bluegill (*Lepomis macrochirus)* were included as a biological marker for environmental genotoxicity. Strand break analysis was included for two important reasons. First, it is a facile assay (Shugart, 1988) and therefore cost effective; and second, an increase in the level of strand breaks is indicative of a nonspecific types of DNA damage that can result from exposure to any of a wide range of genotoxic agents that have the potential to cause transient DNA strand breakage.

Results from the EFPC study obtained over a 14-month period (Shugart, 1990b) indicated that the incidence of DNA strand breakage in sunfish at several EFPC sampling stations was significantly higher than fish from the reference streams at the beginning of the study. However, the temporal and spatial decrease in DNA strand breakage in the fish population (Figure 4) suggested a decrease in genotoxic stress indicative of an improving aquatic environment during a period of intense remedial action. These results assume even greater importance when integrated into a much larger set of data obtained through the BMAP. Analyses of these data are being used to: (1) identify early warning signs of impaired health in fish;

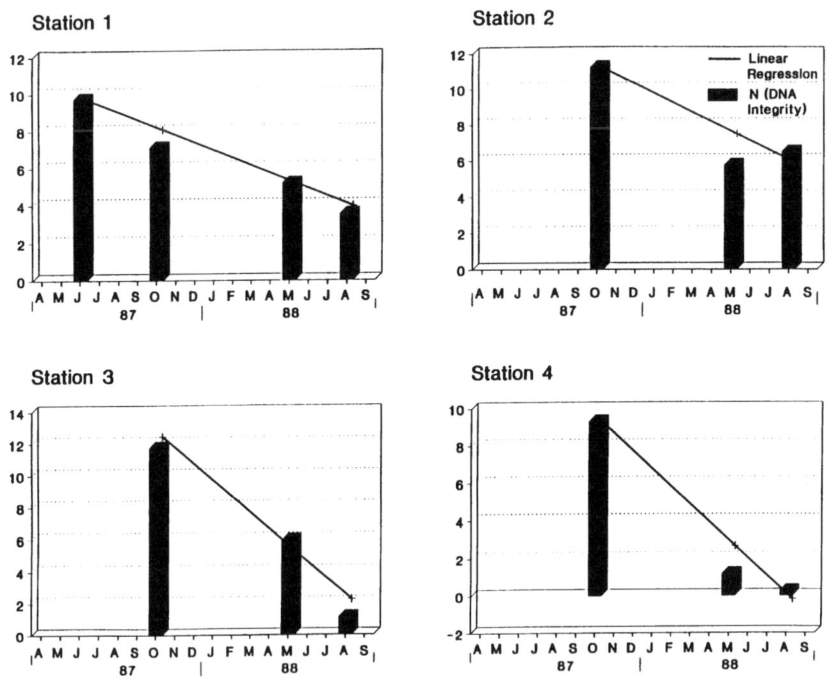

Figure 4 Temporal and spatial relationship of relative DNA strand breakage in bluegill sunfish (*Lepomis macrochirus*) from EFPC sampling stations compared to sunfish from a control, nonpolluted site from June 1987 through August 1988. Station No. 1, 23.4 km; No. 2, 18.2 km; No. 3, 13,8 km; and No. 4, 6.3 km. (From Shugart, L. R., *13th Symposium on Aquatic Toxicology and Risk Assessment: Sublethal Indicators of Toxic Stress,* Landis, W. G. and van der Schalie, W. H., Eds., ASTM, Philadelphia, 1990b, 348. With permission.)

(2) suggest causative relationships between contaminant exposure and effects at the organism level; and (3) document effects of contaminant stress on biota (Adams et al., 1990).

CASE STUDY: BOSTON HARBOR, MASSACHUSETTS

As stated previously, increased occurrences of neoplasia in aquatic species is often correlated with pollution to the environment (Harshbarger and Clark, 1990). Winter flounder (*Pseudopleuronectes americanus*) from Boston Harbor show high incidence of liver lesions that include hepatocellular and cholangiocellular carcinomas in contrast to fish from adjacent, less polluted coastal areas. Sediments in Boston Harbor are heavily contaminated with PAHs and other polychlorinated hydrocarbons, suggesting a possible chemical etiology (McMahon et al., 1990). A study was undertaken to determine whether structural damage to the DNA in the form of adducts could be detected. This was assessed by the ^{32}P-postlabeling assay. The pattern of adduct that were detected in winter flounder from Boston Harbor by the ^{32}P-postlabeling assay (Varanasi et al., 1989) were similar to that found in the

English sole from areas in Puget Sound, Washington, contaminated with similar chemicals. Adduct analyses were not performed on the winter flounder from noncontaminated areas.

Subsequent studies were performed to determine whether another genotoxic response, in the form of oncogene activation, was also occurring in this species of fish (McMahon et al., 1990). To accomplish this, the DNA from the livers of 13 winter flounder obtained from Boston Harbor and exhibiting gross liver lesions was transfected into NIH/3T3 mouse fibroblasts and assayed for the potential to induce subcutaneous fibrosarcomas in athymic nude mice. DNA from 7 to 13 diseased livers were effective in eliciting tumor formation, while DNA from the remaining lesion-bearing fish and 5 control fish gave negative results. It was concluded that the finding of c-K-*raz* gene activation in most of the diseased flounder livers represented a strong correlation between environmental exposure to genotoxic chemicals and a deleterious disease process.

SUMMARY

Cleanup of the environment is one of the most difficult decisions facing society today. The pressure to do something is immense, and a decision to do too little may cause irreversible harm to the environment, while a decision to do more than is necessary may waste large sums of money that could have better use elsewhere (Peakall, 1992). The correct decision will require that society and science address two basic questions. First, how much damage to our health and that of the environment are we prepared to tolerate; and second, how much proof is necessary before action is taken? Pragmatically, the biomarker approach would help to assess the health of the environment, and, in addition, would enable scientists to defend the advice they provide concerning environmental issues. The success of the biomarker approach is based on its predictive and anticipatory capabilities. In this chapter DNA-damaging events are proposed as a class of biomarkers suitable for assessing genotoxic insult. Strategies for the use of and evaluation of this type of biomarker are discussed in detail as well as the documentation of *in situ* applications.

ACKNOWLEDGMENT

The Oak Ridge National Laboratory is managed by Martin Marietta Energy Systems, Inc., under contract DE-AC05-84OR21400 for the U.S. Department of Energy. Environmental Sciences Division Publication No. 3937.

REFERENCES

Adams, S. M., Shugart, L. R., Southworth, G. R., and Hinton, D. E., Application of bioindicators in assessing the health of fish populations experiencing contaminant stress, in *Biological Markers of Environmental Contaminants,* McCarthy, J. F. and Shugart, L. R., Eds., Lewis Publishers, Boca Raton, FL, 1990, 333.

Alberts, B., Bray, D., Lewis, J., Raff, M., Roberts, K., and Watson, J. D., *Molecular Biology of the Cell,* 2nd ed., Garland Publishing, New York, 1989, 220.

Ashby, J., Comparison of techniques for monitoring human exposure to genotoxic chemicals, *Mutat. Res.,* 204, 543, 1988.

Bartsch, H., Hemminki, K., and O'Neill, I., Eds., *Methods for Detecting DNA Damaging Agents in Humans: Application in Cancer Epidemiology and Prevention,* IARC Scientific Publication No. 89, IARC, Lyon, France, 1988.

CBMNRC/Committee on Biological Markers of the National Research Council, Biological markers in environmental health research, *Environ. Health Perspect.,* 74, 3, 1987.

CDC/ATSDR Summary Report, *Biomarkers of Organ Damage or Dysfunction for the Renal, Hepatobiliary and Immune Systems,* Center for Disease Control, Atlanta, 1990.

Depledge, M., The rational basis for detection of the early effects of marine pollutants using physiological indicators, *AMBIO,* 18, 301, 1989.

Ehrenberg, L. E., Moustacchi, S., Osterman-Golkar, G., and Ekman, G., Dosimetry of genotoxic agents and dose/response relationship of their effects, *Mutat. Res.,* 123, 121, 1983.

Groopman, J. D. and Kensler, T. W., Molecular biomarkers for human chemical carcinogen exposures, *Chem. Res. Toxicol.,* 6, 764, 1993.

Harshbarger, J. C. and Clark, J. B., Epizootiology of neoplasms in bony fish of North America, *Sci. Total Environ.,* 94, 1, 1990.

Harvey, R. C., Polycyclic hydrocarbons and cancer, *Am. Scientist,* 70, 386, 1982.

Huggett, R. J., Kimerle, R. A., Mehrle, P. M., and Bergman, H. L., Eds., *Biomarkers: Biochemical, Physiological, and Histological Markers of Anthropogenic Stress,* Lewis Publishers, Boca Raton, FL, 1992.

Jenkins, K. D. and Sanders, B. M., Monitoring with biomarkers: A multi-tiered framework for evaluating the ecological impact of contaminants, in *Ecological Indicators,* McKenzie, D., Hyatt, E., and McDonald, J., Eds., Elsevier, New York, 1992.

Kohn, H. W., The significance of DNA-damaging assays in toxicity and carcinogenicity assessment, *Ann. N.Y. Acad. Sci.,* 407, 106, 1983.

Loar, J. M., Ed., *First Annual Report on the Y-12 Plant Biological Monitoring and Abatement Program,* Draft ORNL/TM, Oak Ridge National Laboratory, Oak Ridge, TN, 1988.

Lyne, T. B., Bickham, J. W., Lamb, T., and Gibbons, J. W., The application of bioassays in risk assessment of environmental pollution, *Risk Analysis,* 12, 361, 1992.

Marnett, L. J., Frontiers in molecular toxicology, *Chem. Res. Toxicol.,* 6, 739, 1993.

McCarthy, J. F., Adams, S. M., Jimenez, B. D., and Shugart, L. R., Environmental monitoring of biological markers in animals and plants, in *Proc 2nd US-USSR Symposium on: Air Pollution Effects on Vegetation,* Noble, R. D., Martin, J. L., and Jensen, K. F., Eds., Northeastern Forest Experiment Station, Broomall, PA, 1989, 187.

McCarthy, J. F. and Shugart, L. R., Eds., *Biological Markers of Environmental Contamination,* Lewis Publishers, Boca Raton, FL, 1990.

McCarthy, J. F., Halbrook, R. S., and Shugart, L. R., *Conceptual Strategy for Design, Implementation, and Validation of a Biomarker-Based Biomonitoring Capability,* ORNL/TM-11783, Oak Ridge National Laboratory, 1992.

McMahon, G., Huber, L. J., Moore, M. J., Stegeman, J. J., and Wogan, G. N., c-K-*Ras* oncogenes: Prevalence in livers of winter flounder from Boston Harbor, in *Biological Markers of Environmental Contaminants,* McCarthy, J. F. and Shugart, L. R., Eds., Lewis Publishers, Boca Raton, FL, 1990, 229.

NRC/CEE. National Research Council/Committee on Environmental Epidemiology, *Environmental Epidemiology: Public Health and Hazardous Wastes,* National Academy Press, Washington, D.C., 1991.

Peakall, D., *Animal Biomarkers as Pollution Indicators,* Chapman and Hall, London, 1992.

Peakall, D. and Shugart, L. R., Eds., *Biomarkers: Research and Application in the Assessment of Environmental Health,* NATO ASI Series H: Cell Biology, Vol. 68, Springer-Verlag, Heidelberg, 1993.

Phillips, D. and Sims, P., PAH metabolites: Their reaction with nucleic acids, in *Chemical Carcinogens and DNA,* Vol. 2, Grover, P. L., Ed., CRC Press, Boca Raton, FL, 1979, 9.

Phillips, D., Fifty years of benzo[a]pyrene, *Nature,* 303, 478, 1983.

Pohl, I. R., An immunochemical approach of identifying and characterizing protein targets of toxic reactive metabolites, *Chem. Res. Toxicol.,* 6, 786, 1993.

Rahn, R. O., Chang, S. S., Holland, J. M., and Shugart, L. R., A fluorometric HPLC assay for quantitating the binding of benzo[a]pyrene metabolites to DNA, *Biochem. Biophys. Res. Commun.,* 109, 262, 1982.

Randerath, K., Reddy, M. V., and Gupta, R. C., ^{32}P-Labeling test for DNA damage, *Proc. Natl. Acad. Sci. U.S.A.,* 78, 6126, 1981.

Sancar, A. and Sancar, G. G., DNA repair enzymes, *Ann. Rev. Biochem.,* 57, 29, 1988.

Sandhu, S. S. and Lower, W. R., *In situ* assessment of genotoxic hazards of environmental pollution, *Toxicol. Ind. Health,* 5, 73, 1989.

Santella, R. M., Gasparo, F., and Hsieh, L., Quantitation of carcinogen-DNA adducts with monoclonal antibodies, *Prog. Exp. Tumor. Res.,* 31, 63, 1987.

Santella, R. M., Application of new techniques for the detection of carcinogen adducts to human population monitoring, *Mutat. Res.,* 205, 271, 1988.

Shugart, L. R. and Kao, J., Examination of adduct formation in vivo in the mouse between benzo[a]pyrene and DNA of skin and hemoglobin of red blood cells, *Environ. Health Perspect.,* 62, 223, 1985.

Shugart, L. R., Quantitation of chemically induced damage to DNA of aquatic organisms by alkaline unwinding assay, *Aquatic Toxicol.,* 13, 43, 1988.

Shugart, L. R., Adams, S. M., Jimenez, B. D., Talmage, S. S., and McCarthy, J. F., Biological markers to study exposure in animals and bioavailability of environmental contaminants, in *Biological Monitoring for Pesticide Exposure: Measurement, Estimation, and Risk Reduction,* Wang, R. G. M., Franklin, C. A., Honeycutt, R. C., and Reinert, J. C., Eds., ACS Symposium Series 382, American Chemical Society, Washington, D.C., 1989, 86.

Shugart, L. R., Biological monitoring: Testing for genotoxicity, in *Biological Markers of Environmental Contaminants,* McCarthy, J. F. and Shugart, L. R., Eds., Lewis Publishers, Boca Raton, FL, 1990a, 205.

Shugart, L. R., DNA damage as an indicator of pollutant-induced genotoxicity, in *13th Symposium on Aquatic Toxicology and Risk Assessment: Sublethal Indicators of Toxic Stress,* Landis, W. G. and van der Schalie, W. H., Eds., ASTM, Philadelphia, PA, 1990b, 348.

Shugart, L. R., McCarthy, J. F., and Halbrook, R. S., Biological markers of environmental and ecological contamination: An overview, *Risk Analysis,* 12, 353, 1992a.

Shugart, L. R., Bickham, J., Jackim, G., McMahon, G., Ridley, W., Stein, J., and Steinert, S., DNA alterations, in *Biomarkers: Biochemical, Physiological, and Histological Markers of Anthropogenic Stress,* Huggett, R. J., Kimerle, R. A., Mehrle, P. M., and Bergman, H. L., Eds., Lewis Publishers, Boca Raton, FL, 1992b, 125.

Shugart, L. R., *Use of Biomarkers in Assessing Health and Environmental Impacts of Chemical Pollutants,* Travis, C. C., Ed., Plenum Press, New York, 1993a, 237.

Shugart, L. R., Hemoglobin adducts, in *Nondestructive Biomarkers in Vertebrates,* Fossi, M. C. and Leonzio, C., Eds., Lewis Publishers, Boca Raton, FL, 1993b, 159.

Shugart, L. R. and Halbrook, R. S., The use of biomarkers to assess environmental health, in *International Conference on Monitoring Toxic Chemicals and Biomarkers,* Vol. 1716, Vo-Dinh, T. and Camman, K., SPIE Publishers, Bellingham, WA, 1993, 199.

Shugart, L. R. and Theodorakis, C., Environmental genotoxicity: Probing the underlying mechanisms, *Environ. Health Perspect.,* 102 (Suppl. 12), 13, 1994.

Stein, J. E., Reichert, W. L., and Varanasi, U., Molecular epizootiology: Assessment of exposure to genotoxic compounds in teleosts, *Environ. Health Perspect.,* in press.

Thilly, W. G. and Call, K.M., Genetic toxicology, in *Third Edition of Casarett and Doull's Toxicology,* Klaassen, C. D., Amdur, M. O., and Doull, J., Eds., Macmillan, New York, 1986, 174.

USEPA/ORD Document, *ORD Health Biomarkers Program: Research Strategy Document.* U.S. EPA/Office of Research & Development, Research Triangle Park, NC, 1991.

Varanasi U., Reichert, W. L., and Stein, J. E., ^{32}P-Postlabeling analysis of DNA adducts in liver of wild english sole (*Paraophrys vetulus*) and winter flounder (*Pseudopleuronectes americanus*), *Cancer Res.,* 49, 1171, 1989.

Weinstein, I. B., Current concepts on mechanism of chemical carcinogenesis, *Bull. N.Y. Acad. Med.,* 54, 336, 1978.

Wogan, G. N. and Gorelick, N. J., Chemical and biochemical dosimetry to exposure to genotoxic chemicals, *Environ. Health Perspec.,* 62, 5, 1985.

CHAPTER 5

In Situ Evaluation of Genetic Hazards of Complex Chemical Mixtures

Shabeg S. Sandhu, William R. Lower, and Yog R. Ahuja

CONTENTS

INTRODUCTION

During the last three decades, increasing concerns have been expressed by the scientific community and the general public about the impact of environmental pollutants on human health and ecosystems. Environmental pollutants such as teratogens, mutagens, carcinogens, and neurotoxins have been detected in a variety of products used in our daily life. In addition to the individual chemicals, large

1-56670-141-4/96/$0.00+$.50

quantities of complex mixtures of chemicals are released into the environment as solid wastes, effluents, and emissions. Potent mutagens and carcinogens have been discovered in these complex mixtures as well (Waters et al., 1990; Sandhu et al., 1989).

The discovery of a large number of industrial waste sites and their potential impact on human and environmental health (see Chapter 1) has further amplified public concerns related to environmental safety. The Comprehensive Environmental Response, Compensation, and Liability Act of 1980 (CERCLA, or Superfund), as amended in 1986 (Superfund Amendment and Reauthorization Act — SARA), mandated the U.S. Environmental Protection Agency (EPA) to establish programs to respond to the release of hazardous substances into the environment. Under this mandate, EPA is required to identify hazardous waste sites (HWS), provide analysis on the toxic potential of these sites for human health and ecology, and suggest remedial alternatives and develop guidelines for evaluating residual toxicity before the HWS are listed in the National Priorities List (NPL), as well as after the remedial action before the sites are considered "clean" and delisted from NPL. As discussed in Chapter 3, each site is a unique entity and different technologies may be applied to list and delist each site on NPL.

The major criteria currently used to determine the potential health hazards from chemical waste sites include considerations of the concentration of certain chemicals and of their toxicological profiles. Although sophisticated technologies are available for a detailed chemical analysis, concentrations and toxicological profiles of individual chemicals of environmental samples alone may not be adequate for determining the hazard potential of chemical mixtures. Under natural environments the exposure to environmental pollutants rarely occurs from individual chemicals, but most often human and other organisms are exposed to toxic chemicals as complex mixtures of chemicals. The toxic potential of a chemical depends on the kind and the proportion of other chemicals in the mixtures (Gill and Sandhu, 1993). It may be misleading to estimate the potential toxic effects of chemical mixtures from the chemical analysis of their individual components (also see Chapter 3). Thus, in addition to chemical analysis, bioanalysis of the environmental samples as well as *in situ* biomonitoring should be an integral part of the process for determining the cleanup standards.

Assessment of the biological hazards from exposure to chemicals can be performed by either bringing the samples into the laboratory to be evaluated or by monitoring the alterations in the biomarkers of concern in the indicator organisms placed or indigenous to the test site. In the laboratory, some of the environmental variables such as temperature, radiation, nutrient stress, pH of the test media, etc., can be controlled and the contribution of each of these factors to the variability in data can be evaluated. However, it is not possible to duplicate in the laboratory the complex interactive effects of chemical and physical agents that exist in nature. Consequently, the extrapolative interpretations of the data from the laboratory to the potential effects under natural environment may have some limitations.

Hazard assessment under field conditions can be accomplished by monitoring the biomarkers of concern in indigenous sentinel populations of plants and animals and by enumerating alterations in specific endpoints in the organisms specifically introduced into a test site. A sentinel species can be any organism that can be used as an indicator of environmental hazard as revealed by body burden analysis for the target chemicals or by bioanalysis for specific biomarker of an individual after the exposure (Lower and Kendell, 1990).

An integrated assessment of the impact of environmental pollutants on human health requires the evaluation of biomarkers of exposure and effects related to several disciplines such as neurotoxicology, genetic toxicology, developmental toxicology, and toxicological evaluations for pulmonary, immune, and cardiovascular functions. However, for this chapter, a brief review pertaining only to one of the disciplines related to health effects research, i.e. genetic toxicology, has been selected. In the following sections a discussion is presented on the potential utility of animal and plant species as sentinels and for *in situ* monitoring of genetic hazards from exposure to chemicals at industrial waste sites. Detailed reviews of *in situ* methods for assessing biological hazards of environmental pollutants have been published recently (McCarthy and Shugart, 1990; Sandhu et al., 1989). For a review of the utility of short-term bioassays performed in the laboratory for the genetic analysis of complex mixtures, the reader is referred to a series of publications by Waters et al. (1979, 1981, 1983, 1990) and Sandhu et al. (1987).

DETECTION OF GENETIC HAZARDS AT INDUSTRIAL WASTE SITES

Mutagens in the environment are present in relatively low concentrations and may be more concentrated in one component of the environment than another. Therefore, a battery of sentinel species and *in situ* bioassays is needed to assess the genetic hazards in a particular environment. In this chapter, we list the potential utility of some of the species that have been used for evaluating exposure to and effects of environmental genotoxins. These species, considered as sentinels, are listed according to the media in which these may be utilized, although some can be used for assays in more than one media. Plants and aquatic animals are uniquely suitable for *in situ* monitoring; the role of these organisms in environmental monitoring is emphasized in this chapter. Since contamination of aquifers through leachate from hazardous wastes and pollution of ambient environment by volatile compounds is a major source of pollution from HWS, *in situ* monitoring of the toxic chemicals in the atmosphere and in the aquatic environment is emphasized.

BIOASSAY FOR GENOTOXICITY EVALUATION OF ATMOSPHERIC POLLUTANTS

A variety of indicator species representing microorganisms, plants, insects, and animals have been used for evaluating the exposure and effects of atmospheric pollutants. The choice of the indicator species for use at a specific site depends upon the purpose of monitoring, available expertise, and location of the test site.

Microbial Organisms

For evaluating the genotoxic potential of volatile organics and airborne particulates, the general practice is to collect the samples from the media, extract these samples with organic solvents, and perform bioanalysis in laboratory using microbial organisms such as *Salmonella typhimurium.* The advantage of this approach is that the experiments can be performed within the confines of a laboratory environment, thus reducing within-laboratory variability. However, during the process of sample collection and preparation, a number of short-lived, intermediate mutants may be lost. A more suitable test system for detecting the levels of mutagens in the ambient environment would be an *in situ* mutagenicity assay that may allow for direct exposure of test organisms to the conditions that closely approximate the real world environment. Ong et al. (1985) developed and partially validated such a test system using *Salmonella typhimurium* for gene mutation and human lymphocytes for sister chromatid exchanges for monitoring mutagenicity of volatile organics in the atmosphere at work places. The technique has potential utility but its application for *in situ* use may be limited for lack of sufficient entrapment of gases and, possibly, the inability to detect mutagens that are strongly bound to particles.

Plant Species

Several plant species listed in Table 1 can be used for monitoring the levels of mutagens in the environment. Among these, *Tradescantia* is perhaps the most suitable species for environmental monitoring. It can be used for monitoring short-term exposure (for less than 1 h) or long-term chronic exposure. It can be used either as an intact plant for assessing multimedia exposure or as stem cuttings for measuring the levels of atmospheric pollutants. *Tradescantia* can be cultured easily in the laboratory and maintained under field conditions. Since it is a perennial species, *Tradescantia* is a suitable test organism for studying the mutational dynamics in ecosystems.

Two endpoints, namely, stamen hair assay for gene mutation in somatic cells and micronucleus assay for chromosome aberrations in pollen mother cells, have been used for monitoring the mutagenicity of vapor phase organics in the atmosphere.

Tradescantia Stamen Hair (Trad. SH) — The Trad. SH test system was developed by Arnold Sparrow and his colleagues at Brookhaven National Laboratory (BNL), Upton, New York, for monitoring levels of ionizing radiation.

The Trad. SH assay measures gene mutation as indicated by a change in pigmentation from blue to pink in stamen hair cells. The most commonly used clone of *Tradescantia,* clone 4430, is a hybrid between two *Tradescantia* species and propagation of this clone is only through vegetative means, ensuring the genetic uniformity of the test material. This species has been used for monitoring the levels of radiation around nuclear power plants in Japan (Ichikawa, 1981; Cebulska-Wasilewska, 1993) and for measuring the genotoxic effects of radiation in the area affected by the failure of the Chernobyl nuclear power plant in 1986 in Ukraine (Shevchenko, 1989). Background information and assay protocol for the *Tradescantia* assay have been reviewed by Ichikawa (1993).

For the *in situ* monitoring of ambient air with *Tradescantia,* a mobile monitoring vehicle (MMV) was used by BNL under contract to the National Institute of Environmental Health Sciences and the EPA (Schairer et al., 1979). In this study, plant cuttings with developing inflorescences were exposed to the test environment for a period of 10–17 days. After exposure at the site, inflorescence were fixed and

Table 1 Bioindicators for in situ monitoring for genotoxic effects in Terrestrial Environment

Assay	Organism	Reference
	1. Point mutation	
Stamen hair	*Tradescantia*	Schairer et al. (1979), Lower et al. (1978, 1983, 1984)
Waxy Pollen	*Zea-maze*	Lower et al. (1978), Gentile et al. (1982), Plewa et al. (1984), Hopke et al. (1982)
Chlorophyll mutants	*Hordeum vulgare*	Constantin and Nilan (1982)
Albino embryo	*Arabidopsis*	Redei (1982)
	2. Micronuclei formation	
Micronuclei	*Tradescantia*	Ma (1981), Ma et al. (1983, 1985, 1987), Lower et al. (1985), Gill et al. (1991)
White-footed mouse	(*Peromyscus leucopus)*	Tice et al. (1987)
	3. Chromosome aberrations	
Royal fern	(*Osmunda regalis)*	Klekowski (1978)
Barley root cells	*Hordeum vulgare*	Constantin & Nilan (1982)
Beans root cells	*Vicia faba*	Ma (1982)
Onions root cells	*Allium cepa*	Grant (1982)
Voles	(*Microtus)*	Rowley et al. (1983)
White-footed mouse	(*Peromyscus leucopus)*	Tice et al. (1987)
Cotton rat	(*Sigmadon hispidus)*	Thompson et al. (1988)
Kangaroo rats	(*Dipodomys merriami)*	Lower et al. (1983)
	(D. deserti)	Schaeffer et al. (1987)
	4. SCEs	
Wild house mouse	(*Mus musculus*)	Nayak and Petras (1985)
White–footed mouse	(*Peromyscus leucopus)*	Tice et al. (1987)
	5. Carcinoma	
Lung cancer	Albino rats and inbred mice	Wang et al. (1984)
Breast cancer	Dogs	Schneider et al. (1969), Schneider (1970)
Osteosarcoma	Dogs	Tjalma (1966)
Mesothelioma	Dogs	Glickman et al. (1983)
Bladder cancer	Dogs	Reif and Cohen (1970), Hays et al. (1981)

transported to the laboratory for enumerating the frequency of mutation from blue to pink cells. The intact plants can also be used for exposure to atmospheric pollutants if they are planted in uncontaminated soil and placed at the test site. The MMV was equipped with plant growth chambers to provide a constant flow of untempered air. The clean air control chamber, equipped with filter train composed of activated charcoal, HEPA particulate, and purafil filters served as a concurrent control. The MMV was used to monitor air quality in 18 cities from 1976–1980 in the United States. Two clean sites, Grand Canyon, Arizona, and Pittsboro, North Carolina, registered no increase in mutation frequencies above the historic laboratory background, whereas Elizabeth, New Jersey, which has a high concentration of petrochemical industries, showed mutation frequencies as much as 90% higher than the concurrent controls. The utility of this assay has encompassed monitoring the mutagenicity of leachate from sewage sledges, water pollution, bottom sediment from a drinking water reservoir, chemical landfills, diesel emissions, air pollution from petrochemical plants, and soil and air pollution from a lead smelter (Lower et al., 1978, 1983a,b; Ma et al., 1983).

Tradescantia Micronucleus Assay (Trad. MCN) — The Trad. MCN assay detects chromosome aberrations in the meiotic cells of *Tradescantia* as micronuclei. The genetic basis and the detailed protocol for this assay has been reported by Ma (1982). The assay is very sensitive and relatively easy to perform. The procedure consists of exposing the cuttings of *Tradescantia* with young inflorescences (contained in a beaker of water) to the ambient atmosphere at a test site. After exposure, the inflorescences are fixed and pollen mother cells are stained and analyzed for micronuclei. The exposure period is usually for a few hours. This assay has been used by Ma and his colleagues for measuring the levels of clastogens at truck stops, indoor environments in households, and at municipal waste sites and by Lower and colleagues to monitor sewage sludge and obscurant smoke used in smoke screens (Lower et al., 1983a,b; Schaeffer et al., 1987).

The Trad. MCN test system was employed by Gill et al. (1991) to assess the mutagenicity of volatile compounds at a Superfund site in North Carolina, contaminated with high levels of agricultural insecticides. Exposure of *Tradescantia* stem cuttings at the site before remediation showed significantly elevated frequencies of micronuclei in the pollen mother cells as compared to the control sites (a laboratory control as well as a "clean site" control). After remediation, chemical analysis showed 90% reduction in the levels of chemical contaminants in the air. Concomitantly, the analysis of the *Tradescantia* inflorescence exposed at the site after remediation also showed no significant differences in the micronuclei frequencies between the concurrent controls and the *in situ* exposure groups, indicating the suitability of this assay system for detecting mutagens in the ambient environment.

The plant assays are simple, cost effective, and easily adaptable for application in laboratory or under field conditions. Their sensitivity for evaluating chemicals for mutagenicity has been reported to be as good as any of the commonly used mammalian cell bioassays (Ennever et al., 1988). However, there is hesitancy on the part of decision-makers to consider data from plant species for use in hazard assessment. Since plants are multicellular organisms, the data from plant assays is perhaps more relevant to humans than the data from prokaryotic cells, but unfortunately the bias against plants as test organisms continues to exist.

Insects

Among the insects, *Drosophila melanogaster* may be potentially useful organism for *in situ* biomonitoring because of an extensive genetic database on this insect. A number of *Drosophila* tests such as sex-linked recessive lethal, sex-linked mosaicism, recessive morphological mutants, and tests for aneuploidy (Valencia et al., 1984; Zimmering et al., 1986) have been used in the laboratory. Application of these assays could be extended for *in situ* assessment of genetic hazards of atmospheric contaminants. In addition, efforts are being made to develop transgenic fruitflies with mammalian receptors, regulatory elements, and detoxication enzymes to parallel human response to environmental health hazards (Harshman, 1989).

Mammalian Species

Domestic and wild species of animals suffer acute as well as chronic effects of exposure to environmental contaminants in ways similar to human beings and can serve as sentinels for human exposure. Most mammals have shorter life spans than humans, therefore, most environmentally induced diseases appear earlier in sentinel animals than in humans. The potential of domestic and wild mammals for *in situ* hazard assessment has been emphasized only recently. A few preliminary studies have been conducted to determine if wild, feral, pet, or laboratory mammals can serve as useful indicators for the presence of genotoxins in the environment. The most appropriate use of the data from sentinel animals would be for hazard assessment and hazard characterization. As is the case with laboratory animals, domestic and wild species of animals differ from humans in metabolism and pharmacokinetics. Therefore, the quantitative extrapolation of data for risk assessment from sentinel species to humans has limitations. An excellent and extensive review on the utility of animals as sentinels of environmental health hazards has been published by the National Research Council (NRC) of the National Academy of Sciences (1991). The NRC review provides an in–depth analysis on the utility of terrestrial and aquatic animals for observational, experimental, analytical, and descriptive epidemiological assessments.

Domestic Animals — Domestic animals such as dogs, cats, and farm animals can potentially provide very useful information on the exposure and effects of contaminants in the environment. Pets share their owners' environment but are free from confounding factors such as alcohol consumption and active smoking. The use of pets for environmental risk assessment has been reviewed by Glickman (1989). Unfortunately, most of the work reported in the literature (Table 1) has been limited to studies on the incidence of carcinomas in domestic pets. Information on the utility of molecular and genetic biomarkers of environmental contaminants in domestic animals is lacking. However, most of the HWS are distant from human habitation and, consequently, the usefulness of the domestic animals for evaluating the safety of the HWS may be rather limited.

Wild and Feral Mammals — Only preliminary studies have been performed for evaluating the utility of wild and feral mammals as indicators of genetic hazards from exposure to chemicals from industrial waste sites or from industrial emissions (Table 1). An excellent review on this topic is presented in Chapter 6. Thompson et al. (1988) and McBee et al. (1987) found elevated frequencies of chromosome aberrations in both the cotton rat (*Sigmodon hispidus)* and the white-footed mouse (*Peromyscus leucopus)* captured near the HWS as compared to the animals captured from the "pristine" sites. Evidence of increase in sister chromatid exchange (SCE) frequencies as compared to the concurrent control sites, was found in kangaroo rats (*Dipodomys deserti)* exposed to obscurant smokes used in smoke screens (Lower et al., 1983a; Schaeffer et al., 1987).

Nayak and Petras (1985) studied SCEs in wild house mice (*Mus musculus)* caught from corn cribs at varying distances from industrial sites in Southern Ontario, Canada. A comparison among SCE frequencies in wild mice vs. inbred mice reared in the laboratory and wild mice after a 9-month quarantine in the laboratory vs. inbred mice kept outdoors for 9 months suggested that environmental factors may be involved in inducing significantly higher levels of SCEs in the animals inhabiting particular areas. Tice et al. (1987) concluded from a preliminary study on the cytogenetic analysis of the bone marrow and germ cell damage of wild populations of the white-footed mouse (*Peromyscus leucopus)* from HWS and from uncontaminated areas that these animals can be employed to monitor the sensitivity and seasonal variability of hazards from exposure to chemicals at specific test sites.

GENOTOXICITY ASSESSMENT IN THE AQUATIC ENVIRONMENT

Green plants and aquatic animals can easily be used for monitoring the levels of genotoxins in the aquatic environment. A summary of the endpoints used for monitoring the genotoxic hazards in aquatic systems is given in Table 2. A few examples on the utility of plant species and aquatic animals for assessing the genotoxic levels in the aquatic environment are discussed below.

PLANTS

Genetic assays utilizing two plant species, i.e., royal fern (*Osmunda regalis)* and *Tradescantia,* will be discussed as an example of how plant assays can be used to monitor mutagens in the aquatic environment.

Klekowski (1978) used royal ferns to show a high incidence of postzygotic mutational damage in a population growing in a river heavily polluted with effluents from a paper mill, whereas no genetic damage was detected in the population of ferns growing in the unpolluted environment. Since these organisms are long-lived perennials with an ontogenetic system that stores mutagenic damage, they can be used to give an integrated estimate of genetic damage over a period of time. A number of developmental, genetic, and cytogenetic endpoints are available in ferns that render them as very useful and inexpensive *in situ* test systems for detecting levels of mutagens in the aquatic environment. Unfortunately, no subsequent use has been made of this test system.

Table 2 Bioindicators of Exposure to Genotoxic Agents in Aquatic Animals

	Organism	Reference
	1. Chromosome aberrations	
Central mud minnow	(*Umbra limi*)	Kligerman and Bloom (1975)
Eastern mud minnow	(*U. pygmeae*)	Prein et al. (1978), Hooftman and Vink (1981)
Oyster toadfish	(*Opsanus tau*)	Maddock et al. (1986)
Marine worm	(*Neanthes arenaceodentata*)	Pesch et al. (1981)
Marine worm	(*Pomotoceres triqueter*)	Dixon (1986)
	2. Micronuclei	
Eastern mud minnow	(*U. pygmeae*)	Hooftman and de Raat (1982)
Amphibian	(*Caudiverbera caudiverbera*)	Venegas et al. (1987)
Algae and grasses		Fang et al. (1983)
	3. SCEs	
Central mud minnow	(*Umbra limi*)	Kligerman and Bloom (1976)
Eastern mud minnow	(*U. pygmeae*)	Alink et al. (1980), Hooftman and Vink (1981)
Oyster toadfish	(*Opsanus tau*)	Maddock et al. (1986)
Mussel	(*Mytilus edulis*)	Dixon and Clarke (1982), Harrison and Jones (1982)
Marine mussels	(*M. galloprovincialis*)	Brunetti et al. (1986)
	4. DNA perturbations	
Marine mussels	(*M. galloprovincialis*)	Kurelec et al. (1990)
Redbreast sunfish	(*Lepomis auritus*)	Adams et al. (1990)
	5. DNA damage and repair	
Marine sponge	(*Thethya lyncurium*)	Zahn et al. (1983)

Tradescantia — micronucleus assay was used by Ma et al. (1987) to study water quality in a reservoir and shallow wells in a rural area over a period of years. The results from these studies indicated that Trad.MN was an efficient tool for detecting the seasonal variation in the levels of mutagenicity in the reservoir and the shallow wells which collected runoff water from nearby farms.

Recently, Grant et al. (1992) employed three plant bioassays, including Trad. MN assay, for *in situ* detection of genotoxins in a creek and an area of Lake Superior contaminated with raw effluent from a paper mill. The same bioassays were used to evaluate water samples from the test sites in laboratory for their genotoxic potentials. Their data indicated a significant decrease in genotoxic effects when the samples from a test site are tested in the laboratory vs. *in situ* measurements. The reduction in mutagenicity observed in samples evaluated in the laboratory, as compared to the *in situ* evaluations, may be due to the dissipation of active agents in the water after removal from the test site, their conversion to inactive metabolites, and the removal of the atmospheric component.

AQUATIC ANIMALS

A number of biomarkers in various species of aquatic animals (Table 2) have been used for evaluating the mutagenicity of water samples in the laboratory and as sentinels for detecting genetic hazards in the aquatic environment (de Raat et al., 1990). All of these bioindicators were developed initially for evaluating the genotoxic potential of individual chemicals or complex chemical mixtures in laboratory. However, these endpoints can also be used for the *in situ* assessment of genetic hazards in aquatic and terrestrial environments. For example, SCEs have been used as an indicator of exposure to genotoxins in the heavily polluted Rhine River using Eastern mud minnows (*Umbra pygmaea)* (Prein et al., 1978; Alink et al., 1980; Hooftman and Vink, 1981). Similarly, SCEs have been utilized in marine animals such as polychete worms (*Neanthes arenaceandentata)* (Pesch and Pesch, 1980; Pesch et al., 1981), mussel (*Mytilu edulis)* larvae (Harrison and Jones, 1982), mussel (*M. galloprovincialis)* eggs (Brunetti et al., 1986), and oyster toadfish (*Opanus tau)* (Maddock et al., 1986) for evaluating exposure to genotoxins.

Other cytogenetic endpoints such as micronuclei and chromosome aberrations have also been observed in aquatic animals as indicators of exposure to pollutants in the aquatic environment (Prein et al., 1978; Hooftman and de Raat, 1982; Hose et al., 1987). The micronucleus test is convenient because it does not depend on the quality of karyotype and is relatively easy to perform. However, chromosome effects, in general, require relatively higher concentrations of exposure to pollutants. Thus, consideration of negative results for chromosomal effects may be somewhat misleading.

Increased prevalence of neoplasms in native fish has been detected in the areas affected by municipal and industrial effluent and emissions (Couch and Harshberger, 1985; Mix, 1986). Because of the antagonistic and synergistic effects of the chemicals in mixtures, it is not always possible to correlate the concentrations of chemical pollutants in the aquatic systems and the incidence of neoplasm in the aquatic animals. Also, there is sufficient lag time between the exposure of animals to carcinogens and the appearance of carcinoma. In spite of these limitations, the increase in incidence of neoplasia in aquatic animals is a reliable indicator of environmental exposure.

TRANSGENIC FISH

Recent progress in recombinant DNA and gene transfer technologies has made it possible to produce transgenic animals harboring vectors with well-defined markers that can be rescued from test animals after exposure for enumerating induced genetic alterations (Dubridge and Carlos, 1988). The utility of transgenic mouse models containing bacteriophage *lamda* or *phi*X174 is now being evaluated for identifying genetic hazards of environmental chemicals (Malling and Burkhart,

1992). In addition to transgenic mouse, the gene transfer technology has been utilized to produce a variety of other transgenic animals such as *Drosophila,* sea urchins, farm animals, and fish (see review by Chen and Powers, 1990).

In order to produce transgenic fish models for detecting and quantifying the mutagens in an aquatic environment, efforts are under way to develop transgenic Japanese medaka *(Oryzias latipes)* and teleost fish *Fundulus heteroculitis* containing *phi*X174 shuttle vector stably integrated in their genome. The basic principle entails the introduction of recombinant genes with a well-defined mutation target into the chromosomal genome of a host. Following the exposure of the host to a mutagen, the target gene is recovered and serves as an indicator of mutagenesis. The Japanese medaka is a freshwater fish, whereas *Fundulus* is an estuarine animal. The genetics, developmental biology and specific developmental stages for both of these species are well documented (Powers, in press).

The transgenic fish can be used to study the mutation index of an aquatic ecosystem by exposing these animals in specially designed cages to evaluate the effects of the total environment or the test can be performed in the biomonitoring trailer described by Gardner et al. (1989).

INTEGRATION, INTERPRETATION, AND LIMITATIONS

In situ bioassays, as a component of environmental monitoring program, may be applied as an initial part of a cost effective and efficient way of identifying the levels of genotoxic hazards in the environment. If no biological hazard is identified, the concern may be downgraded without costly chemical analysis. *In situ* monitoring may be conducted in concert with both biological and chemical analysis of environmental samples in the laboratory. If *in situ* monitoring indicates the presence of genotoxins at a test site, laboratory analysis of the samples may be necessary to identify the chemicals, or the classes of chemicals, responsible for the observed effects. Frequently, the environmental samples have to be fractionated to identify the chemical fraction responsible for the genotoxic effects. However, in the process of sample preparations, many artifacts may be introduced in the samples that may lead to misinterpretation of the results.

Factors such as availability of nutrients, pH of soil, temperature, age, sex, disease, etc., have an impact on the data collected under field conditions. Therefore, the influence of these factors on the variability of data should be analyzed under laboratory conditions. Under field conditions it is difficult to control environmental variables, and the need for careful planning of experimental design, sampling techniques, and statistical analysis cannot be overemphasized.

Concurrent positive and negative field and laboratory controls are necessary for the proper interpretation of the data from *in situ* monitoring. A field reference site should be geographically close and as similar as possible to the test site, except for the variable under investigation. The selection of concurrent field controls usually is difficult due to widespread application of agricultural chemicals, wind directions relevant to a chemical plant, radiation, etc., which can impact on the variability of the data and its interpretation. Therefore, for evaluating the environmental consequences of exposure to chemicals at HWS, more than one concurrent field control may be necessary.

The selection of the test organism for *in situ* monitoring will depend upon the media to be monitored. Preference in the selection should be given to a test organism that may have the following attributes: (1) can be used in the laboratory as well as under field conditions; i.e., *Tradescantia,* transgenic animals or plants; (2) provides ease for karyotypic analysis; (3) has short generation time; e.g., *Arabidopsis;* (4) has clearly characterized multiple genetic markers; (5) provides wide geographical distribution, and should provide samples for most of the time during the year; (6) in the case of plants, perennial species should be preferable; and (7) requires short exposure periods.

Since it is not possible to control environmental variables under field conditions, there is generally a considerable variability in the data collected under such conditions. In addition, the complex chemical mixtures may invoke response in more than one biomarker in more than one species. Consequently, the interpretation of the data from any single biomarker may lead to erroneous conclusions. A suite of biomarkers may be required for the proper understanding of the mechanisms of toxicity and the interpretation of the results.

The utility of *in situ* assays for monitoring environmental mutagens is especially relevant for the developing countries where resources are relatively scarce. Results of a recent international collaborative study on the utility of plant test systems (Sandhu et al., 1994) in genetic toxicology show that test systems such as Tradescantia micronucleus and Arabidopsis embryo assay are sensitive and potentially useful for assessing the mutational dynamics of an ecosystem.

With a few exceptions, the genetic markers for environmental monitoring have been limited to gene mutations and chromosomal aberrations. During the last two decades, revolutionary advances have been made in the biological sciences as a results of spectacular discoveries in molecular biology and sophisticated biological instrumentation. Although the medical community has readily incorporated these technologies into medical research, environmental scientists have been slow to adopt these approaches for assessing the consequences of anthropogenically induced environmental stresses in ecosystems.

Excellent reviews have been published recently on the availability and potential utility of molecular and cellular biomarkers for environmental monitoring (Huggett et al., 1992; McCarthy et al., 1990). The list of available cellular and subcellular biomarkers include a wide spectrum ranging from those designed to measure general and physiological response of organisms to environmental insults, to those that measure chemical or organ specific effects (Huggett et al., 1992). However, these biomarkers have been developed and calibrated with model chemicals in laboratory and their potential utility under field conditions need to be validated. We propose the further development and *in situ* validation of the following molecular biomarkers;

DNA Perturbations — Interaction of environmental chemicals with DNA may induce structural alterations to DNA molecule resulting in the formation of adducts, strand breaks or of chemically modified bases. If these damages are not repaired they may be passed to the next generations of cells and the individuals inheriting these altered cells may exhibit subclinical manifestation of genotoxic disease. Two biomarkers of DNA perturbation are proposed for validation for *in situ* environmental monitoring.

DNA Adducts — Excellent reviews on the utility of DNA adducts for detecting genotoxins in the environment have been published (Kurelec and Gupta, 1993; Shugart et al., 1992). Studies reported by Dunn et al. (1987) and Varanasi et al. (1989), with benthic fish and Kurelec and Gupta (1993) with muscles have shown the usefulness of DNA adducts (^{32}P-postlabeling assay) as a biomarker of exposure to genotoxic compounds. These studies show that in several species of fish from U.S. and European waters, substantially higher DNA adduct levels were detected in fish from contaminated sites than in fish from relatively uncontaminated sites.

DNA Strand Breaks — A novel microgel electrophoresis assay has been developed for directly evaluating in individual cells the frequency of single-strand DNA breaks and/or alkali labile sites (Singh et al., 1988). This technique, called single cell gel (SCG) electrophoresis assay, requires only a few cells (1 to 10,000 cells) for analysis. This small amount of cells makes it possible to evaluate the level and intercellular variability of DNA damage induced by genotoxic agents in virtually any eukaryote cell population. For the detection and quantification of DNA strand breaks, a small number of eukaryotic cells are embedded in an agarose gel on a microscope slide, lysed by detergents and high salt concentrations at pH 10, and electophored for a short time under alkaline conditions. Cells with increased DNA damage display increased migration of the DNA from the nucleus toward the anode. The importance of this technique lies in its ability to detect intercellular differences in DNA damage in virtually any eukaryotic cell population.

The SCG assay is highly sensitive, rapid (results can be obtained within a few hours), and relatively inexpensive. The assay has been partially validated with a variety of model physical and chemical agents and complex mixtures using different tissues from a variety of animals (Tice et al., 1991).

The endpoints of DNA perturbation, i.e., DNA adducts and DNA strand breaks, both provide estimates of nonspecific exposure to environmental pollutants. Whereas DNA adducts analysis (32P-postlabeling) has been well validated with model compounds in laboratory and has been applied for detecting the levels of mutagens/carcinogens in the aquatic environments, the SCG technique for measuring DNA strand breaks has not been used under field conditions. On the other hand, analysis for DNA adducts is relatively more expensive as compared to SCG.

Mitosis Promoting Factor — The measurement of growth in the field is difficult because one has to follow the mass of known cohorts over time. Many molecular correlates of growth have been measured, i.e., metabolic enzyme activity, fatty acid analysis, and RNA-DNA ratio, but none of these have been found to be a reliable indicator of growth of organisms in an ecosystem. Thus, a specific indicator of the actual rate of cell proliferation is needed to give an index of growth in the stressed environment.

Recent advances in biochemical regulation of cell cycle indicate that a protein kinase, mitosis promoting factor (MPF), acts as "master switch" of the cell cycle (Powers, 1993). Its function is absolutely conserved and its activity is required for initiation, maintenance, and exit from mitosis for all eukaryotic cells from yeast to man. In teleosts, MPF is active over a period of 3–6 minutes prior to mitosis and can easily be measured by the histone H1 assay. Thus, MPF's characteristics of universality, absolute specificity of function, abbreviated temporal activity, and a well-characterized assay for its measurement suggest that MPF has a tremendous potential as a highly specific measure of cellular proliferation and growth.

Vitellogenin —Classically, gametogenic capacity is estimated by counting changes in gametic number or gonad weight. While these measures have been useful for some species, these procedures are time consuming and do not always demonstrate the impact of subtle environmental changes on reproductive capacity. A possible alternative method for determining the impact of environmental variables on reproductive capacity is the assessment of specific reproductive genes that affect vitellogenin in vertebrate females.

During oogenesis in fish, the egg-yolk protein (vitellogenin) is synthesized in the liver, secreted into the vascular system and then deposited in the developing oocytes as lipovitellin and phosvitine (Chen, 1983). Therefore, the reproductive status of female fish in a population might be determined by measuring the levels of vitellogenin in the serum, the rate of vitellogenin synthesis in the liver or the accumulation of vitellogenin mRNA in the liver. Serum vitellogenin can easily be detected both in hormone-induced and reproductively active female fish. When these fish are in the impacted environment, the levels of vitellogenin are significantly altered, making the measurements of vitellogenin an excellent indicator of sublethal environmental stress. The levels of vitellogenin in the serum of female fish from pristine and impacted environments can be measured by using an antivitellogenin antibody either by radioimmunoprecipitation method or by RNA-DNA hybridization ratio.

Stress Proteins — When ambient temperature is raised beyond a critical point, a family of heat shock proteins genes is expressed in all organisms (Atkinson and Waldon, 1985). The heat shock proteins are actually stress proteins and are responsive to a variety of environmental stresses, including changes in temperature, oxygen, several classes of dissolved organics, and even some physical stresses.

The cDNA for heat shock proteins from several terrestrial and aquatic species have already been isolated and have been shown to cross-hybridize with heat shock mRNA from closely related species. This fact allows for the rapid and quantitative measure of the heat shock response from a large number of individuals through northern blot and/or dot blot hybridization. Comparison of northern or dot blots of individuals from suspected areas with those of individuals from pristine environments may provide sensitive and reliable measure of environmental pollutants in the marine environment.

CONCLUSION

Only a few studies have been performed on the utility of plants and animals for *in situ* environmental monitoring. From the limited data that has been analyzed, it is evident that indigenous and introduced species of plants as well as feral animals can be used effectively for detecting levels of genotoxins in ecosystems. However, for most of the studies reported in the literature, the measurements for the exposure and effects of environmental pollutants have been made at organismal level (i.e., carcinoma in aquatic animals) or cytogenetic effects (i.e., chromosome breaks or micronuclei formations). In general, measurements at organismal level or even chromosomal levels provide useful information but these endpoints are not sensitive enough to serve as early warning systems for the potential impact of environmental agents on ecosystems. Biochemical and cellular events tend to be more responsive to environmental insults than the manifestation of biological effects at higher levels of biological organization. Therefore, in order to enhance the effectiveness of *in situ* biological monitoring and assessment, it is imperative to integrate cellular and subcellular biomarkers of stress into indicator species.

ACKNOWLEDGMENT

Y. R. Ahuja acknowledges the support of ICMR (No.74/7/90–Pers/EMS).

REFERENCES

Adams, S. M., Shugart, L. R., and Southworth, G. R., Application of bioindicators in assessing the health of fish populations experiencing contaminant stress, in *Biomarkers of Environmental Contamination,* McCarthy, J. F. and Shugart, L. R., Eds., Lewis Publishers, Boca Raton, FL, 1990, 333.

Alink, G. M., Frederix–Wolters, E. M. H., Van der Gaag, M. A., van de Kerkhoff, J. F., and Poels, C. L. M., Induction of sister chromatid exchanges in fish exposed to Rhine water, *Mutat. Res.,* 78, 369, 1980.

Atkinson, B. G. and Waldon, D. B., *Changes in Eukaryotic Gene Expression in Response To Environmental Stress,* Academic Press, Orlando, FL, 1985, 379.

Brunetti, R., Gola, J., and Majone, F., Sister chromatid exchange in developing eggs of *mytilus galloprovlinchialis* Bivalvia, *Mutat. Res.,* 174, 207, 1986.

Cebulska-Wasilewska, A., *Tradescantia* stamen–hair mutation bioassay on the mutagenicity of radioisotope–contaminated air following the Chernobyl nuclear accident and one year later, *Mutat. Res.,* 270, 23, 1992.

Chen, T. T., Identification and characterization of estrogen-responsive gene products in the liver of the rainbow trout, *Can. J. Biol.,* 61, 802, 1983.

Chen, T. T. and Power, D. A., Transgenic fish, *Trends Biotechnol.,* 8, 209, 1990.

Constantine, M. J. and Nilan, R. A., The chlorophyll-deficient mutant assay in barley *(Hordeum vulgare):* A report of the U. S. Environmental Protection Agency Gene-Tox Program, *Mutat. Res.,* 99, 37, 1982.

Constantine, M. J. and Nilan, R. A., Chromosome aberration assays in barley (Hordeum vulgare): A report of the U.S. Environmental Protection Agency Gene-Tox program, *Mutat. Res.*, 99, 13, 1982.

Couch, J. A. and Harshberger, J. C., Effect of carcinogenic agents on aquatic animals: An environmental and experimental overview, *Environ. Carcin. Rev.,* 3, 63, 1985.

de Raat, W. K., Vink, G. J., and Hanstveit, A. C., The significance of mutagenicity as a criterion for ecotoxicological evaluations, in *Genetic Toxicology of Complex Mixtures,* Waters, M. D., Daniel, F. B., Lewtas, J., Moore, M. M., and Nesnow, S., Eds., Plenum Press, New York, 1990, 249.

Dixon, D. R., *Promotoceres* triqueter: A test system for environmental mutagenesis, in *The Effects of Stress and Pollution on Marine Animals,* Praeger Press, New York, 115, 1986.

Dixon, D. R. and Clark, K. R., Sister chromatid exchange: A sensitive method for detecting damage caused by exposure to environmental mutagens in the chromosomes of adult *Mytilus edulis, Mar. Biol. Lett.,* 3, 163, 1982.

DuBridge, B. R. and Carlos, M. P., Recombinant shuttle vectors for the study of mutation in mammalian cells, *Mutagenesis,* 3, 1, 1988.

Dunn, B., Black, J., and Maccubbin, A., ^{32}P–postlabeling analysis of aromatic DNA adducts in fish from polluted areas, *Cancer Res.,* 47, 6543, 1987.

Ennever, F. K., Andreano, G., and Rosenkranz, H. S., The ability of plant genotoxicity assays to predict carcinogenicity, *Mutat. Res.,* 205, 99, 1988.

Fang, T., Ma, T.-H., Lin, G., Ho, J., Dai, J., Jhou, R., Chen, D., Ou, Y., and Cui, J., Preliminary report on the development of a bioassay for detection on mutagens in sea water, *Environ. Exp. Bot.,* 23, 303, 1983.

Gardner, H. S., Van der Schalie, W. H., Wolfe, M. J., and Finch, R. A., New methods for on-site biological monitoring of effluent water quality, in *In Situ Evaluation of Biological Hazards of Environmental Pollutants,* Sandhu, S. S., Lower, W. R., de Serres, F. J., Suk, W. A., and Tice, R. R., Eds., Plenum Press, New York, 1989, 277.

Gentile, J. M., Gentile, G. J., Bultman, J., Sechriest, R., Wagner, E. D., and Plewa, M. J., An evaluation of the genotoxic properties of insecticides following plant and animal activation, *Mutat. Res.,* 101, 19–22, 1982.

Gill, B. S. and Sandhu, S. S., Application of the *Tradescantia* micronucleus assay for the genetic evaluation of chemical mixtures in soil and aqueous media, *Mutat. Res.,* 270, 65, 1993.

Gill, B. S., Sandhu, S. S., Baker, L. C., and Casto, B. C., Application of a plant test system in the identification of potential genetic hazards at chemical wastes sites, in *Plants for Toxicity Assessment, Volume 2,* ASTM STP 1115, Gorsuch, J., Lower, W. R., and St. John, K. R., Eds., ASTM, Philadelphia, 1991, 309.

Glickman, L. T., Domanski, L. M., Marquire, T. G., Dubielzig, R. R., and Chung, A., Mesothelioma in dogs associated with exposure of their owners to asbestos, *Environ. Res.,* 32, 305, 1983.

Glickman, L., Analytical epidemiology in pet population for environmental risk assessment, in *In Situ Evaluation of Biological Hazards of Environmental Pollutants,* Sandhu, S. S., Lower, W. R., de Serres, F. J., Suk, W. A., and Tice, R. R., Eds., Plenum Press, New York, 1989, 133.

Grant, W. F., Chromosome aberration assays in *Allium:* A report of the U. S. EPA's Gene-Tox Program, *Mutat. Res.,* 99, 273, 1982.

Grant, W. F., Lee, H. G., Logan, D. G., and Salamone, M. F., The use of *Tradescantia* and *Vicia faba* bioassays for the in situ detection of mutagens in an aquatic environment, *Mutat. Res.,* 270, 53, 1992.

Harrison, F. L. and Jones, J. M., An in vitro sister chromatid exchange assay in the larvae of the mussel *mytailus edulis:* Response to three mutagens, *Mutat. Res.,* 105, 235, 1982.

Harshberger, J. C., Cullen, L. J., Calabrese, M. J., and Spero, P. M., Epidermal, hepatocellular and cholangiocellular carcinomas in brown bullheads, *ictalurus nebulosus* from industrially polluted Black River, Ohio, *Mar. Environ. Res.,* 14, 535, 1984.

Harshman, L. G., The use of *Drosophila melanogaster* for in situ biomonitoring, in *In Situ Evaluation of Biological Hazards of Environmental Pollutants,* Sandhu, S. S., Lower, W. R., de Serres, F. J., Suk, W. A., and Tice, R. R., Eds., Plenum Press, New York, 1989, 191.

Hayes, H. M., Jr., Hoover, R., and Tarone, R. E., Bladder cancer in pet dogs: A sentinel for environmental cancer?, *Am. J. Epidemiol.,* 114, 229, 1981.

Hooftman, R. N. and Vink, G. J., Cytogenetic effects on the Eastern mud minnow *(Umbra pygmeae)* exposed to ethyl methanesulphonate, benzo-a-pyrene and river water, *Ecotox. Environ. Safety,* 5, 261, 1981.

Hooftman, R. N. and de Raat, W. K., Induction of nuclear anomalies (micronuclei) in the peripheral blood erythrocytes of the eastern mud minnow, *umbra pygmeae*, by ethyl methanesulphonate, *Mutat. Res.,* 104, 147, 1982.

Hope, P. K., Plewa, M. J., Johnston, J. B., Weaver, D., Wood, S. G., Larson, R. A., and Hinsley, T., Multitechnique for screening Chicago municipal waste sludge for mutagenic activity, *Environ. Sci. Technol.,* 16, 140, 1982.

Hose, J. E., Cross, J. N., Smith, S. G., and Diehl, D., Elevated circulating erythrocyte micronuclei in fishes from contaminated sites off southern California, *Mar. Environ. Res.,* 22, 167, 1987.

Huggett, R. J., Kimerle, R. A., Mehrle, P. M., and Bergman, H. L., *Biomarkers, Biochemical, Physiological and Histological Markers of Anthropogenic Stress,* Lewis Publishers, Boca Raton, FL, 1992.

Ichikawa, S., In situ monitoring with *Tradescantia* around nuclear power plants, *Environ. Health Perspect.,* 37, 145, 1981.

Ichikawa, S., *Tradescantia* stamen-hair system as an excellent botanical tester of mutagenicity: Its response to ionizing radiation and chemical mutagens and some synergistic effects found, *Mutat. Res.,* 270, 3, 1992.

Klekowski, E. D., Screening aquatic ecosystems for mutagens with fern bioassays, *Environ. Health Perspect.,* 37, 99, 1978.

Kligerman, A. D. and Bloom, S. E., A cytogenetic model for the study of chromosome aberrations in fishes, *Mutat. Res.,* 31, 334, 1975.

Kligerman, A. D. and Bloom, S. E., Sister chromatid differentiation and exchange in adult mud minnows *(Umbra limi)* after in vivo exposure to 5-bromodeoxyuridine, *Chromosoma,* 56, 101, 1976.

Kurelec, B., Garg, A., Krca, S., and Gupta, R. C., DNA adducts in marine mussel *Mytilus galloprovincialis* living in polluted and unpolluted environments, in *Biomarkers of Environmental Contamination,* McCarthy, J. F., and Shugart, L. R., Eds., Lewis Publishers, Boca Raton, FL, 1990, 217.

Kurelec, B. and Gupta, R. C., Biomonitoring of aquatic systems, IARC-SCi-Publ. (124), 365, 1993.

Lower, W. R., Rose, P. S., and Drobney, V. K., In situ mutagenic and other effects associated with lead smelting, *Mutat. Res.,* 54, 65, 1978.

Lower, W. R., Drobney, V. K., Aholt, B. J., and Politte, R., Mutagenicity of environments in the vicinity of oil refinery and petrochemical complex, *Teratog. Carcin. Mutagen.,* 3, 65, 1983a.

Lower, W. R., Thompson, W. A., Drobney, V. K., and Yanders, A. K., Mutagenicity in the vicinity of a lead smelter, *Teratog. Carcin. Mutagen.,* 3, 231, 1983b.

Lower, W. R. and Kendell, R. J., Sentinel species and sentinel bioassays, in *Biomarkers of Environmental Contamination,* McCarthy, J. F. and Shugart, L. R., Eds., Lewis Publishers, Boca Raton, FL, 1990.

Ma, T.-H., *Tradescantia* cytogenetic tests (root-tip mitosis, pollen mitosis, pollen mother cell meiosis): A report of the U. S. EPA's Gene-Tox Program, *Mutat. Res.,* 99, 257, 1982.

Ma, T.-H., Lower, W. R., Harris, F. D., Paku, J., Anderson, V. A., Harris, M. M., and Bare, J. L., Evaluation by the *Tradescantia*-micronucleus test of the mutagenicity of internal combustion engine exhaust fumes from Diesel and Diesel-soybean oil mixed fuels, in *Short-Term Assays in the Analysis of Complex Mixtures III,* Waters, M. D., Sandhu, S. S., Chernoff, J., and Nesnow, S., Eds., Plenum Press, New York, 1983.

Ma, T.-H. and Harris, M. M., In situ monitoring of environmental mutagens, in *Hazardous Assessment of Chemicals, Current Developments,* Vol. 4, Academic Press, New York, 1985, 77.

Ma, T.-H., Neas, R. E., Harris, M. M., Xu, Z., Cook, C., and Swofford, D., In vivo tests (*Tradescantia*- and mouse-micronucleus) and chemical analyses on drinking water of rural communities, in *Short-Term Assays in the Analysis of Complex Mixtures II,* Sandhu, S. S., DeMarini, D. M., Mass, M. J., Moore, M. M., and Mumford, J. L., Eds., Plenum Press, New York, 1987, 409.

Maddock, M. B., Northrup, H., and Ellingham, T. J., Induction of sister chromatid exchanges and chromosomal aberrations in hematopoietic tissue of a marine fish following in vivo exposure to genotoxic carcinogens, *Mutat. Res.,* 172, 165, 1986.

Malling, H. V. and Burkhart, J. G., Comparison of mutation frequencies obtained using transgenes and specific-locus mutation systems in male mouse germ cells, *Mutat. Res.,* 279, 149, 1992.

McBee, K., Buckman, J. W., Brown, K. W., and Donnelly, K. C., Chromosomal aberrations in native small mammals (*Permyscu leucopus* and *Sigmodon hispidus)* at a petrochemical waste disposal site. I. Standard karyology, *Arch. Environ. Contam. Toxicol.,* 16, 681, 1987.

McCarthy, J. F. and Shugart, L. R., Eds. *Biomarkers of Environmental Contamination,* Lewis Publishers, Boca Raton, FL, 1990.

Mix, M. C., Cancerous diseases in aquatic animals and their association with environmental pollutants: A critical literature review, *Mar. Environ. Res.* (Special Issue), 20, 1, 1986.

National Research Council, *Animals as Sentinels of Environmental Health Hazards,* National Academy Press, Washington, D.C., 1991.

Nayak, B. N. and Petras, M. L., Environmental monitoring for genotoxicity: In vivo sister chromatid exchange in house mouse (*Mus musculus), Can. J. Genet. Cytol.,* 27, 351, 1985.

Ong, T.-M., Stewart, J. D., Tucker, J. D., and Whong, W. Z., Development of an in situ test system for detection of mutagens in the workplace, in *Short-Term Bioassays in the Analysis of Complex Environmental Mixtures IV,* Waters, M. D., Sandhu, S. S., Lewtas, J., Claxton, L., Strauss, G., and Nesnow, S., Eds., Plenum Press, New York, 1985, 25.

Pesch, G. G. and Pesch, C. E., *Neanthes arenaceodentata* (polychaeta, Annelida), a proposed model for marine genetic toxicology, *Can. J. Fish. Aquat. Sci.,* 37, 1225, 1980.

Pesch, G. G., Pesch, C. E., and Malcolm, A. R., *Neanthes arenaceodentata,* A cytogenetic model for marine genetic toxicology, *Aquat. Toxicol.,* 1, 301, 1981.

Plewa, M. J., Plant genetic assays to evaluate complex environmental mixtures, in *Short-Term Bioassays in the Analysis of Complex Environmental Mixtures IV,* Waters, M. D., Sandhu, S. S., Lewtas, J., Claxton, L., Strauss, G., and Nesnow, S., Eds., Plenum Press, New York, 45, 1984.

Powers, D. A., Applications of molecular and biotechnological techniques to the study of large marine ecosystems, in *Stress, Mitigation and Sustainability of Large Marine Ecosystems,* Vol. 5, Sharman, K., Alexander, L. M., and Gold, B. D., Eds., American Association for the Advancement of Science Press, Washington, D.C., in press.

Prein, A. E, Thie, G. M., Alink, G. M., Koeman, J. H., and Poeles, C. L. M., Cytogenetic changes in fish exposed to water of the river Rhine, *Sci. Total Environ.,* 9, 287, 1978.

Rawley, M. H., Christian, J. J., Basu, D. K., Pawlikowski, M. A., and Paign, B., Use of small animals (voles) to assess a hazardous waste site in Love Canal, Niagara Falls, NY, *Arch. Environ. Contam. Toxicol.,* 12, 383, 1983.

Redei, G. P., Mutagen assay with *Arabidopsis:* A report to the U. S. Environmental Protection Agency's Gene-Tox Program, *Mutat. Res.,* 99, 243, 1982.

Reif, J. S. and Cohen, D., Retrospect radiographic analysis of pulmonary disease in rural and urban dogs, *Arch. Environ. Health,* 20, 684, 1970.

Sandhu, S. S., DeMarini, D. M., Mass, M. J., Moore, M. M., and Mumford, J. L., Eds., *Short-Term Assays in the Analysis of Complex Mixtures II,* Plenum Press, New York, 1987, 409.

Sandhu, S. S., Lower, W. R., de Serres, F. J., Suk, W. A., and Tice, R. R., Eds., *In Situ Evaluation of Biological Hazards of Environmental Pollutants,* Plenum Press, New York, 1989.

Sandhu, S. S., de Serres, F. J., Gopalan, H. N. B., Grant, W. F., Svendsgaard, D., Veleminsky, J., and Becking, G. C., Results and recommendations (IPCS Collaborative Study on Plant Test Systems), *Mutat. Res.,* 301, 257, 1994.

Schaeffer, E. W., Novak, E. W., Lower, W. R., Yanders, A. F., Kapila, S., and Wang, R., Effects of chemical smokes on flora and fauna under field and laboratory exposures, *Ecotoxicol. Environ. Safety,* 13, 301, 1987.

Schairer, L. A., Van't Hof, J., Hays, C. G., Burton, R. M., and de Serres, F. J., Measurement of biological activity of ambient air mixtures using a mobile laboratory for in situ exposures: Preliminary results from *Tradescantia* plant test system, in *Application of Short-Term Bioassay in the Fractionation of Complex Mixtures,* Waters, M. D., Nesnow, S., Huisingh, J. L., Sandhu, S., and Claxton, L., Eds., Plenum Press, New York, 1979, 421.

Schneider, R., Dorn, C. R., and Taylor, D. O. N., Factors influencing canine mammary cancer development and postsurgical survival, *J. Natl. Cancer Inst.,* 43, 1249, 1969.

Schneider, R., Comparison of age, sex and incidence rates in human and canine breast cancer, *Cancer,* 26, 419, 1970.

Shevchenko, V. A., Some aspects of the genetic consequences of the Chernobyl disaster, Trans. 10th Int. Conf. Structural Mechanisms in Reactor Technology, Los Angeles, Vol. D, 1989, 245.

Singh, N. P., McCoy, M. T., Tice, R. R., and Schneider, E. L., A simple technique for quantitation of levels of DNA damage in individual cell, *Exp. Cell Res.,* 175, 184, 1988.

Shugart, L., Bickam, J., Jackim, G., McMahon, G., Ridley, W., Stien, J., and Steinert, S., DNA alterations, in *Biomarkers, Biochemical, Physiological, and Histological Markers of Anthropologic Stress,* Huggett, R. J., R. A., Kimerle, P. M., Mehrle, and H. L., Bergman, Eds., Lewis Publishers, Boca Raton, FL, 1992, 125.

Thompson, B. A., Schroder, G. D., and Connor, T. H., Chromosomal aberration in the cotton rat (*Sygmodo hispidus*) exposed to hazardous waste, *Environ. Mol. Mutagen.,* 11, 359, 1988.

Tice, R. R., Ormiston, B. G., Boucher, R., Luke, C. A., and Paquette, D. E., Environmental biomonitoring with feral rodent species, in *Short-Term Assays in the Analysis of Complex Mixtures II,* Sandhu, S. S., DeMarini, D. M., Mass, M. J., Moore, M. M., and Mumford, J. L., Eds., Plenum Press, New York, 1987, 175.

Tice, R. R., Andfrew, P. W., Hirari, O., and Singh, N. P., The single cell (SCG) assay: An electrophoretic technique for the detection of DNA damage in individual cells, in *Biologically Reactive Intermediates. IV. Molecular and Cellular Effects and Their Impact on Human Health,* Winter, C. M., et. al., Eds., Plenum Press, New York, 1991, 182.

Tjalma, R. A., Canine bone sarcoma: Estimation of relative risk as a function of body size, *J. Natl. Cancer Inst.,* 36, 1137, 1966.

Valencia, R., Abrahamson, S., Lee, W. R., Von Holle, E. S., Woodruff, R. C., Wurglers, F. E., and Zimmering, S., Chromosome mutation test for mutagenesis in Drosophila melanogaster, *Mutat. Res.,* 134, 61, 1984.

Varanasi, U., Reichert, W. L., and Stein, J., ^{32}P–postlabeling analysis of DNA adducts in liver of wild English sole (*Parophrys vetulus*) and winter flounder (*Pseudopleuronectes americanus*), *Cancer Res.,* 49, 1171, 1989.

Venegas, W., Hermosilla, I., Gavilan, J. F., Naveas, R., and Carrasco, P., Larval stages of the anuran amphibian *Caudiverbera caudiveverbea:* A biological model for studies of genotoxic agents, *Bol. Soc. Biol. Concepcion,* 58, 171, 1987.

Wang, H., You, X., Qu, Y., Wang, W., Long Y., and Ni, J., Investigation of cancer epidemiology and study of carcinogenic agents in the Shanghai rubber industry, *Cancer Res.,* 44, 3101, 1984.

Waters, M. D., Nesnow, S., Huisingh, J., Sandhu, S. S., and Claxton, L., Eds., *Application of Short-Term Assays in the Fractionation and Analysis of Complex Mixtures,* Plenum Press, New York, 1979.

Waters, M. D., Sandhu, S. S., Huisingh, J., Claxton, L., and Nesnow, S., Eds., *Short-Term Assays in the Analysis of Complex Mixtures II,* Plenum Press, New York, 1981.

Waters, M. D., Sandhu, S. S., Lewtas, J., Claxton, L., Chernoff, N., and Nesnow, S., Eds., *Short-Term Assays in the Analysis of Complex Mixtures III,* Plenum Press, New York, 1983.

Waters, M. D., Daniel, F. B., Lewtas, J., Moore, M. M., and Nesnow, S., Eds., *Genetic Toxicology of Complex Mixtures,* Plenum Press, New York, 1990.

Wurgler, F. E. and Kramers, P. G. N., Environmental effects of genotoxins (eco–genotoxicology), *Mutagenesis,* 7, 321, 1992.

Zimmering, S., Mason, J. M., and Osgood, C., Current status of aneuploidy testing in Drosophila, *Mutat. Res.,* 167, 71, 1986.

Zahn, R. K., Zahn-Daimler, C., Muller, W. E. G., Michaelis, M. L., Kurelec, B., Rijavec, M., Batel, R., and Bihari, N., DNA damage by PAH and repair in a marine sponge, *Sci. Total Environ.,* 26, 137, 1983.

CHAPTER 6

Wildlife Toxicology in Biomonitoring and Bioremediation: Implications for Human Health

Karen McBee and Robert L. Lochmiller

CONTENTS

INTRODUCTION

Much of the research in ecotoxicology has been directed toward identification and characterization of toxic effects of pollutants in laboratory models followed by extrapolation or modeling of potential effects on exposed biota. Investigation of effects of chemical pollutants in "real-life" situations has often been neglected because such studies can be lengthy, cumbersome, and difficult to conduct with appropriate rigor and necessary controls. However, a number of people working in an area broadly known as wildlife toxicology are developing ways to address concerns about just such "real-life" situations. Wildlife toxicology draws on the skills and knowledge of ecologists, chemists, physiologists, toxicologists, biometricians, and systems ecologists, making it a truly multidisciplinary field (Hoffman et al., 1990).

1-56670-141-4/96/$0.00+$.50

Recorded observations of the effects of environmental contaminants on exposed wildlife date at least from the mid-19th century (Hoffman et al., 1990; McBee and Bickham, 1990). With the rapid proliferation of synthetic agricultural pesticides after World War II, accounts of negative impacts on wildlife became much more common. Several early studies of the effects of modern pesticides such as DDT indicated cause for concern for potential hazard to exposed wildlife. These early efforts have expanded during the last 50 years to include a battery of innovative toxicological methods adapted for use with wildlife species. Almost all this work has been conducted with wild birds and mammals; however, very little is known about the toxic effects of pollutants on wild populations of amphibians and reptiles. Among the techniques that have been successfully adapted for use with avian and mammalian wildlife are chemical analytical, behavioral, genetic, morphological, immunological, and biochemical methods. All may have direct relevance to environmental remediation and human health.

A major component of any remediation program for restoring contaminated areas to acceptable levels of toxicity should include surveillance and monitoring activities to evaluate and assess program effectiveness and determine when program objectives have been achieved (Hartig and Zarull, 1991). Realization that the more traditional approaches to estimating risks of exposure to humans (e.g., analysis of air, water, and soil for specific contaminants) have several severe shortcomings has led to the development of alternative methods of biomonitoring. Among the most recent advances has been an increased interest in employing resident wildlife species for in situ assessments of health risks associated with contaminated terrestrial environments to aid in biomonitoring activities (Talmage and Walton, 1991; McBee and Bickham, 1990; Lamb and Kenaga, 1981; NRC, 1991; Peterle, 1991).

Several strong arguments exist to support the use of wild animals in biomonitoring programs. Wild animals naturally reside in a very stressful environment where extremes in temperature (seasonal stressors), competition (behavioral stressors), and nutritional deficiencies (physiological stressors) are common; laboratory-derived toxicity data may not be completely applicable under such natural environmental conditions. Populations and communities of wild animals may frequently encounter a wide variety of environmental pollutants via a variety of exposure routes. Analytical assays for specific toxicants typically recognize only those toxicants for which specific procedures were performed. Toxicity in the natural environment is usually the result of exposure to complex mixtures of toxicants, as with petroleum hydrocarbons from fuels and solvents, or their metabolites or degradation products. Additive insults of low levels of multiple toxicants may cause deleterious effects (Rowley et al., 1983). Toxicants that have been "aged" in the environment may be more toxic than corresponding quantities of technical mixtures (which are used in most laboratory toxicity experiments); this has been demonstrated with polychlorinated biphenyls (PCBs) (Hornshaw et al., 1983).

Another very strong argument in support of *in situ* biomonitoring is that resident vertebrates should provide a more accurate model system for evaluating the risks associated with more "realistic" exposure routes and dosages. Wild animals residing in contaminated terrestrial environments are undoubtedly exposed to a variety of xenobiotics via multiple avenues such as ingestion and inhalation; laboratory toxicity assessments rarely duplicate such systems. Lastly, wild animals afford the opportunity to examine the effects of multiple stressors (temperature, nutritional, physiological stress) on toxicity.

Only a fraction of the wildlife species available have been examined in sufficient detail to warrant their use in biomonitoring programs. Due to the vast array of wildlife species and contaminants available, there is no way all species will ever be tested for their suitability as biomonitors. To be suitable, a significant database of physiological and population information is necessary to establish normal or expected responses given the daily, seasonal, and annual cycles in demographics and physiological condition in most wild animals. Without such information, it becomes extremely difficult to separate toxicant effects from normal changes in life processes of animals in their natural environment. Only a handful of wildlife species currently meet these information criteria.

The working hypothesis in *in situ* biomonitoring is that exposure to chemicals that may have adverse metabolic, immunosuppressive, mutagenic and clastogenic, or pharmacologic properties could greatly alter normal physiology of resident wildlife species within an ecosystem. Consequently, such physiological alterations could have profound effects on intrinsic population characteristics such as survival and recruitment rates, and structural characteristics of communities, including species richness, diversity, and abundance. Thus, technologies or approaches to *in situ* biomonitoring can range from routine assessments of body burdens of selected toxicants in the individual organism to measuring indices of vertebrate community structure. Development of efficacious means of evaluating effects of toxic chemicals to higher vertebrates residing within terrestrial environments would undoubtedly provide additional predictive capabilities for assessing risks to human populations as well.

POPULATION DYNAMICS

The structure and composition of populations and communities are of interest to toxicologists because this structure should theoretically reflect the overall impact of environmental contaminants within a terrestrial ecosystem. Unfortunately, studies regarding the effects of contaminants on wild mammals have mainly been limited to monitoring uptake (McBee and Bickham, 1990). However, a few studies have attempted to elucidate how population dynamics of resident wildlife species can potentially change in response to exposure to toxic environmental contaminants (Anderson and Barrett, 1982; Maly, 1984; Elangbam et al., 1989a; Batty et al., 1990). Two of the potential endpoints that have been considered for ecological risk assessment at the population level are local pseudoextinction (probability of a population falling below a given density) and decline in temporal mean density (Emlen, 1989).

Batty et al. (1990) demonstrated significant increases in the age structure of natural populations of white-footed mice (*Peromyscus leucopus),* suggesting decreased reproduction, in an area heavily contaminated with PCBs. Similar changes were reported in wild populations of cotton rats (*Sigmodon hispidus)* inhabiting a toxic industrial-waste disposal site (Elangbam et al., 1989a). Pomeroy and Barrett (1975) demonstrated that pesticide contamination can inhibit reproduction of cotton rats, resulting in a lower peak population density. Similar findings have been shown with voles (*Microtus pennsylvanicus)* in the Love Canal Emergency Declaration Area, which was considered safe for humans and animals (Rowley et al., 1983). Population studies on voles from this area suggested a significant decline in the population density because of the loss of mature females and increased *in utero* mortality.

Considerably less is known about the impact of environmental contaminants on wild animal communities, which should reflect a culmination of effects on all populations within the community. As populations change, relationships among other populations could change as well as a result of modification in food chains and competition (Scanlon, 1987). For example, Batty et al. (1990) observed two species of shrew on control areas but none on PCB-polluted sites.

Population studies tend to be tedious and time consuming, and must be rigidly controlled (Rowley et al., 1983). Even when coupled with toxicologic and pathologic evaluation of tissues, potential long-term effects, especially carcinogenesis, may be difficult to assess from population studies. Demographic estimates and measurements are inherently difficult to acquire accurately, as myriad other environmental variables often complicate their interpretation (Krebs, 1989). Hence, it may be misleading to assume that all observed alterations of the structure and composition of animal populations and communities is largely due to the direct effect of contaminants on the immunology, metabolism, and genetics of resident organisms, resulting in changes in survival and reproductive activity. The careful selection of ecologically similar, replicated, reference (uncontaminated) sites for comparison with contaminated areas undergoing remediation is desirable. Additionally, logistical and financial commitments can be extremely high when attempting to monitor changes in demographic and community endpoints.

Conceivably, the degree of site contamination could be severe enough to prevent population of the area by wild animals, or prolonged exposure of populations to contaminants on selected sites could result in selection for "resistant" physiological phenotypes. Although largely unexplored as a tool in terrestrial biomonitoring, the use of small enclosures or mesocosms to confine selected wild animal models to specific sites of contamination for monitoring remediation efforts may provide a solution. Such "controlled" *in situ* systems may effectively minimize the limitations discussed above, as well as providing control of the type of animals used (age, sex, physiological status, etc.) and flexibility to choose the location and duration of exposure (thus permitting short- or long-term exposures on sites of choice). Biologists routinely have used field enclosures to study the dynamics of wild rodent populations under natural conditions (Lidicker, 1979; Tamarin et al., 1984), and have also examined their use in assessing impacts of municipal sludge on populations of meadow voles (*Microtus pennsylvanicus)* (Tilghman et al., 1990; Maly, 1984).

REPRODUCTION

Ratcliffe's (1967) classic work on the effects of contaminants on eggshell-thinning in birds was instrumental in stimulating a wealth of studies on reproductive toxicology in wild birds and mammals; these research efforts continue today (Newton and Wyllie, 1992). The majority of this work has focused on the consequences of pesticide applications on reproductive activity of birds (Grue and

Christian, 1981; Stromborg, 1981), primarily predaceous species such as raptors and fish-eating species. In comparison, efforts to elucidate the effects of nonpesticide contaminants on terrestrial mammalian reproduction have been few. Research efforts in both avian and mammalian reproductive toxicology have been recently reviewed by Peterle (1991). As with most wildlife toxicology studies, few attempts have been made toward relating observed reproductive dysfunction in wildlife to risks for humans.

MIXED FUNCTION OXIDASE ENZYME SYSTEM

The mixed function oxidase (MFO) enzyme system has been of considerable interest to toxicologists because of its importance in detoxification of certain foreign chemicals and its ability to produce carcinogens or reactive toxic metabolites. The MFO enzyme system has been found in the hepatic endoplasmic reticulum of every animal species so far examined (Klaassen et al., 1986) and is induced by exposure to a wide variety of toxicants such as dioxin, polyhalogenated biphenyls (PCBs and polybrominated biphenyls), alkanes, olefins, aromatic carbon systems (monocyclic and polycyclic hydrocarbons and halobenzenes), nitrogen-based systems (nitrosamines, aromatic amines, azo dyes), sulfur-based systems, and metal-based systems (Conney, 1986; Guengerich and Liebler, 1985; Payne et al., 1987). This enzyme system is considered by far the most important group of enzymes involved in the process of biotransformation of xenobiotic compounds. This oxidative enzyme system acts to add a hydroxyl moiety to a xenobiotic compound. This is followed by cleavage of either an ester or an amide bond, which produces a carboxylic acid plus either an amine or an alcohol for further biotransformation. Basically, induction or enhanced activity of this enzyme system is the result of increased rates of synthesis (Klaassen et al., 1986). Overall induction of the MFO system is assayed by measuring cytochrome P-450; however, the system is highly specific with as many as 10 different isozymes, which can be selectively induced by specific xenobiotics, represented in some mammalian liver models.

The selectivity of the MFO system of wild animal models provides a potentially useful means of indicating exposure and toxicity of a variety of chemicals in our environment (Elangbam et al., 1989b; Knight and Walker, 1982). MFO induction has been widely used to assess toxicity in a wide variety of aquatic organisms under laboratory and field conditions (see review by Payne et al., 1987); although a number of laboratory sensitivity trials have been conducted on several wild avian and mammalian models, field validation trials are largely unavailable.

Some of the first attempts at using MFO induction as an index of toxicity in a wild animal model involved indirectly measuring *in vivo* reductions in pentobarbitol-induced (metabolized via MFO system) sleeping times. Laboratory sensitivity trials were used to examine sleeping times of white-footed mice (*Peromyscus leucopus;* Sanders and Kirkpatrick, 1975) and cottontail rabbits (*Sylvilagus floridanus;* Zepp et al., 1974) exposed to PCB and cottontails exposed to mirex (Warren et al., 1978). Reproductive responses of wild animals have also been examined in the laboratory because of the physiological correlation between MFO enzyme activity and metabolism of steroid hormones. Lustick et al. (1972) demonstrated significant correlations between DDT exposure and reproductive steroid metabolism and testis size in wild bobwhite quail (*Colinus virginianus)* fed dietary doses ranging from 10 to 150 ppm. Concern has also been expressed that elevated levels of MFO enzymes in a variety of seabirds and waterfowl species may decrease egg production and overall recruitment in populations inhabiting oil-contaminated environments (Payne et al., 1987).

More recent approaches to assessing metabolic alterations in wild animals have involved measurement of overall cytochrome P-450 induction as well as induction of specific isozymes with resorufin ethers, benzo(a)pyrene (aryl hydrocarbon hydroxylase), or other substrates. Rattner et al., (1989) recently reviewed the literature regarding the measurement and use of MFO enzymes to monitor xenobiotic exposure in avian and mammalian wildlife. A list of relevant studies demonstrating a relationship between environmental contaminant exposure and MFO induction in wildlife species can be found in Table 1 of Rattner et al., (1989).

Wild small mammals may be good subjects for assessing induction of MFO enzymes. The cotton rat shows marked inducibility (Elangbam et al., 1989a).

Lubet et al., (1990) suggested that induction of alkoxyresorufin metabolism could be a potential indicator of environmental contamination; we evaluated these metabolic assays in the cotton rat under laboratory (Novak and Qualls, 1990) and field conditions (Elangbam et al., 1989a). Induction of CYP1A-mediated *o*-dealkylase activity has recently been demonstrated as an effective biomarker for exposure to Aroclor® 1254 (a PCB mixture) in the harvest mouse (*Reithrodontomys fulvescens)* (Lubet et al., 1992). These field results revealed highly significant differences between control and contaminated sites, suggesting that the *o*-dealkylation of alkoxyresorufins by various cytochrome P-450 isoenzymes has the potential to be a very sensitive assay to indicate pathophysiologic alteration in wild small mammals and potentially to be used as an indicator of human health risks. These assays have the potential to be streamlined and made feasible as laboratory assays through the development and validation of new techniques in the indicator species.

IMMUNOTOXICITY

The effects of a wide variety of environmental toxicants on the immune system of laboratory animals are extensively described in the literature. Many industrial chemicals, pesticides, petroleum-derived hydrocarbons and mixtures (benzene, benzo(a)pyrene, toluene, methylcholanthrene), as well as PCBs and heavy metals such as lead and cadmium, are common environmental pollutants and can be potent immunotoxicants, capable of causing changes in lymphoid organs, impairment of immunological responsiveness, or susceptibility to disease agents (for several excellent reviews see Descotes, 1988; Luster et al., 1982; Vos, 1977). The immune system provides a sensitive assay system for detecting immunotoxicity in animals and assessing risks to humans because (1) many cell types of the immune system undergo rapid proliferation and differentiation, and (2) the immune system is a highly organized and regulated system where imbalances result in measurable immune system alterations. Immunotoxicity testing usually involves the screening of both morphological characteristics and functional responses of cell-mediated and humoral arms of the immune system in animals (Table 1) (Dean et al., 1982; Vos et al., 1990).

Some of the most widespread environmental contaminants (benzene, PCBs, and heavy metals) are also potent immunosuppressants to both laboratory rodents

Table 1 Procedures for Defining Immune System Alterations in Wild Animal Species Exposed to Immunotoxicants

Immune system component	Procedure performed
Pathotoxicology	Hematology profile—hemoglobin, white blood cell count with differential lymphocyte counts, red blood cell counts
	Clinical chemistry—total protein, albumin, total globulins, gamma globulins
Immune organs	Weights—spleen, thymus, Peyer's patches, liver, kidney
	Cellularity—total splenocyte counts, total thymocyte counts
	Histology—liver, thymus, Peyer's patch, adrenals
Humoral immunity	Immunoglobulins—quantitation of IgG levels
	Responsiveness—IgM antibody responses to sheep red blood cell challenge
	Complement—complement fixation test
Lymphocytes	Lymphoproliferation—response to mitogenic stimulation (Con-A, PWM)
	Subpopulations—enumeration of T and B lymphocytes in blood/spleen
Cell-mediated	Hypersensitivity—delayed type hypersensitivity responsiveness to dermal oxazalone applications
	Cytotoxic T-cell activity
	Natural killer cell activity

Adapted from Dean et al., *Pharmacol. Rev.,* 34, 137, 1982. With permission.

and humans. Considerable progress has been made in the study of pollutant-disease relationships and in the development of biomonitoring techniques for assessing immunotoxicity in fish inhabiting aquatic environments (reviews by Zeeman and Brindley, 1981; Anderson et al., 1984). However, similar advances in the use of wild avian and mammalian species to evaluate immunotoxicity risks associated with contaminated terrestrial environments have been slow to develop. The effects of various toxicants on the development of immunocompetence have been studied in wild deer mice *(Peromyscus maniculatus),* cotton rats *(Sigmodon hispidus),* bobwhite quail *(Colinus virginianus),* and mallards *(Anas platyrhynchos)* in the laboratory. Only limited attempts have been made toward adapting these assay systems for use in biomonitoring programs under field conditions, primarily because of the technical difficulty associated with assay development and unfamiliarity with immune systems of wild animals.

Because many environmental contaminants have the potential to suppress immunity, several investigators have examined the susceptibility of wild animals to common disease organisms following exposure to these contaminants. Zeaks et al. (1981) explored whether consumption of sublethal doses (levels that might be encountered under natural conditions) of a broad-spectrum insecticide (Sevin) by bobwhite quail *(Colinus virginianus)* would influence their susceptibility to a normally nonpathogenic protozoan parasite *(Histomonas meleagridis).* As little as 10 Êg Sevin in the presence of histomonads was successful at inducing pathological evidence of histomoniasis; histomoniasis resulted in death at dosages of >20 Êg Sevin but not in controls. Similar relationships between duck hepatitis virus and pollutants have been explored in mallard ducks *(Anas platyrhynchos)* exposed under controlled conditions to dieldrin (Friend and Trainer, 1974a), DDT (Friend and Trainer, 1974b), and PCBs (Friend and Trainer, 1970). An interesting study by Rocke et al. (1984) examined the effect of sublethal doses of crude oil, fuel oil, and chemical dispersants on resistance of mallard ducks *(Anas platyrhynchos)* to infection with *Pasteurella multocida,* which causes avian cholera in wild waterfowl and is responsible for thousands of deaths among ducks each year. Waterfowl typically ingest oil mixtures as a result of preening or consuming contaminated food. Intake of 2.5 or 5.0 ml/kg body weight daily for 28 days resulted in significant reductions in immune resistance, which is primarily cell-mediated, to avian cholera. Although significant reductions in spleen weights were observed, there was no indication of depressed antibody-mediated immunity [measured as plaque-forming cell (PFC) response to sheep red blood cell (SRBC)] challenge in mallards consuming oil. Goldberg et al. (1990) examined further the role of immunosuppressants in predisposing waterfowl to duck plague virus (DPV) infection, but reported no increase in susceptibility to DPV challenge following crude oil consumption (28 days). Several investigators are presently developing *in vivo* assay systems [antibody response to erythrocytes and cell-mediated response to phytohemagglutinin (PHA)] for assessing immunocompetence of wild waterfowl for use in biomonitoring programs under field conditions (Schrank et al., 1990).

Increased susceptibility to a variety of infectious disease organisms in waterfowl exposed to lead poisoning from ingestion of spent lead shot has been postulated (Franson, 1986). Spent shot is a significant source of lead in many wetland environments that have received heavy hunting pressure over many years. Disease outbreaks in waterfowl populations occur frequently and are often associated with a high incidence of lead poisoning. Trust et al. (1990) examined the effects of lead ingestion on *in vitro* and *in vivo* indices of immune function in mallards (*Anas platyrhynchos)* confined to outdoor enclosures on a hunted waterfowl marsh. Mallards dosed with one No. 4 lead shot had elevated blood lead levels and lowered hemagglutination titers to SRBC (a T-cell-dependent humoral immune response) by day 7; no measurable effect on lymphoproliferative responsiveness to lipopolysaccharide (LPS) and PHA was observed. Biomonitoring programs employing immunocompetence assessments of wild waterfowl collected from lead-contaminated wetlands may prove useful in evaluating the success of remediation programs such as steel-shot only hunting zones and soil tillage.

Although immunotoxicity studies have not been conducted for wild rodents in the field, recent research by Porter et al. (1984), Olson and Hinsdill (1984), and Fairbrother et al. (1986) using wild deer mice *(Peromyscus maniculatus)* and McMurry et al. (1991) using cotton rats (*Sigmodon hispidus)* in the laboratory indicates that *in situ* immunotoxicity testing may be extremely useful in biomonitoring and assessing risks to humans in terrestrial environments contaminated with complex mixtures of immunotoxicants.

Plant growth regulators used in agricultural chemicals were found to have dose-dependent immunomodulatory effects in deer mice (*Peromyscus maniculatus),* influencing lymphocyte viability, white blood cell counts, plasma proteins, PFC responses to SRBC challenge, thymus weights, and hemolysin titers (Olson and Hinsdill, 1984). Low-level consumption of plant growth regulators can also influence mortality in deer mice (*Peromyscus maniculatus)* challenged with arboviruses such as Venezuelan equine virus (VEEV) without significant alterations in the magnitude and duration of viremias (Fairbrother et al., 1984, 1986). Age (Fairbrother et al., 1986) and nutritional state (Porter et al., 1984) can greatly influence how the immune system of deer mice will respond to virus infections after exposure to plant growth regulators or PCBs. These studies demonstrate the feasibility of using wild mammalian biomonitors, which are subjected to multiple stressors under natural conditions, to more realistically assess risks to humans that may be stressed due to malnutrition or other causes.

Similar approaches toward developing a sensitive *in situ* method for evaluating immunotoxicity hazards to wildlife and humans have been explored by McMurry et al. (1991) using wild cotton rats (*Sigmodon hispidus*). Benzene exposure in animals is associated with suppression of all marrow-derived lymphocytes, which can easily result in death due to overwhelming infection (Hadden, 1985). Adult cotton rats (*Sigmodon hispidus*) exposed intraperitoneally to graded doses of benzene showed increasing spleen weights, splenocyte yields, white blood cell counts, and PFC responses to SRBC, but no cell-mediated immune suppression was documented (McMurry et al., 1991).

GENOTOXICTIY

Pervasive among environmental toxicants are mutagens that act by altering the normal DNA complement of a cell. Chemically induced mutations can be either alterations of specific DNA sequences or alterations of normal chromosome structure and number. Chemicals that cause structural alterations in chromosomes are a special class of mutagens known as clastogens.

Clastogenic mutagens can have two possible major deleterious effects on the health of exposed wildlife. Germ cell mutations can lead to lowered fertility and fecundity through gamete, zygote, and fetal death, or they can be passed on through heritable, genetic defects that affect future generations (Leonard, 1976; Hook, 1982; Hsu, 1983). Somatic cell mutations have a more immediate effect on the health of an organism. These mutations are involved in tumor formation (Miller and Miller, 1981; Marx, 1982), cancer induction and progression (Farber and Cameron, 1980; Sasaki, 1982), and formation of tumor metastases (Fidler and Hart, 1982). Through the induction of focal lesions, somatic mutations are also implicated in such broad physiological effects as heart disease, cataracts, and aging (Hartman, 1983). Thus, the health effects of exposure to clastogens may not be expressed until years after the initial exposure event.

Although studies of genotoxicity using wildlife species as monitors of mutagenic pollutants are limited, initial research is promising. Ellenton and McPherson (1983) examined sister chromatid exchange (SCE) rates in unincubated herring gull (*Larus argentatus*) eggs collected from five breeding colonies in polluted areas around the Great Lakes and from a colony in a relatively pristine area on the Atlantic Coast. The average number of SCEs per chromosome for the colonies ranged from 0.069 to 0.101. No significant differences between any of the colonies were found. Nayak and Petras (1985) monitored levels of SCEs in house mice (*Mus musculus*) collected from corn cribs at varying distances from industrial sites in Canada. The SCE frequency in wild-caught mice was inversely proportional to distance between the site of capture and the nearest industrial site. After 9 months in captivity, wild-caught mice showed levels of SCE (3.46 ± 0.05) similar to that of laboratory strains (3.42, 3.62, and 3.97). Laboratory mice maintained in outdoor enclosures near industrial sites showed increased SCE frequencies (5.07 ± 0.22) compared to laboratory controls and newly collected wild mice showed similar elevated SCE frequencies (6.02 ± 0.16).

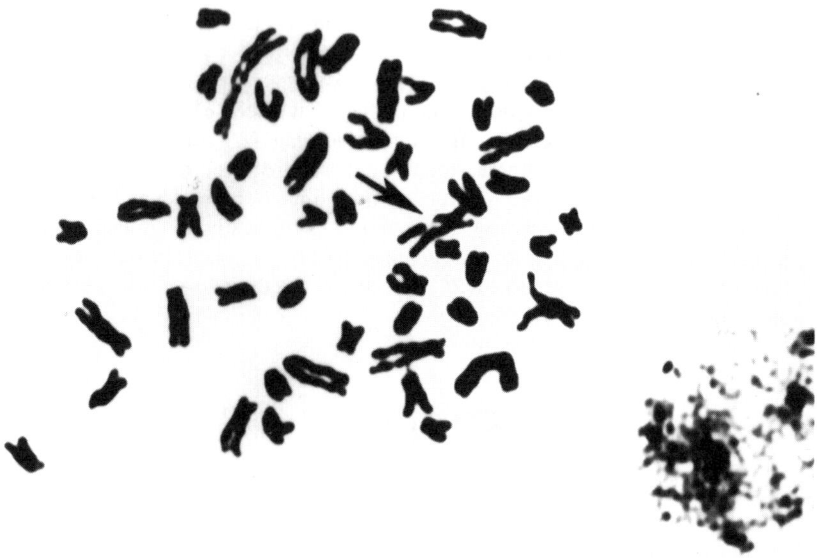

Figure 1 Chromatid break in an acrocentric chromosome (arrow) of *Peromyscus leucopus* (white-footed mouse) from a hazardous waste site in Texas.

Tice et al. (1987) used white-footed mice (*Peromyscus leucopus)* collected from a hazardous waste site (HWS) and from pristine sites in the northeastern United States in a series of genotoxicity assays. These workers found significantly elevated levels of SCEs in bone marrow and testicular tissue, decreased mitotic indices, increased levels of micronucleated polychromatic erythrocytes, and increased frequencies of sperm head abnormalities in animals from the waste site compared to animals from reference sites. Thompson et al. (1988) demonstrated a significant increase in frequencies of bone marrow chromosome aberrations in cotton rats (*Sigmodon hispidus)* from two Superfund sites in south Texas compared to animals collected from a matched reference site.

Examination of standard metaphase chromosome preparations from bone marrow of white-footed mice (*Peromyscus leucopus)* and cotton rats (*Sigmodon hispidus)* trapped over a two year period at a site polluted with a complex mixture of petrochemical waste products, heavy metals, and PCBs, and at two matched reference sites indicated significant increases in levels of chromosomal aberrations in animals collected at the contaminated site (Figures 1 and 2). Increases in lesions per cell and aberrant cells per individual were shown for both species at the contaminated site compared to the reference sites; however, animals from the two reference sites were not significantly different for either lesions per cell or aberrant cells per individual. Additionally, white-footed mice had significantly higher frequencies of lesions per cell at both the contaminated site and one reference site than did cotton rats, although the two species did not differ in aberrant cells per individual (McBee et al., 1987).

A subset of these animals were examined using G-banding analysis of bone marrow metaphase chromosomes (McBee, 1991). This is a differential staining technique that results in a unique banding pattern for each pair of homologs. Several classes of aberrations not detectable in the standard metaphase preparations

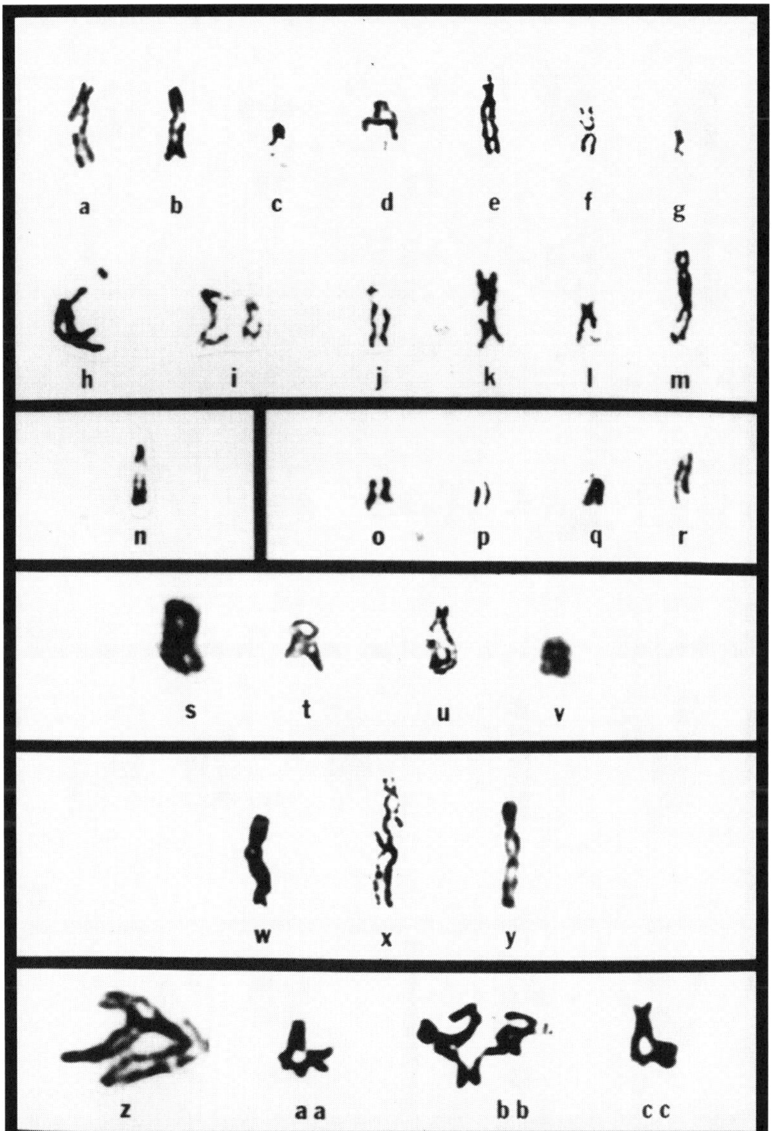

Figure 2 Examples of several classes of chromosomal aberrations found in *Sigmodon hispidus* (cotton rat) and *Peromyscus leucopus* (white-footed mouse) inhabiting an area contaminated with a complex mixture of petrochemical pollutants. a–m: chromatid breaks; n: chromosome break; o–r: acentric fragments; s–v: ring chromosomes; w–y: dicentric chromosomes; z–cc: translocation figures.

were identified, including Robertsonian fusions, pericentric and paracentric inversions, and interstitial deletions (Figure 3). Four classes of these previously unidentified aberrations were significantly higher in animals from the contaminated site compared to animals from reference sites. Several animals also carried apparently

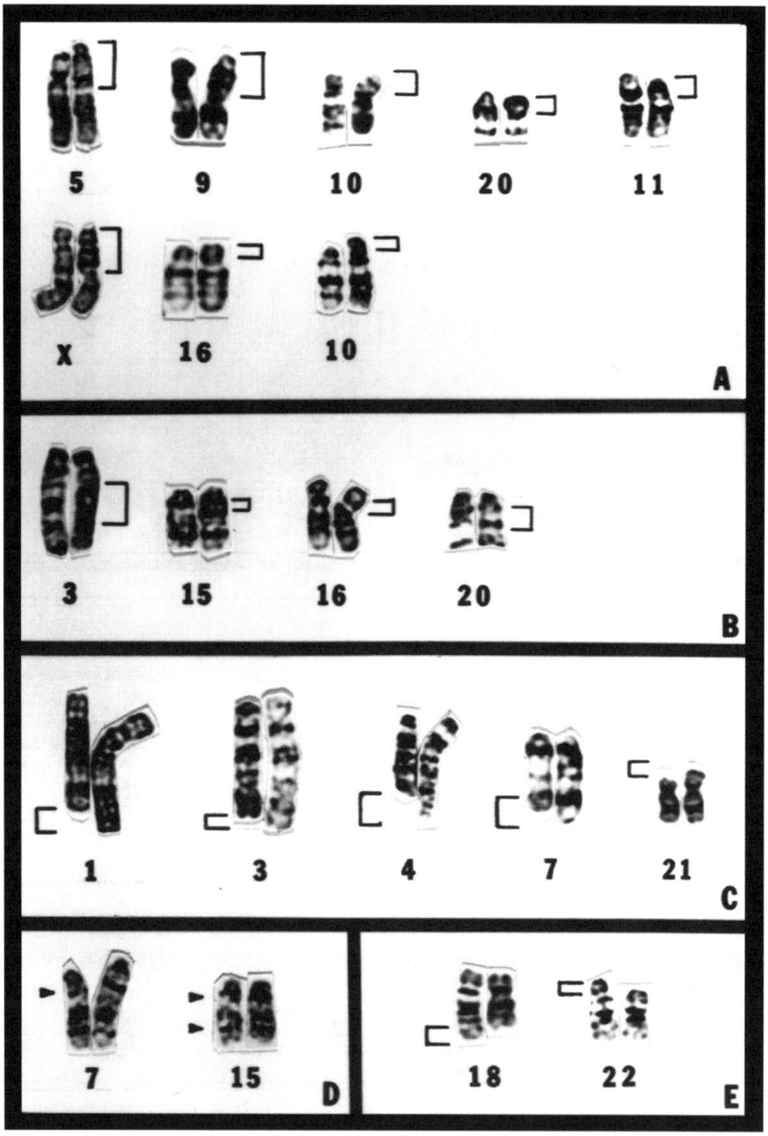

Figure 3 Aberrations observed in G-banded karyotypes of *Peromyscus leucopus* (white-footed mouse) from a site contaminated with petrochemical pollutants. A and B: pericentric inversions, and paracentric inversions respectively; the normal chromosome is on the left in each pair and lines indicate the inverted segment of the aberrant chromosome. C: terminal deletions; the normal homolog is on the right in each pair and lines mark deleted segments. D: interstitial deletions; the normal homolog is on the right in each pair and triangles mark sites of deletions. E: duplications; the normal chromosome is on the right; lines indicate duplicated regions (from McBee, 1991).

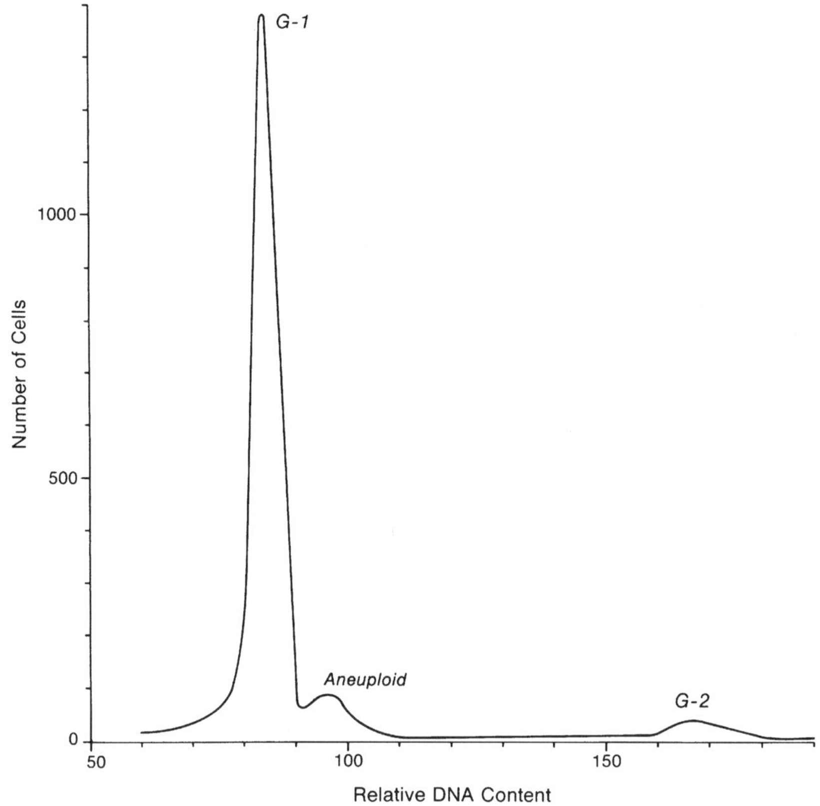

Figure 4 DNA histogram of spleen cells from *Peromyscus leucopus* (white-footed mouse) from a hazardous waste site. The small subpeak at the base of the G_1 peak represents a subpopulation of aneuploid cells. The area under the G1 peak includes 10,000 cells, the coefficient of variation for mean DNA content in this cell sample was 3.36 (From McBee, K. and J. W. Bickham, *Bull. Environ. Contam. Toxicol.* 40, 347, 1988. With permission.)

inherited aberrations in three chromosomes, suggesting that aberrations induced in a parent may be transmitted to offspring (McBee, 1991).

Flow cytometric analysis of nuclear DNA content variation in white-footed mice from this site showed significantly increased coefficients of variation (CV) around mean DNA content within individuals compared to animals from one of the uncontaminated sites. Flow cytometry also identified individuals that apparently had stable aneuploid cell populations, and individuals in which DNA replication apparently had been blocked resulting in accumulation of cells in the S region of DNA histograms (Figure 4) (McBee and Bickham, 1988).

Flow cytometry has also been used to examine red blood cell DNA in mallards (*Anas platyrhynchos)* exposed to environmental ^{137}Cs (Cesium). George et al. (1991) released 100-day-old male mallards, previously unexposed to ^{137}Cs, on an abandoned nuclear reactor cooling tower reservoir. Blood samples were periodically collected over a 1-year period from these animals and from two reference populations and analyzed for nuclear DNA content. Significant differences in CV between exposed individuals and reference individuals were detected after 9 months of exposure. Two individuals from the reservoir also showed evidence of aneuploidy after 9 months but none of the reference animals showed the typical dual peak pattern indicative of aneuploidy.

Shaw-Allen and McBee (1993) examined bone marrow metaphase chromosomes from white-footed mice, cotton rats, and harvest mice collected from a site contaminated with the PCB mixture Aroclor® 1254 and two uncontaminated reference sites and found no evidence of statistical differences among these groups. This substantiates results from laboratory cell culture assays of clastogenicity of this PCB mixture.

EXTRAPOLATION FROM WILDLIFE MODELS TO POTENTIAL HUMAN EFFECT

Any time data from one group of organisms is used to extrapolate to a possible response or effect in another group of organisms, several important considerations must be kept in mind. This is especially critical when the effort is to extrapolate from a wild animal model to a potential human health effect. Many of these concerns are addressed in Chapter 4 of this volume and will not be reiterated here. Effective extrapolation depends, in part, on an understanding of the underlying cellular and molecular mechanisms involved in the toxic endpoints measured in animal models (Brusick, 1980). Depending on the amount of information available for understanding basic mechanisms and the level of similarity in mechanisms assumed between model species and humans, some types of endpoints can be much more reliably extrapolated than others. For instance, based on the similarity of the genetic code and the similarity in mechanisms of action involved in interaction of a clastogenic chemical and the DNA molecule, extrapolation of an observed genetic effect in an animal model to a potential effect in humans is appropriate. However, extrapolation from endpoints such as behavioral responses may be much more difficult. Additionally, it is important to stress that it is not necessary for a wildlife model to show a "one-to-one" relationship in toxic response with humans. These animals serve as bioindicators of the potential for human health risk, not as exact measures of human disease levels.

Much previous work involving extrapolation of an observed effect in an animal model to a potential health effect in humans has been done using laboratory-reared strains of a few rodent taxa. These strains have often been developed to be genetically highly homogeneous. This high level of genetic homogeneity is valuable in the context of elucidating basic mechanisms of action and in gaining information on extremes of response. However, these laboratory models cannot provide information on the range of response possible under varying conditions of age, season, disease, nutrition, and typical genetic heterogeneity that is characteristic of natural populations of both wild animal models and humans. The same factors such a similarity of underlying mechanisms that argue for the use of laboratory-reared animal models holds true for wildlife species. In spite of the complexity and sometimes great variability of response inherent in using wildlife species as models for potential human health effects, these models may have far greater potential, both economically and biologically, than some laboratory models for advancing our understanding of the complex array of responses to environmental pollutants possible in human populations.

REFERENCES

Anderson, D. P., van Muiswinkel, W. B., and Robertson, B. S., Effects of chemically induced immune modulation on infectious diseases of fish, in *Chemical Regulations of Immunity in Veterinary Medicine,* Kende, M., Gainer, J., and Chirigos, M., Eds., Alan R. Liss, New York, 187, 1984.

Anderson, T. J. and Barrett, G. W., Effects of dried sewage sludge on meadow vole *(Microtus pennsylvanicus)* populations in two grassland communities, *J. Appl. Ecol.,* 19, 759, 1982.

Batty, J., Leavitt, R. A., Biondo, N., and Polin, D., An ecotoxicological study of a population of the white-footed mouse *(Peromyscus leucopus)* inhabiting a polychlorinated biphenyls-contaminated area, *Arch. Environ. Contam. Toxicol.,* 19, 283, 1990.

Brusick, D., *Principles of Genetic Toxicology,* Plenum Press, New York, 1980.

Conney, A. H., Induction of microsomal cytochrome P-450 enzymes: The first Bernard B. Brodie lecture at Pennsylvania State University, *Life Sci.,* 39, 2493, 1986.

Dean, H. J., Lester, M. I., Boorman, G. A., and Lauer, L. D., Procedures available to examine the immunotoxicity of chemicals and drugs, *Pharmacol. Rev.,* 34, 137, 1982.

Descotes, J., *Immunotoxicology of Drugs and Chemicals,* Elsevier, New York, 1988.

Elangbam, C. S., Qualls, C. W., Lochmiller, R. L., and Novak, J., Development of the cotton rat *(Sigmodon hispidus)* as a biomonitor of environmental contamination with emphasis on hepatic cytochrome P-450 induction and population dynamics, *Bull. Environ. Contam. Toxicol.,* 42, 482, 1989a.

Elangbam, C. S., Qualls, C. W., and Baudy, M., Induction of hepatic cytochrome P-450 in wild cotton rats *(Sigmodon hispidus)* by phenobarbital and 3-methylcholanthrene, *Bull. Environ. Contam. Toxicol.,* 42, 716, 1989b.

Ellenton, J. A. and McPherson, M. F., Mutagenicity studies of herring gulls from different locations on the Great Lakes. I. Sister-chromatid exchange rates in herring gull embryos, *J. Toxicol. Environ. Health,* 12, 317, 1983.

Emlen, J. M., Terrestrial population models for ecological risk assessment: A state-of-the-art review, *Environ. Toxicol. Chem.,* 8, 831, 1989.

Fairbrother, A., Yuill, T. M., and Olson, L. J., Effects of ingestion of chlorocholine chloride and cyclophosphamide on Venezuelan equine virus infection in deer mice *(Peromyscus maniculatus), Toxicology,* 31, 67, 1984.

Fairbrother, A., Yuill, T. M., and Olson, L. J., Effects of three plant growth regulators on the immune response of young and aged deer mice *Peromyscus maniculatus, Arch. Environ. Contam. Toxicol.,* 15, 265, 1986.

Farber, E. and Cameron, R., The sequential analysis of cancer development, *Adv. Cancer Res.,* 31, 125, 1980.

Fidler, I. J. and Hart, I. R., Biological diversity in metastatic neoplasms: Origins and implications, *Science,* 217, 998, 1982.

Franson, J. C., Immunosuppressive effects of lead, in *Lead Poisoning in Wild Waterfowl — A Workshop,* Feirabend, J. S. and Russell, A. B., Eds., National Wildlife Federation, Washington, D.C., 106, 1986.

Friend, M. and Trainer, D. O., Polychlorinated biphenyl: Interaction with duck hepatitis virus, *Science,* 170, 1314, 1970.

Friend, M. and Trainer, D. O., Experimental dieldrin–duck hepatitis virus interaction studies, *J. Wildl. Manage.,* 38, 896, 1974a.

Friend, M. and D. O., Trainer, Experimental DDT–duck hepatitis virus interaction studies. *J. Wildl. Manage*., 38, 887, 1974b.

George, L. S., Dallas, C. E., Brisbin, I. L., and Evans, D. L., Flow cytometric analysis of red blood cell DNA of mallards accumulating environmental ^{137}Cs over time, *Ecotoxicol. Environ. Safety,* 21, 337, 1991.

Goldberg, D. R., Yuill, T. M., and Burgess, E. C., Mortality from duck plague virus in immunosuppressed adult mallard ducks. *J. Wildl. Dis.,* 26, 299, 1990.

Grue, C. E. and Christian, C. L., Use of captive starlings to determine effects of pollutants on passerine reproduction, in *Avian and Mammalian Wildlife Toxicology: 2nd Conference,* ASTM STP 757, Lamb, D. W. and Kenaga, E. E., Eds., ASTM, Philadelphia, 5, 1981.

Guengerich, F. P. and Liebler, D. C., Enzymatic activation of chemicals to toxic metabolites, *CRC Crit. Rev. Toxicol.,* 14, 259, 1985.

Hadden, J. W., Immunotoxicology and immunorestoration, in *Immunotoxicology and Immunopharmacology,* Dean, J. H., Luster, M. I., Munson, A. E., and Amos, H., Eds., Raven Press, New York, 229, 1985.

Hartig, J. H. and Zarull, M. A., Methods of restoring degraded areas in the Great Lakes, *Rev. Environ. Contam. Toxicol.,* 117, 127, 1991.

Hartman, P. E., Mutagens: Some possible health impacts beyond carcinogenesis, *Environ. Mutagen.,* 5, 139, 1983.

Hoffman, D. J., Rattner, B. A., and Hall., R. J., Wildlife toxicology, *Environ. Sci. Technol.,* 24, 276, 1990.

Hook, E. B., Contribution of chromosome abnormalities to human morbidity and mortality, *Cytogenet. Cell Genet.,* 33, 101, 1982.

Hornshaw, T. C., Aulerich, R. J., and Johnson, H. E., Feeding Great Lakes fish to mink: Effects on mink and accumulation and elimination of PCBs by mink, *J. Toxicol. Environ. Health,* 11, 933, 1983.

Hsu, T. C., Genetic instability in the human population: A working hypothesis, *Hereditas,* 98, 1, 1983.

Klaassen, C. D., Amdur, M. O., and Doull, J., *Casarett and Doull's Toxicology: The Basic Science of Poisons,* 3rd ed., Macmillan, New York, 1986.

Knight, G. C. and Walker, C. H., A study of the hepatic microsomal monooxygenase of sea birds and its relationship to organochlorine pollutants, *Comp. Biochem. Physiol.,* 73C, 211, 1982.

Krebs, C. J., *Ecological Methodology,* Harper & Row, New York, 1989.

Lamb, D. W. and Kenaga, E. E., *Avian and Mammalian Wildlife Toxicology: 2nd Conference,* ASTM 757, ASTM, Philadelphia, 1981.

Leonard, A., Heritable chromosome aberrations in mammals after exposure to chemicals, *Radiat. Environ. Biophys.,* 13, 1, 1976.

Lidicker, W. Z., Analysis of two freely growing enclosed populations of the California vole, *J. Mammal.,* 60, 447, 1979.

Lubet, R. A., Guengerich, F. P., and Nims, R. W., The induction of alkoresorufin metabolism: A potential indicator of environmental contamination, *Arch. Environ. Contam. Toxicol.,* 19, 157, 1990.

Lubet, R. A., Nims, R. W., Beebe, L. E., Fox, S. D., Issaq, H. J., and McBee, K., Induction of hepatic CYP1A activity as a biomarker for environmental exposure to Aroclor® 1254 in feral rodents, *Arch. Environ. Contam. Toxicol.,* 22, 339, 1992.

Luster, M. I., Dean, J. H., and Moore, J. A., Evaluation of immune function in toxicology, in *Principles and Methods of Toxicology,* Hayes, A. W., Ed., Raven Press, New York, 1982, 561.

Lustick, S., Voss, T., and Pertle, T. J., Effects of DDT on steroid metabolism and energetics in bobwhite quail *(Colinus virginianus), Proc. 1st Bobwhite Quail Symp.,* 1, 213, 1972.

Maly, M. S., Survivorship of meadow voles, *Microtus pennsylvanicus,* from sewage sludge-treated fields, *Bull. Environ. Contam. Toxicol.,* 32, 724, 1984.

Marx, J. L., Change in cancer gene pinpointed, *Science,* 218, 667, 1982.

McBee, K., Chromosomal aberrations in native small mammals *(Peromyscus leucopus)* at a petrochemical waste disposal site: II. Cryptic and inherited aberrations detected by G-band analysis, *Environ. Toxicol. Chem.,* 10, 1321, 1991.

McBee, K. and Bickham, J. W., Petrochemical-related DNA damage in wild rodents detected by flow cytometry, *Bull. Environ. Contam. Toxicol.,* 40, 343, 1988.

McBee, K. and Bickham, J. W., Mammals as bioindicators of environmental toxicity, in *Current Mammalogy,* Vol. 2, Genoways, H. H., Ed., Plenum Press, New York, 1990, 37.

McBee, K., Bickham, J. W., Brown, K. W., and Donnelly, K. C., Chromosomal aberrations in native small small mammals *(Peromyscus leucopus* and *Sigmodon hispidus)* at a petrochemical waste disposal site. I. Standard karyology, *Arch. Environ. Contam. Toxicol.,* 16, 6881, 1987.

McMurry, S. T., Lochmiller, R. L., Vestey, M., Elangbam, C. S., and Qualls, C. W., Jr., Acute effects of benzene and cyclophosphamide exposure on cellular and humoral immunity of cotton rats *(Sigmodon hispidus), Bull. Environ. Contam. Toxicol.,* 46, 937, 1991.

Miller, E. D. and Miller, J. A., Mechanisms of chemical carcinogenesis, *Cancer,* 47, 1055, 1981.

National Research Council, *Animals as Sentinels of Environmental Health Hazards,* National Academy Press, Washington, D.C., 1991.

Nayak, B. N. and Petras, M. L., Environmental monitoring for genotoxicity: In vivo sister chromatid exchange in house mouse *(Mus musculus), Can. J. Genet. Cytol.,* 27, 351, 1985.

Newton, I. and Wyllie, I., Recovery of a sparrowhawk population in relation to declining pesticide contamination, *J. Appl. Ecol.,* 29, 476, 1992.

Novak, J. and Qualls, C. W., Effects of phenobarbital and 3-methylcholanthrene in the hepatic cytochrome P-450 metabolism of various alkoxyresorufin ethers in the cotton rat *(Sigmodon hispidus), Comp. Biochem. Physiol.,* 94, 543, 1990.

Olson, L. J. and Hinsdill, R. D., Influence of feeding chlorocholine chloride and glyphosine on selected immune parameters in deer mice, *Peromyscus maniculatus, Toxicology,* 30, 103, 1984.

Payne, J. F., Fancey, L. L., Rahimtula, A. D., and Porter, E. L., Review and perspective on the use of mixed-function oxygenase enzymes in biological monitoring, *Comp. Biochem. Physiol.,* 86C, 233, 1987.

Peterle, T. J., *Wildlife Toxicology,* Van Nostrand Reinhold, New York, 1991.

Pomeroy, S. E. and Barrett, G. W., Dynamics of enclosed small mammal populations in relation to an experimental pesticide application, *Am. Midl. Nat.,* 93, 91, 1975.

Porter, W. P., Hinsdill, R., Fairbrother, A., Olson, L. J., Jaeger, J., Yuill, T., Bisgaard, S., Hunter, W. B., and Nolan, K., Toxicant-disease-environment interactions associated with suppression of immune system, growth and reproduction, *Science,* 224, 1014, 1984.

Ratcliffe, D. A., Decrease in eggshell weight in certain birds of prey, *Nature,* 215, 208, 1967.

Rattner, B. A., Hoffman, D. J., and Marn., C. M., Use of mixed-function oxygenases to monitor contaminant exposure in wildlife, *Environ. Toxicol. Chem.,* 8, 1093, 1989.

Rocke, T. E., Yuill, T. M., and Hinsdill, R. D., Oil and related toxicant effects on mallard immune defenses, *Environ. Res.,* 33, 343, 1984.

Rowley, M. H., Christian, J. J., Basu, D. K., Pawlikowski, M., and Paigen, B., Use of small mammals (voles) to assess a hazardous waste site at Love Canal, Niagara Falls, New York, *Arch. Environ. Contam. Toxicol.,* 12, 383, 1983.

Sanders, O. T. and Kirkpatrick, R. L., Effects of polychlorinated biphenyl (PCB) on sleeping times, plasma corticosteroids, and testicular activity of white-footed mice, *Environ. Physiol. Biochem.,* 5, 308, 1975.

Sasaki, M., Role of chromosomal mutation in the development of cancer, *Cytogenet. Cell Genet.,* 33, 160, 1982.

Scanlon, P. F., Heavy metals in small mammals in roadside environments: Implications for food chains, *Sci. Total Environ.,* 59, 317, 1987.

Schrank, C. S., Cook, M. E., and Hansen, W. R., Immune response of mallard ducks treated with immunosuppressive agents: Antibody response to erythrocytes and in vivo response to phytohemagglutinin-P, *J. Wildl. Dis.,* 26, 307, 1990.

Shaw-Allen, P. L. and McBee, K., Chromosome damage in wild rodents inhabiting a site contaminated with Aroclor® 1254, *Environ. Contam. Toxicol.,* 12, 677, 1993.

Stromborg, K. L., Reproductive tests of diazinon on bobwhite quail, in *Avian and Mammalian Wildlife Toxicology: 2nd Conference,* ASTM STP 757, Lamb, D. W. and Kenaga, E. E., Eds., ASTM, Philadelphia, 1981, 19.

Talmage, S. S. and Walton, B. T., Small mammals as monitors of environmental contaminants, *Rev. Environ. Contam. Toxicol.,* 119, 47, 1991.

Tamarin, R. H., Reich, L. M., and Moyer, C. A., Meadow vole cycles within fences, *Can. J. Zool.,* 62, 1796, 1984.

Thompson, R. A., Schroder, G. D., and Conner, T. H., Chromosomal aberrations in the cotton rat (*Sigmodon hispidus)* exposed to hazardous waste, *Environ. Mol. Mutagen.,* 11, 359, 1988.

Tice, R. R., Ormiston, B. G., Boucher, R., Luke, C. A., and Paquette, D. E., Environmental biomonitoring with feral rodent species, in *Short-Term Assays in the Analysis of Complex Mixtures II,* Sandhu, S. S., DeMarini, D. M., Mass, M. J., Moore, M. M., and Mumford, J. L., Eds., Plenum Press, New York, 175, 1987.

Tilghman, A., Taylor, D. H., and Woods, P. E., Effects of municipal sludge on locomotor activity and exploratory behavior of meadow voles, (*Microtus pennsylvanicus), Environ. Toxicol. Chem.,* 9, 31, 1990.

Trust, K. A., Miller, M. W., Ringelman, J. K., and Orme, I. M., Effects of ingested lead on antibody production in mallards (*Anas platyrhynchos*), *J. Wildl. Dis.*, 26, 316, 1990.

Vos, J. G., Immune suppression as related to toxicology, *CRC Crit. Rev. Toxicol.*, 5, 67, 1977.

Vos, J. G., Krajnc, E. I., and van Loveren, H., Immunotoxicity and altered host resistance in rats: Tributyltin oxide as an example, in *Basic Science in Toxicology*, Volans, G. N., Sims, J., Sullivan, F. M., and Turner, P., Eds., Taylor & Francis, New York, 1990, 354.

Warren, R. J., Kirkpatrick, R. L., and Young, R. W., Barbiturate-induced sleeping times, liver weights, and reproduction of cottontail rabbits after Mirex ingestion, *Bull. Environ. Contam. Toxicol.*, 19, 223, 1978.

Zeaks, S. J., Hansen, M. F., and Robel, R. J., Increased susceptibility of bobwhites (*Colinus virginianus*) to Histomonas meleagridis after exposure to Sevin insecticide, *Avian Dis.*, 25, 981, 1981.

Zeeman, M. G. and Brindley, W. A., Effects of toxic agents upon fish immune systems: A review, in *Immunologic Considerations in Toxicology*, Vol. 2, Sharma, R. P., Ed., CRC Press, Boca Raton, FL, 1981, 1.

Zepp, R. L., Sanders, O. T., and Kirkpatrick, R. L., Reduction of pentobarbitol-induced sleeping times in PCB-treated cottontail rabbits, *Bull. Environ. Contam. Toxicol.*, 12, 518, 1974.

CHAPTER 7

Environmental Bioremediation: Approaches and Processes

Celia Bonaventura, Joseph Bonaventura, and Donovan F. Bodishbaugh

CONTENTS

1-56670-141-4/96/$0.00+$.50

BIOREMEDIATION DEFINED

Bioremediation is the application of living organisms or products, or the application of biological principles, to ameliorate adverse environmental conditions. The first question one might ask is "Why bioremediation?" One answer is quite simple. Bioremediation is a fundamental mechanism by which living systems have accommodated to changes in environmental conditions throughout the course of life on Earth. Through bioremediation, living systems have altered the face of the Earth in a profound way. Changes in the biosphere, initiated by the activities of living organisms, have actually molded the geologic history of our home planet. The primitive, abiotic Earth was baked in an anoxic, reducing atmosphere (Walker, 1977). The first life forms (which were, of course, obligatory anaerobes) probably made use of incident sunlight and the highly reducing conditions to drive the first biological engines. The evolutionary process eventually led to a photosynthetic mechanism, driven by the splitting of water molecules, that "polluted" the planet with a toxic by-product, free oxygen (O_2). The eventual global buildup of O_2 led, in turn, to the evolution of enzymatic remediation of oxygen toxicity (Fridovich, 1975), and eventually to global domination by life forms that actually utilized the oxidative potential of O_2 in their metabolism. Thus began the era of aerobic life forms. In a sense, "the rest of the story" is history. This brief description demonstrates the ever-present self-correcting nature of living systems, i.e., their ability to adapt to change, and thereby bring into balance a system that has been perturbed and has lost balance.

The biosphere is not a system in equilibrium *per se*. Life is maintained at a (nearly) steady-state condition, with a continuous throughput of energy. In order to maintain the steady state, a number of controlled systems and mechanisms for recycling are in operation. In even the simplest single-cell organisms, metabolic by-products have to be treated in a number of ways in order for them not to build up to toxic levels. Accordingly, an elegant metabolic machinery has evolved in order to transform and excrete toxic materials. In many instances, transformed wastes are maintained within the cell and used as building blocks for resynthesis. A similar situation occurs in multicellular organisms that have, in addition, organ-based systems (kidneys, liver, etc.) that function in ways to allow the organism to regulate body chemistry, and maintain a homeostatic steady state. Cycling at the geochemical level is now recognized as an important function of the overall scheme of the planet. The crust of the Earth itself is recycled, as evidenced by the relatively recently discovered lithospheric subduction zones, where tectonic plates meet, and continental margins are subducted deep into the Earth's mantle (Press and Siever, 1986).

Human activity appears to be increasingly disruptive to the global steady state by the production of either amounts or types of waste materials that do not easily enter into the natural process of degradation and cycling. As a species, we are becoming aware of this looming problem, and we are beginning to attempt to remediate the situation, to bring our waste levels and output back within the stability thresholds of the global waste cycle. Given the fact that biological systems have evolved in a way that cycling and self-remediation are natural functions of life, it makes a great deal of sense for us to examine closely the natural biological processes in an effort to learn sound remediation techniques with which to ameliorate environmental damage.

As presently used, the term bioremediation usually refers to some means of harnessing bacterial metabolism of waste although fungi and fungal products are being increasingly exploited, and both algae and higher plants are sometimes employed (see Chapter 8). Long before the existence of microbes was recognized, man learned how to take advantage of the biodegradative ability of the microbial world. From observations of natural decay came the technology of composting wastes. What we know of today as "bioaugmentation" (enhancement of pollutant degradation by augmentation of bacterial growth) has been practiced for centuries in small- and large-scale agricultural communities by the use of various composting schemes. Thus, agricultural waste is transformed into a valuable resource by reducing it to a state that actually contributes to new crop growth.

In a parallel fashion, mankind made use of biodegradation of human and animal waste in developing sewage treatment facilities of various sorts. Microbial biodegradative processes are the key components of the degradation of these wastes. With an increasing awareness of the role played by microorganisms in these processes, engineering techniques for handling large quantities of human and animal wastes are continually improving. Although microbial action is recognized as a normal, even essential part of sewage treatment, public and regulatory acceptance of microbial degradation of anthropogenic environmental pollutants has been handicapped by the fear that "microbes on the loose" can bring about unexpected, adverse consequences. The fact that mankind has successfully harnessed the biodegradative ability of microbes for thousands of years has done little to lessen these fears. Perhaps public awareness of the dangers of biological warfare, along with some sensationalized treatment by the press, has contributed to the hesitancy to accept microbes as partners in dealing with environmental pollutants (for a detailed discussion of this phenomenon, see McCabe and Fitzgerald, 1991). In considering this issue it is important to distinguish between naturally occurring and genetically engineered microbes. In the case of the former, the microorganisms utilized in bioremediation systems typically are already present in the contaminated soil, water, etc., and are a natural feature of the ecosystem. It is only the microbial density and/or growth rate that is manipulated by man. Genetically engineered microbes are a special case, with special concerns. Considerations related to their use are treated in Chapter 8.

Aside from the "natural" role of bioremediation, there are distinct advantages to a biological approach over "conventional" or nonbiological remediation techniques. Bioremediation offers clear advantages over some techniques that merely contain or repackage waste materials. In general, bioremediation leads to actual reduction and detoxification of wastes. Ideally, wastes can be completely degraded or "mineralized" to their simple inorganic constituents: carbon dioxide (CO_2) and water in the casc of pure organic compounds. In some instances, notably heavy metal contamination, biological systems can be adapted to concentrate or "bioaccumulate" toxic materials. Concentration and subsequent removal of these materials may enhance environmental quality or, potentially, make the metals available in a form suitable for recycling. Another great advantage of bioremediation is the potential for *in situ* remediation. This is especially important in situations where large areas and/or large volumes need remediation as in the case of entire ecosystems (wetlands, lakes, fields, etc.) that are environmentally impacted by anthropogenic pollutants. Approaches that rely on "mucking, sucking, and trucking" unwanted materials, putting them out of sight, out of mind, with a vain hope of somehow hiding them from natural food webs and chemical distribution systems, have tremendous long-term expenses, are labor intensive, and are destructive to sensitive environments.

Before embarking on a detailed treatment of the subject of bioremediation, a general disclaimer is in order. There are some potential caveats that are best kept in mind when looking to nature for solutions. For example, we know from studies of natural detoxification processes that, in some instances, the cellular mechanisms of toxicant metabolism can create chemical species that are more toxic than the parent compounds. The situation is complicated by the fact that biochemical reactions rarely proceed by a single pathway. Enzymatic catabolism (degradative metabolism) of a given substrate (compound to be degraded) is likely to be by multiple, competing pathways, each of which may produce a different set of metabolites (chemical breakdown products). For a given set of conditions, one pathway may dominate, resulting in nearly uniform biodegradation. If conditions change, however, the dominant reaction, or even the dominant actively-degrading organism in a mixed culture, may change, resulting in a change in primary metabolite identity. As will be seen, one of the greatest difficulties in assessing the success of bioremediation efforts is verification of the exact fate of the waste, especially for complex waste mixtures, and for uncontained *in situ* remediation. The applicability of bioremediation must be individually assessed for every contaminated site. No single bioremediation approach (or conventional remediation approach, for that matter) will work in every case, and many cases may be wholly unsuitable for bioremediation. Much of the technical literature related to bioremediation is generated by workers in the biotechnology industry, and it seems prudent to examine critically any dramatic claims in such cases. In the face of sometimes overwhelming environmental problems, it is tempting to bill promising techniques as miracle cures. The folly of this may be illustrated by returning to the ultimate example of the adaptive capacity of the biosphere described at the beginning of this chapter. While the introduction of O_2 into the primitive Earth's atmosphere and the subsequent evolution of aerobic organisms comprise as powerful an example of nature's adaptive response to perturbation as could be imagined, some sobering lessons should be drawn relative to modern perturbations. One is the time course of change. The transformation from reducing to oxidizing atmosphere is believed to have taken literally billions of years (Schidlowski, 1980). As detailed in other chapters, human activities are creating geochemical changes at a rate several orders of magnitude faster today. A second obvious point concerns the mechanism by which the imbalance was corrected. The evolution of aerobic life forms ensured the survival of life on Earth, but was of little benefit to the extant anaerobic life forms that had polluted their world with a toxic gas. In fact, it assured the end of their dominance! Those today who would suggest that the biosphere is capable of absorbing any insult thrown at it, or even that human-directed bioremediation as described in the following pages is the cure-all for the Earth's pollution woes, may not be entirely wrong. On the other hand, the full implications to mankind of such a suggestion may not be as palatable as they first appear. Accordingly, bioremediation should not be looked on as the panacea for all of our environmental problems. No form of remediation can supplant the need to reduce production of waste if the quality of life on Earth as we know it is to be maintained.

SCOPE OF THIS CHAPTER

There has been an explosion of bioremediation-related literature in recent years. Many of these articles and books are narrowly focused on specific waste-related problems or technical systems. Our objective here (and in the following application-oriented, companion chapter) is a general and reasonably brief overview of the principles and practice of waste bioremediation. While some readers will undoubtedly be motivated by an immediate need for a working system, it is not our intention to produce a comprehensive review of available technology. Considering the rapid pace of development in the field, such an undertaking could fill volumes, and would almost certainly be out of date before publication. This work is intended to be a primer for those interested in the subject, and an introduction to those about to dive into the sea of research papers and vendor-supplied technical documentation. New, more efficient bioremediation systems will continue to be developed, and capacities beyond those described here may soon be achieved. There are some basic limitations and potential problems inherent in any living system, however, and hopefully this treatment of the subject will equip the reader with the knowledge needed to understand and anticipate their occurrence.

BIOREMEDIATION AS A NATURAL PROCESS

BIOREMEDIATION AND BIODEGRADATION

The terms bioremediation and biodegradation are sometimes erroneously treated as being synonymous. Two distinct concepts are involved, and separation of these terms is beneficial. Environmental bioremediation refers to a process of environmental improvement, in which biological organisms or products play a key role. A human value judgment is implicit in this concept, since two conditions are being compared and one judged to be an improvement. The improvement may not involve degradation of waste at all, as in the case of waste bioaccumulation, although degradation is usually the desired action. Biodegradation specifically refers to chemical breakdown or mineralization of materials (not necessarily waste) facilitated by biological organisms or products. The results of this action may or may not be judged to be remediative, although biodegradation of waste often leads to bioremediation. Many microorganisms can adapt their catabolic machinery to make use of certain environmental pollutants as food sources. Thus, the microbes bioremediate the environment as they biodegrade the pollutant to obtain energy. As alluded to above, biodegradation of organic chemicals is accomplished enzymatically. Catabolic enzymes are proteins that catalyze specific degradation reactions of organic molecules and that usually display high affinity for certain substrates. By this process, organic wastes (along with a few inorganics) may be absorbed by the microbe and broken down internally, or wastes may be degraded externally by secreted enzymes, after which the metabolites are absorbed and utilized. Alternatively, microbial catabolic enzymes, produced for breaking down normal food sources, may fortuitously degrade certain wastes that are present as well. This process, termed "co-metabolism," obviously requires an ample supply of the preferred food substrate and also requires that the degrading enzymes come into contact with the waste by waste absorption or by enzyme secretion. Either direct metabolism or co-metabolism can be enhanced in rate and extent over that which occurs naturally by measures that boost the microbial population or increase the availability of food and/or growth-limiting nutrients. The basic information required to enhance natural biodegradative processes is knowledge of the microorganisms present in a given site and the food substrates they prefer to use. As described in more detail in the following pages, the objective of bioremediation is to detoxify environmental contaminants constructively and harmlessly. With respect to biodegradation, two potential caveats are best kept in mind: (1) it is possible for pollutant mixtures to be metabolized into more harmful mixtures by biological action; and (2) even relatively harmless wastes can cause environmental problems if production rates exceed the capacity of natural cycles of breakdown and renewal (the same is true for waste metabolites). Therefore the occurrence of biodegradation, in and of itself, does not necessarily constitute bioremediation.

NATURAL CYCLES TO MIMIC OR UTILIZE

The global carbon and nitrogen cycles, as presently understood (Figures 1 and 2), represent the *ne plus ultra* of bioremediation. Biological organisms, especially microbes, play a central role in these grand cycles of renewal. Various organisms have co-evolved to play complementary roles in perpetuating the cycle. Average cycling times for different elements and compounds, e.g., O_2, N_2, H_2O, CO_2, vary from years to thousands of years, and also differ depending on the environmental compartment being examined. Atmospheric CO_2, for example, is thought to turn over about every 3 years on average, while CO_2 in the hydrosphere cycles much more slowly, with a mean residence time for the whole ocean being about 350 years (Schlesinger, 1991). The global carbon cycle in particular is now under intense scientific scrutiny, as man struggles to understand the complex phenomenon of global warming, and the contribution of man's increasing use of fossil fuels that generate CO_2 (reviewed by Hileman, 1992).

In a balanced system, the waste of one organism is utilized by another to the eventual benefit of the whole ecosystem. Nutrient cycling is thus an integral part of a continuously functioning living system. The concept of balance enters here. When input and output of any substance remain unequal for a prolonged period of time, the result will either be a shortage or a glut of that substance. Shortages of substances essential for ecosystem function, or gluts of substances that impair normal function can be expected to have serious adverse effects. This basic principle of mass balance cannot be avoided in any system. Ecosystem damage has resulted when this principle has been disregarded. There have been many cases where mankind has ignored the basic laws of thermodynamics by dumping harmful substances of a quality or quantity that exceeds natural degradative capacities.

Beyond simple mass balances, the chemical form and environmental distribution of wastes has a profound effect on their susceptibility to biodegradation. This introduces the concept of "bioavailability." To be metabolized, wastes must either come into contact with or be brought in very close proximity to the degrading microbes in a chemical form and under environmental conditions that are conducive

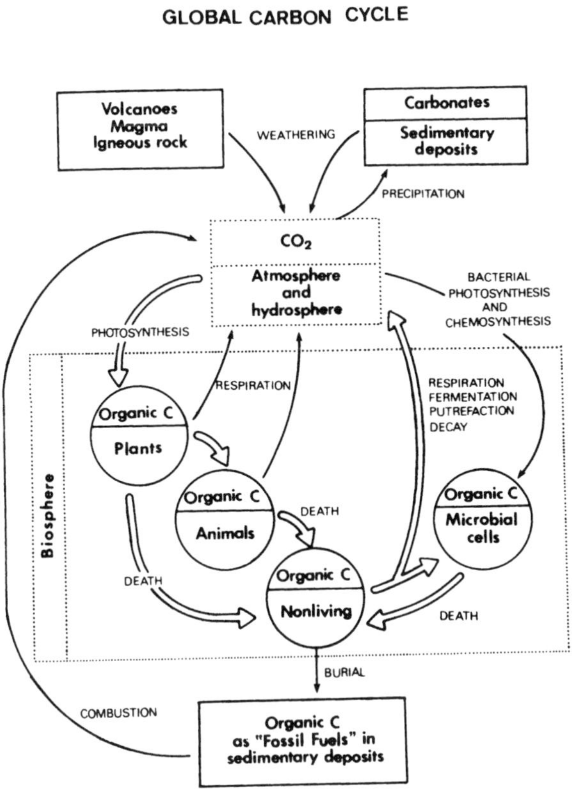

Figure 1 Biogeochemical cycling of carbon. (From Evans, W. C., *Nature,* 270, 17, 1977. Reprinted with permission from *Nature,* Copyright 1977, Macmillan Magazines Ltd.)

to degradation. For example, a simplified soil ecosystem might be characterized as a mixed, two-phase system: the dry soil particles, and the groundwater that fills some spaces between and coats some surfaces of the particles. A nonsoluble soil contaminant, associated exclusively with the dry particles, would be bioavailable to liquid-phase organisms only to the extent that it contacted the soil-water interface. Bioavailability would therefore largely depend upon soil water content, which may change dramatically with season and weather events. In another scenario, even though degrading microbes have physical access to a given waste, they may not be able to metabolize it if the waste is chemically bound to some other substance. In real ecosystems, compounds are always something less than completely bioavailable to a given organism. Furthermore, bioavailability is dynamic, and may change with time and environmental conditions. Precise knowledge of waste bioavailability may not be possible, but an understanding of potential problems and actions that may improve bioavailability is essential to bioremediation efforts.

Thus, two separate yet linked goals of bioremediation are (1) to biotransform toxic materials into nontoxic ones; and (2) to make accumulating anthropogenic waste enter natural biogeochemical cycles more quickly.

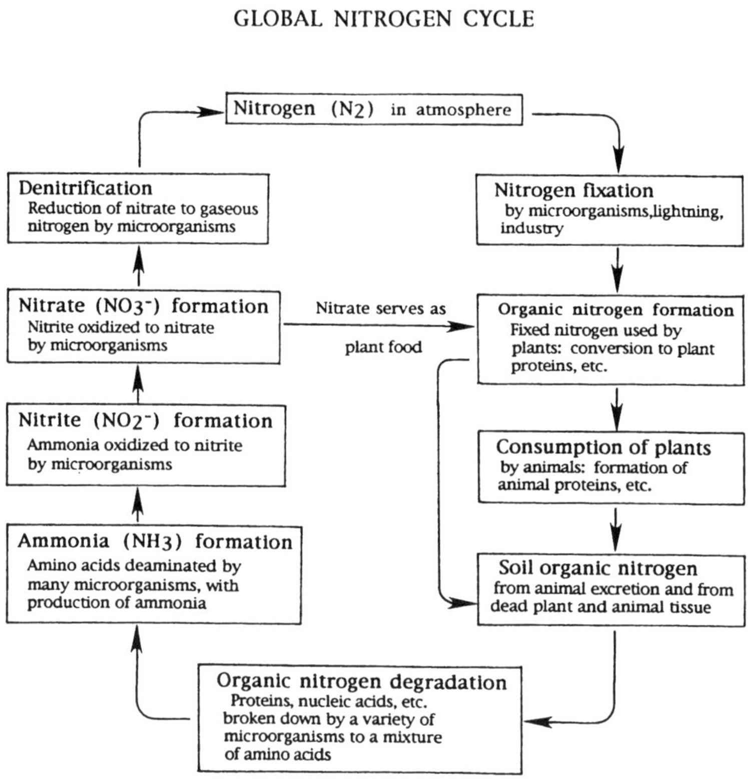

Figure 2 Biogeochemical cycling of nitrogen. (From Keeton, W. T., *Biological Science,* 3rd ed., W. W. Norton, New York, 1980; Pelczar, M., Jr., Reid, R. D., and Chan, E. C. S., *Microbiology,* 4th ed., McGraw-Hill, New York, 1977. Reprinted with permission from McGraw-Hill.)

AEROBIC VS. ANAEROBIC METABOLISM OF WASTES

For purposes of bioremediation, aerobic microbial metabolism has traditionally been the exclusive focus of attention. Aerobic degradative pathways in microbes, as in animals, break down organic molecules oxidatively, by using divalent oxygen (O_2) or other active oxygen species, such as hydrogen peroxide (H_2O_2), as electron acceptors. Aerobic catabolism of organics ultimately results in the familiar mineral products CO_2 and H_2O. Aerobes are capable of degrading most organic wastes, provided enough oxygen is available. Some compounds, notably the organohalogens, are highly resistant to aerobic biodegradation (termed recalcitrant or persistent wastes). Resistance of most aromatic and aliphatic compounds to degradation is dramatically increased by halogenation (most commonly chlorination), and further halogenation results in increased resistance (Stroo, 1992).

Anaerobic microbes degrade organics reductively, eventually resulting in the mineral end-product methane (CH_4), with CO_2 and free hydrogen (H_2) also produced in the case of carbohydrate compounds. While not usually utilized for routine waste degradation, some anaerobes are very adept at dechlorination of common recalcitrant organochlorine compounds, notably polychlorinated biphenyls (PCBs), organochlorine pesticides, such as DDT, and chlorinated aliphatics, such as the industrial solvent trichloroethylene (TCE) (reviewed by Abramowicz, 1990; Long-de Vallière et al., 1989; Neilson, 1990). Thus, anaerobic microbial catabolism (sometimes called fermentation) offers a bioremediation option for dealing with persistent wastes. Complete anaerobic degradation of wastes, however, may be very slow.

The major problem with anaerobic digestion of organochlorine wastes is that biodegradation is often incomplete (at least on a practical time scale), and may result in toxic metabolites. The use of mixed cultures, containing both aerobes and anaerobes, has been shown to facilitate mineralization of many organochlorines (reviewed by Rochkind et al., 1986). In practice, a sequential bioreactor system utilizing both anaerobic and aerobic reactors could be employed. For example, PCBs or chlorinated aromatics could be dechlorinated anaerobically, then fed into an aerobic bioreactor to be fully mineralized to CO_2 and H_2O. Similarly, TCE and perchloroethylene (PCE) may be reductively metabolized to vinyl chloride (a toxic chemical), which can then be subjected to aerobic biodegradation. Commercial versions of such two-stage hybrid bioreactor systems are currently under development (Thayer, 1991). Isolation and characterization of dehalogenases (dehalogenating bacterial enzymes) for possible development of immobilized enzyme reactors and biofilters is also being conducted (reviewed by Janssen et al., 1990).

BIOREMEDIATION AND PRODUCT DESIGN

A proactive step toward minimizing environmental damage caused by waste imbalance is the utilization of biodegradable materials wherever possible. Effective bioremediation can be advanced and facilitated by making products and containers that are readily susceptible to biodegradation. To achieve long-term success in the implementation of any environmental remediation policy, prevention of future contamination, and compatibility between wastes and available remediation technologies must be addressed. Keeping the principles of bioremediation in mind during product and process development is both prudent and profitable in the long term. Eventual benefits include partial elimination of the need for future remediation and ensuring that methods will exist to deal with inevitable accidental contaminations.

ENVIRONMENTAL HEALTH AND HUMAN HEALTH ASPECTS OF CONTAMINATION

ENVIRONMENTAL QUALITY AND HUMAN HEALTH: PUBLIC PERCEPTIONS AND AWARENESS

A growing impetus for environmental remediation efforts is recognition of the inextricable link between environmental health and human health. Historically, these two fields have operated out of different camps and have not been in close communication. The adverse consequences of this separation of efforts have been considerable. Human health is obviously a very old, broad, and deep field of science. In contrast, the very concept of "environmental health," in a formal sense, may be considered a modern one. While anyone may have an intuitive sense of what constitutes a healthy environment, precise definition is difficult. Ultimately, environmental health is measured as the viability of the inhabitants of a given ecosystem, as affected by ambient environmental factors (Shields, 1990). In practice, this definition requires assessing the health of indigenous organisms, and correlating observed changes in health with changes in environmental conditions. While evidence of deteriorating environmental health as a direct result of human activity is widespread and generally accepted, the human health implications of changes in environmental health remain controversial (reviewed in Anon., 1992). Many human health researchers and clinicians have been reluctant to accept nonhuman environmental health data as indicators of human health risk (although traditional medical science is largely based on laboratory animal work). Realization of the linkage between the two fields is rapidly increasing, however. Rather than treating human and environmental health as independent, many now view human health as a subset (by definition dependent) of overall environmental health. In recent years there has been an extraordinary allocation of research resources toward expanding our understanding of the human health implications of animal health (see NRC, 1991a). The utility of nonhuman environmental health indicators as predictors of human health risk is especially dependent on a thorough understanding of the possibilities and limitations of various environmental biomarkers and sentinel species, as discussed previously in Chapter 6 (see also NRC, 1991a, b, c; Shields, 1990).

There are cases where solid, if indirect, evidence exists for a connection between environmental pollution, animal health, and human health, especially as relates to food species (reviewed by Ahmed, 1991). As a pertinent example, organohalogen levels in nature are increasing, due both to direct deposition of industrial emissions, and to biological production from anthropogenic organic material (Asplund and Grimvall, 1991). There is fairly widespread appreciation that contamination of aquifers, surface waters, and fish populations with these toxicants leads directly or indirectly to exposure of the human population and a concomitant decrease in human health. Epidemiological studies have reported that humans who regularly consume fish from PCB-contaminated waters have highly elevated body burdens of PCBs (Fiore et al., 1989; Kreiss, 1985). Toxicity studies have shown that animals fed fish contaminated with PCBs and other organohalogens develop a variety of behavioral, neurological, and biochemical defects (Cleland et al., 1989; Hertzler, 1990). Even more attention has been given to genotoxic materials. Cancer-bearing fish are now beginning to be recognized as good systems in which to study human disease. As laboratory models of vertebrate toxicity, fish offer unique advantages over traditional rodent species, including low cost, larger numbers, ready observation of developmental effects, and amplification of genetic defects to thousands of individuals in a single generation (reviewed by Bailey et al., 1989). In addition, fish and other aquatic animals are ideally suited for use as indicators of environmental stress that may sound the alarm that human health is at risk in adjacent areas (reviewed by Contie, 1991; Dawe, 1990; Murchelano, 1990). Incontrovertible scientific evidence for direct human health links to sickness in aquatic animals is still sometimes minimal or altogether absent, but searching for these connections is currently an area of intense scientific study (see EPA, 1989 for guidelines and references). Other risks and implications of environmental contamination to human health are thoroughly reviewed in several other chapters of this book.

Public recognition of the intrinsic connections between environmental contamination and hazardous wastes appears to be increasing. A growing number of people are seeking assurance that the organisms they eat are healthful. Red tide occurrences that cause diseased or stressed animals to be washed up on shore in coastal areas result in public avoidance of suspect organisms as food sources. Media coverage and articles on pollution problems such as marine oil spills are also drawing public attention to the need for environmental remediation (Holloway, 1991). Red tides and oil spills are cases where the public is not waiting for all of the scientific assurances to be gathered. To quote Dawe (1990), "It is a self-evident truism that when aquatic animal populations experience ill health from exposure to anthropogenic chemicals, human health is in jeopardy as well."

In the past, there has been little if any resistance to claiming more and more of the planet and its resources as being solely for mankind's use. Beyond that position, our increasingly industrial civilization has resulted in an alteration in the Earth's air, soil, and waters that extends far beyond the city and farm environments that we have claimed for our exclusive use. The interdependence and connectedness with the living and nonliving aspects of nature that were part of the philosophical view of many indigenous new- and old-world cultures seem greatly at odds with current practices of expansion and resource utilization. There is now emerging a call for reconsideration of these fundamental values. Scientists are increasingly recognizing the need for a fully integrated approach to environmental research and resource management. The Sustainable Biosphere Initiative of the Ecological Society of America is a prime example of this approach. It outlines needs for basic research, education, and implementation of policy in the field of ecology, and suggests priorities and strategies for addressing these needs (Lubchenco et al., 1991). This growing recognition of the need for change in present practices and for interdisciplinary cooperation is not limited to the scientific community. Twenty years ago, the notion of preserving the cultural and ecological integrity of jungles and forests was the exclusive domain of radical conservationists. Today, heads of state of developing and industrial nations meet, amid media fanfare and hype, to draft international treaties on the issue. Good stewardship of our planet is an increasingly popular cry of the public. In the long term, it is proving to be not only environmentally advantageous, but also politically and economically sound to clean up polluted air and water systems. The loss of biodiversity resulting from degradation of environmental health threatens multibillion-dollar global industries, such as agriculture, biotechnology, pharmaccuticals, and nature tourism (Reid, 1992). A new emphasis on "sustainable development" has focused global attention on the need to adopt environment-friendly industrial approaches, not only to maintain the Earth's life-support systems, but to ensure the future of natural resources and the economies that they support.

Bioremediation is a technological attempt to exploit the connection between human and environmental health in a positive way, by exploiting the abilities of microbes and other natural members of the biosphere to improve environmental quality for all species, including *Homo sapiens*. The ultimate key to the success of measures designed to improve environmental health and quality, including bioremediaton, is education. In order to develop fully the potential of harnessed biological systems and to prevent misconceptions and misuse, a high level of understanding of the complex and subtle roles of various elements of the biosphere is required of the consumer, of industry, and of the regulatory community. The Senegalese conservationist Baba Dioum was recently quoted as observing, "In the end we will conserve only what we love, we will love only what we understand, and we will understand only what we are taught" (Cortese, 1992).

ANTHROPOGENIC POLLUTANTS

Over 70,000 commercial chemicals are used in the U.S., with up to 3000 being added to the list annually (Hoffmann, 1982). Each year the U.S. EPA compiles a Toxics Release Inventory (TRI) of major toxic chemicals released by U.S. industry. There are over 320 chemicals and chemical categories on this list. While emissions of some chemicals are decreasing, total release of toxicants into the environment by U.S. manufacturers in 1990 was some 4.8 billion pounds (Ember, 1992). In addition, large quantities of a number of toxic products are released into the environment by end-users in more or less unaltered form. These include materials such as fuels, dielectric fluids, flame retardants, heat transfer fluids, lubricants, protective coatings, propellants, pesticides, refrigerants, and many other chemicals. Such materials or their breakdown products often accumulate in soil and aquifers near landfills and dumpsites, or in surface lakes and streams. The breakdown of major chemical wastes found in listed waste sites has already been reviewed (see Chapter 1, Table 2). It should be recognized that these and many other pollutants are present, not only in concentrated waste sites, but throughout the environment, albeit in concentrations too low to trigger regulatory action.

Many constituents of common pollutant classes are present in the environment to some extent due to natural causes, such as crude oil seeps, natural pesticides produced by certain plants, heavy metal ore deposits, etc. Other anthropogenic waste compounds have structural similarities to naturally occurring substances. Either way, microorganisms that have evolved in the presence of natural "pollutants" have often developed the ability to utilize them as food substrates, and can readily metabolize those of anthropogenic origin as well. This ability can be exploited for bioremediation purposes. Nutrient enrichment of a waste-contaminated site improves growth conditions for the entire microbial population in a nonspecific manner, but the microbes adapted to metabolize the waste will be the ones that thrive the most (assuming that the waste concentrations are not so high as to be toxic to the microbes).

Fortunately, it is becoming clear that there exist microbial organisms that can biodegrade almost any organic chemical. By encouraging their growth in contaminated soil or water and their utilization of the contaminants that are present, *in situ* bioremediation can occur. This process takes place without human intervention to an appreciable degree although its time course may be extremely slow. Environmental chemists and microbial ecologists have extensively characterized the natural biodegradation pathways of a number of anthropogenic pollutant classes, and recent reviews have been published for many, including polycyclic aromatic hydrocarbons (PAHs) (Cerniglia and Heitkamp, 1989), PCBs (Abramowicz, 1990; Bedard,

1990), and pesticides (MacRae, 1989; Thorn et al., 1991). Rapid screening assays are being developed by researchers to identify organisms capable of degrading specific wastes (Krieger, 1992).

Organisms that have the capacity to utilize an environmental contaminant as a food source often grow more abundantly in areas where the contaminant is present. Consequently, organisms with the highly developed capability to metabolize specific contaminants can be found in contaminated sites. These can be isolated from the microbiological population and cultured for inoculation of other sites or for use in bioreactors. This approach has been used successfully to isolate bacterial strains capable of metabolizing marine crude oil spills (Rambeloarisoa et al., 1984), jet fuel, and organohalogens in groundwater aquifers (Aelion and Bradley, 1991; Criddle et al., 1990; Henry and Grbic-Galic, 1991), among many others. Inoculation of contaminated sites with such specialized microbes can present additional technical problems, however. For example, a microorganism that degrades marine oil spills may not thrive in oil-contaminated soil, or even in a marine environment that differs much from the one of origin. Nevertheless, organisms isolated in such a way have been used in a number of land treatment and bioreactor systems (see Chapter 8). Most cultures have been developed by private remediation service companies and remain proprietary products. Again, it should be stressed that this process is a selection of genetically able microorganisms from mixed cultures that are found in a natural environment. In contrast is the use of genetically engineered strains of microorganisms, where deliberate manipulation of DNA sequences yields capabilities that the organisms did not previously possess.

Two classes of unusual pollutants that are increasingly produced by human activities merit special mention here. The first is the inorganic pollutant category of heavy metals, such as lead, mercury, and cadmium. While they are natural elements found in the Earth's crust, heavy metals are utilized in industrial processes and released in purer form and higher concentrations than would normally occur in natural systems. The second category is that of radioactive waste materials. These two pollutant classes share a common problematic characteristic, namely that they cannot be detoxified by microbial catabolism. The toxicity of metals is inherent in their atomic structure, and they cannot be further mineralized. Incorporation of heavy metals into organometallic complexes may change their biological activity; but such a process is as likely to increase toxicity, due to increased bioavailability, as decrease it. Radioactive compounds, while they may be biodegraded, will yield radioactive metabolites, which may well be as hazardous as the parent compound. In order to remediate an environment contaminated with either of these pollutants, the objective is concentration and containment of the toxicant.

The methods most appropriate for concentration and containment of pollutants depend greatly on water content. Traditional approaches have involved incineration and wet oxidation with controlled emission to capture the contaminants. When the water content is high, these methods do not work well; and contaminant concentration by bioaccumulation may be an attractive alternative. *In situ* bioaccumulation by certain plants is a promising technology for dealing with slowly leaching metal sources, such as mine drainage. Obviously, the bioaccumulating organisms themselves become loaded with toxic substances; and the resulting biomass, whether it is bacterial or plant, must be collected and disposed of before it decays. Specific examples of the use of biological systems to deal with heavy metal contaminants are given in the next chapter. Bioremediation of radioactive pollutants is practically a nonexistent field, although an exhaustive review of interactions between stored radioactive waste and microorganisms has been published (Bachofen, 1990, 1991). Radioactive metals may potentially be concentrated in the same ways as ordinary heavy metals.

WHY BIOREMEDIATION NOW?

In addition to the generic benefits of waste bioremediation described above, there are some current event-driven factors that make such approaches especially attractive today. While it is unlikely that bioremediation techniques that can handle all types of waste and cleanups will ever be developed, a dramatic increase in the use of biological methods in waste remediation seems likely in the near future.

Enactment of federal legislation, such as the Comprehensive Environmental Response, Compensation, and Liability Act of 1980 (CERCLA) and the Superfund Amendments and Reauthorization Act of 1986 (SARA), has made responsible parties liable for the costs of waste-site cleanup, costs that can be enormous. In addition to actual Superfund remediation, efforts to avoid liability for EPA-administered cleanups under the terms of these statutes have led to large-scale remediation efforts by private industrial concerns in the U.S. One obvious motivation to remediate is, therefore, legal compliance. The immediate stimulus may be direct regulatory-agency action or cleanup requirements related to real estate transactions. Property titles in many states cannot be legally transferred unless the liability for any contamination discovered and cleanup required is covered. The increased need for remediation of diverse classes of waste and waste sites has created a demand for improved remediation techniques and for techniques that are applicable to a wider variety of cases. As an emerging technology, bioremediation is poised for rapid development.

A second current trend is a dramatic increase in the cost of traditional waste-treatment methods, especially bulk disposal. Bioremediation is, in many situations, a more cost-effective approach than containment or treatment by traditional chemical, physical, or thermal processes. For complex mixtures of waste, biodegradative treatment offers typical savings of 60 to 90% over landfill-disposal costs, for example (see Chapter 1). Accurate financial comparisons among waste-treatment techniques require analysis of specific processes and specific waste compositions. Whether the method is biological or nonbiological, the possibility of incomplete remediation exists. Toxic by-products or residues of primary treatment (incinerator fly ash, chemical sludge, spent filters and scrubbers, etc.) may themselves be subject to a final disposal step. Such costs must be included in comparisons. Since bioremediation can ideally destroy organic wastes without residue, such considerations may favor bioremediation over nonbiological alternatives.

GENERAL STEPS AND METHODS IN BIOREMEDIATION

BIOREMEDIATION METHODS

Given that microbes that can biodegrade various wastes exist, and that methods exist for stimulating, identifying, and/or culturing these organisms, the practice of bioremediation largely reduces to techniques of applying the microbes to the waste, and manipulating conditions to maximize the desired activity. For purposes of this discussion, three theoretical categories of such techniques have been identified. Detailed physical descriptions and examples of technical systems that fall into these categories will be presented in Chapter 8. The present discussion is restricted to general and theoretical descriptions. Treatment of contaminated material in place is called *in situ* remediation. Usage of this term varies and can cause confusion. Some workers consider *in situ* treatment to mean treatment of undisturbed material only, with no handling, containment, or transport allowed. Others apply the term more liberally to include any treatment in the immediate environs of the contaminated site. The meaning of "*in situ* land treatment" is clear enough: contaminated soil is treated without excavation. Yet "*in situ* groundwater treatment" usually refers to techniques that involve pumping, treating, and reinjecting contaminated groundwater (see Chapter 8). Obviously, some treatments are more *in situ* than others. We, therefore, use the more liberal definition here.

The simplest method of getting microbes and waste together is generally referred to as land treatment. For the purpose of discussion, we will divide this category into two subcategories: *in situ* land treatment and land farming. *In situ* land treatment is a method for bioremediation of contaminated soil and, to some extent, of associated groundwater. The action may be as simple as nutrient enrichment (fertilization) or may involve further manipulation of site conditions, such as inoculation of the contaminated site with selected microorganisms, mixing and aeration of surface soils, or pH modification. Obviously, *in situ* treatment involves a minimum amount of effort and cost to begin since the contaminated material does not have to be contained or even extensively handled. The expensive and potentially dangerous job of transporting waste is eliminated, and further disturbance of the site itself is minimized. Paradoxically, the chief attraction of this approach is also the chief drawback. Remediation of uncontained toxic waste makes degradation monitoring difficult, and concerns about toxicant dispersal may rule out such methods altogether. The difficulty inherent in assaying the effectiveness of biological vs. abiotic elements in pollutant degradation is addressed in the final section of this chapter. Of course, very limited control over ambient conditions is possible with *in situ* land treatment.

The other version of land treatment, which we have termed land farming (although this term is sometimes used as a synonym for generic land treatment), is applicable to solids, sludges, liquids, or contaminated soil. Wastes are sprayed onto, plowed into, or otherwise mixed with surface soils at the land farm site; and aeration, water, and nutrient enrichment are applied to achieve the best conditions for microbial growth. This technique can be adapted for routine treatment of wastes as well as remediation of contaminated sites. Following waste application, land farming is identical in practice to *in situ* treatment of contaminated soil, except that treatment takes place at a different location than the site of contamination. This transport of waste allows use of a selected, dedicated remediation facility where various considerations, such as containment, monitoring, accessibility, security, etc., can be optimized. Also, the contaminated material is treated exclusively in the surface soil layer where nutrient enrichment, aeration, and monitoring are most easily accomplished. In most cases, however, land farming as a final treatment for wastes is effectively banned in the U.S. by the terms of the 1984 Hazardous and Solid Waste Amendments (HSWA) to the 1976 Resource Conservation and Recovery

Act (RCRA). The so-called "RCRA land ban" prohibits any form of land disposal of waste unless it has been treated to prescribed standards. Thus, land farming can currently be used only as an intermediate treatment step, to be followed by further treatment or disposal. It is also possible to blur the distinction between *in situ* treatment and land farming by excavation of contaminated soil, and redistribution into the surface layer at the contaminated site. This is a gray area with respect to the law, and is currently the source of legal debate (see discussion by Stroo, 1992). As with *in situ* land treatment, land farming requires large amounts of dedicated land, and months to years for bioremediation to be completed.

The second approach to bioremediation consists of biofiltration methods for treating gaseous streams. Thus, this technology, too, is an area in which remediation in the classical sense and normal emissions control overlap. The usual biofilter design is a permeable, organic filter bed, which serves as a culture medium for microorganisms. Nonliving immobilized enzyme biofilters are also possible. Biofilters may concentrate pollutants in a manner analogous to conventional stack gas scrubbers, or be used in some cases to biodegrade contaminants to a nontoxic form that is compatible with natural environmental processes. The latter holds the advantage of actual reduction of waste, rather than simply repackaging for later disposal. Such a biofilter can be thought of as a type of biological catalytic converter. As a general rule, operator intervention is minimal. When accumulating waste levels become toxic to the microbes, or nutrients from the organic medium are depleted, the filter bed is simply changed. To U.S. readers biofilter technology will be less familiar than the other bioremediation approaches discussed here. More established in Europe, this method is well worth examining in that it is the only biological technique currently available to remediate airborne pollutants. Application of biofilter technology is at present limited to removing volatile organic compounds (VOCs) from point sources only. Waste remediation technologies for treatment of ambient air pollution are essentially nonexistent.

Bioreactors are the third and most technologically sophisticated approach to environmental bioremediation. Bioreactors offer a much faster means of waste biodegradation than land treatment and more control over reaction conditions and effluent quality than simple biofilters. In contrast to the months or years required for land treatment, bioreactors may require only days to weeks for effective degradation of specific pollutants. In bioreactors, bacterial growth is optimized in a well-mixed aqueous phase contained in a lagoon, tank, or other reactor vessel into which slurries of waste or contaminated material are introduced. Control over mixing, aeration, temperature, nutrient levels, water content, etc., is increased dramatically over that possible with land treatment techniques. Degradation monitoring is obviously made easier since the system is contained and outputs are regulated. Furthermore, control over release of nonindigenous organisms to the environment is possible. The bioreactor approach is intrinsically more expensive than *in situ* or land treatment methods because specialized reactors and equipment are utilized, and materials must be more extensively handled and sorted. Operation of bioreactors requires a tremendous amount of attention and expertise to achieve the rapid degradation that offsets the higher costs. Maintenance of optimum conditions for bacterial activity is far from trivial. Additionally, variations in reactor conditions, e.g., O_2 concentration, temperature, and pH, can affect not only the rate of degradation, but the identity of the metabolites produced. In contrast to land treatment, bioreactors require better knowledge of both the exact composition of

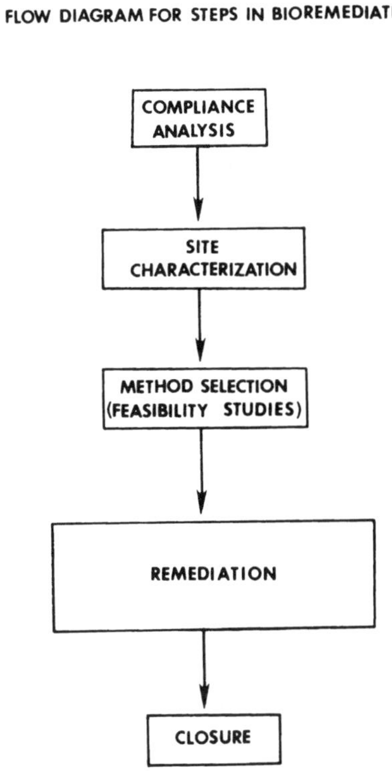

Figure 3 Logical structure of a waste remediation project.

the waste to be degraded and the degradation pathways catalyzed by the microbes in use. Portable bioreactor units permit a type of *in situ* bioreactor treatment with minimal transportation costs. This technique has perhaps the greatest potential for use as a routine treatment method for industrial wastes. Waste effluent streams can be treated at the site of production by directing them into a bioreactor process line.

For comparative purposes, bioremediation has been contrasted with conventional remediation methods. It should be pointed out, however, that the two are not incompatible. Nonbiological methods have been effectively combined with biological ones to enhance degradation of recalcitrant pollutants. For example, chemical preoxidation of organohalogens has been demonstrated to dramatically increase bacterial degradation (Carberry and Benzing, 1991). Furthermore, most sites in need of remediation will be contaminated by mixtures of toxicants. No single method can be expected to work equally well on all constituents of a diverse mixture. In some cases, a combination of remediation approaches may represent the most cost-effective solution among several practicable ones.

STEPS IN BIOREMEDIATION

The steps in a bioremediation project are like those in other approaches to environmental remediation. The process can be divided into at least five stages involved in dealing with the complex assortment of variables found at polluted sites and the regulatory standards that pertain to the project (Figure 3).

The first step, which might be called compliance analysis, requires examination of the contaminated site in light of the regulations that govern it and the actions that need to be taken. Minimum remediation goals based on the level at which contamination is considered unacceptable may be set.

The second stage, site characterization, is one of the more challenging and difficult aspects of a bioremediation project. Site evaluation involves analysis of the physical characteristics of the site, the source of the pollutant(s), the nature of the contaminated soil or water, and the distribution and concentration of contaminants. In addition, the type and volume of material to be treated have to be evaluated in light of the growth and biological activity of microorganisms either known to exist or capable of degrading the pollutant(s) if introduced there.

In the next stage, the feasibility of a possible bioremediation approach, usually involving assays of pollutant degradation carried out under controlled laboratory conditions, has to be determined. It is at this stage that the suitability of a bioremediation approach falls under severe scrutiny. Time considerations are often involved in determining whether or not the bioremediation approach, even if cost effective, is suitable for the situation or problem at hand.

The fourth stage in a bioremediation project is the process of remediation itself, which may involve multiple steps, such as pretreatment, or anaerobic followed by aerobic treatment of contaminated soil or material, as discussed above. Monitoring systems are employed to evaluate the progress and extent of contaminant degradation or conversion.

The last stage of a bioremediation project involves final monitoring of the site for the presence of contaminants and closure of the project. The difficulties discussed in the following section with regard to evaluation of the efficacy of bioremediation processes pertain to the monitoring stage. Also the troublesome issue of "how clean is clean?" arises here. Generally, cleanup will be declared complete when the minimum goals set in Stage 1 are met. However, the regulations that apply and public perception of whether the site is ever "clean" may be at odds.

ASSESSMENT OF BIOREMEDIATION AND BIODEGRADATION

PROBLEM DEFINITION

As mentioned above, mankind learned to make use of microbes to recycle waste products and facilitate crop growth long before his awareness of their existence. Because methods of evaluation of *in situ* biodegradation and bioremediation are lagging far behind the need for such approaches, it is reasonable to expect that mankind will develop an art of bioremediation to deal with environmental contamination prior to fully understanding the science of its activities. Problems can therefore be expected to arise.

With bioreactor remediation of waste contaminated material, assessment of the degree of degradation of waste is relatively straightforward. The system is contained and typically well mixed, and the outlets can be regulated and monitored. Still, identification of all major metabolites may be difficult, and extensive trial and error may be required to establish optimum conditions for degradation of a given waste composition by a given biological system. With land treatment or any other uncontained form of *in situ* bioremediation, however, assessment of bioremediation becomes much trickier. Difficulties and issues related to this task have recently been reviewed (Madsen, 1991) and are partially restated here. Part of the problem lies in assessing the extent to which biological organisms contribute to pollutant degradation. Whether dealing with a contaminated field site or a land farm for waste degradation, it is far from trivial to quantify fully the various transformations and concomitant fates of chemicals released on the open ground. This point can readily be illustrated by envisioning the effects of spilling a complex mixture, such as motor oil, on an open field. Volatile components will partially dissipate into the air. Others will partially disappear into the ground, where they are subject to both chemical and biological action, and may interact with groundwater. The mass balance involved in tracing the disappearance of the volatilized and absorbed fractions may be beyond the scope of all but the most determined effort. Keeping tabs on microbial activity by monitoring the microbes is also problematic. Microbial populations in the soil are complex and heterogeneous. It has been estimated that there are typically 10^9 and 10^6 indigenous microorganisms per gram of surface soil and aquifer sediment, respectively (Alexander, 1977; Beloin et al., 1988). In addition to the abundance and species diversity question, these microorganisms exist in various biological states, so that their presence *per se* does not directly indicate their level of biological activity (Madsen et al., 1991).

LABORATORY FLASKS VS. FIELD PLOTS

In spite of the difficulties listed above, one must establish some measure of the effectiveness of bioremediation at a given contaminated site in order to determine what approach to remediation of the site is most desirable and cost effective. The approach that is usually taken is to initiate a "flask assay" under controlled, laboratory conditions. A pollutant compound is added to a small, contained sample of soil, sediment, or water gathered from the field site under examination. The flask is inoculated with the selected microorganism(s); nutrient levels and aeration, along with other parameters, are adjusted, and the evolution of various metabolites is measured. If a radiolabeled parent compound is used, then the rate of production of radioactive CO_2 can give direct mineralization rates for organic compounds (Klecka et al., 1990). Control flasks of identical field material provide information about abiotic degradation and background biodegradation rates. Both aerobic and anaerobic metabolism of many compounds have been studied in this way (reviewed by Neilson, 1990; Evans, 1977; Long-de Vallière et al., 1989). More elaborate "model ecosystem" tests have also been developed. The results from the latter have been used to assess the partitioning of chemicals between various physical and biological components of a given habitat (Metcalf et al., 1971). While such bench scale tests have frequently been shown to predict accurately the biodegradation rates observed in the field (Stroo et al., 1992), extrapolation of laboratory results to field scale sites is always a concern. In considering the differences between the laboratory and field conditions, it has been pointed out that environmental modelers are confronted with their own version of the Heisenberg uncertainty principle (Gibbs, 1976). Accordingly, the closer a given process is examined, the more likely it is that artifacts will be imposed on measurements of that process. Furthermore, extensive laboratory testing is expensive; and the number of options tested will

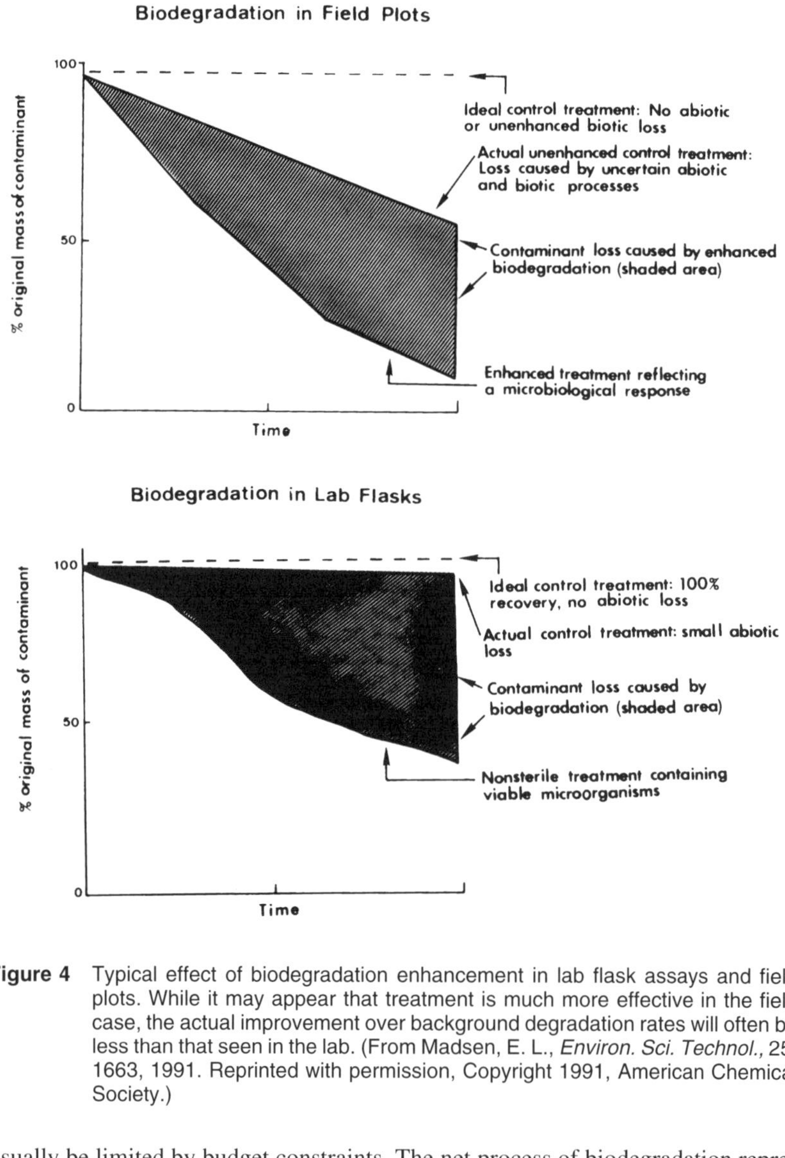

Figure 4 Typical effect of biodegradation enhancement in lab flask assays and field plots. While it may appear that treatment is much more effective in the field case, the actual improvement over background degradation rates will often be less than that seen in the lab. (From Madsen, E. L., *Environ. Sci. Technol.,* 25, 1663, 1991. Reprinted with permission, Copyright 1991, American Chemical Society.)

usually be limited by budget constraints. The net process of biodegradation represents a complex balance of interrelated chemical, physical, and biological parameters. These relationships are sensitive to a wide number of environmental factors, the alteration of which can bring about either synergistic or antagonistic effects. A sample of soil or sediment from a field site, when removed from the field, does not often provide exactly the same environment for living organisms that it did in its original location, either qualitatively or quantitatively. Figure 4 is a graphical comparison of typical laboratory flask data with an estimate of what realistically occurs in an open field environment. The applicability of laboratory tests for predicting bioreactor performance is more direct, but, as with any pilot project, scale-up problems are potentially a major source of concern. Thorough laboratory simulations of bioremediation techniques are essential to the development of practicable systems, but the results should not be blindly extrapolated to the scale of a large field site and expected to predict performance accurately. Field validation of predicted degradation rates and pathways is required.

In a controlled laboratory setting, biodegradation or modification of a given pollutant can be readily demonstrated. It has been much more difficult to establish firmly the role of biological systems in degradation of pollutants in less controlled field settings. Verification of the efficacy of *in situ* field site remediation is most difficult since the least control is exercised by the operator. Also, the lack of "hard numbers" in this arena has contributed to slow public and regulatory acceptance of *in situ* bioremediation as a viable approach to the cleanup of contaminated sites. Nevertheless, one can point to several examples where the disappearance of a known pollutant is intimately associated with biological activity. Table 1 provides examples of investigations suggesting *in situ* biodegradation, and Table 2 gives examples where investigators have successfully proven *in situ* biodegradation. Proof of biodegradation ideally takes the form of mass balance data, although this is usually not possible to obtain. Alternatively, reproducible waste disappearance to metabolite appearance ratios consistent with flask assays may provide convincing evidence (see Madsen, 1991 for references). Confirmatory support of biodegradation is given by the identity, number, and biological activity of microorganisms present. As discussed in the next chapter, molecular biology techniques, including DNA and RNA gene probes, can also provide useful mechanistic information about biodegradation pathways.

SUMMARY AND CONCLUSIONS

Bioremediation is an approach to handling human waste problems that relies on harnessing biological methods of dealing with naturally occurring waste. The increasing human population density and the associated increase in waste generated by human activity pose immediate and long-term threats to human and environmental health. The growing awareness of the connections between human health, environmental health, and economic health mandates the development of new methods of reducing wastes that go beyond traditional repackaging and disposal techniques. In a balanced natural system, many biochemical waste products are kept from accumulating to disruptive levels by biodegradation and biogeochemical cycling. Bioremediation uses living systems (usually microbial) or biological products (such as immobilized enzymes) to biodegrade anthropogenic waste, with the objective being reduction of waste to chemical forms that can be assimilated into natural cycles.

In many cases, bioremediation methods can actually lead to mineralization of organic wastes, i.e., reduction to simple inorganic constituents. In addition, biological systems can be used to bioaccumulate nondegradable wastes, such as heavy metals, leading to waste-volume reduction and possibly to recovery of useful products. A biological approach can be used not only to remediate waste-contaminated sites, but also as part of normal emission control and waste treatment. It is imperative to address both of these parts of the waste problem if the biosphere is to continue to exist as a sustainable resource. It is also necessary to reduce the

Table 1 Examples of Investigations Suggesting *In Situ* Biodegradation

Type of site	Contaminant	Methods	Reference
		Soil habitat	
Agricultural soil	Dinoseb (2-*sec*-butyl-4,6-dinitrophenol), a pesticide	Loss of extractable Dinoseb from soil in laboratory-incubated flasks; enumeration of bacteria	Stevens et al., 1990
Prairie soil	Dicamba (2-methoxy-3,6-dichlorobenzoic acid), a pesticide	Production of $^{14}CO_2$ and intermediary metabolites from radiolabeled Dicamba added to soil in laboratory-incubated flasks	Smith, 1974
		Surface water habitat	
Seawater	Oil	Production of CO_2 in sealed laboratory-incubated flasks receiving oil and various fertilizers; loss of oil from tubes floating on seawater in open tanks	Atlas and Bartha, 1973
Lake water	Quaternary ammonium compounds	Production of $^{14}CO_2$ and incorporation of ^{14}C into bacterial biomass in sealed laboratory-incubated flasks receiving ^{14}C- labeled compounds; addition of compounds to vessels incubated *in situ* prior to initiating the biodegradation assay; enumeration of bacteria	Ventullo and Larson, 1986
		Groundwater habitat	
Shallow aquifer and wastewater pond sediments	Surfactant chemicals	Analysis of field site waters for organic and inorganic constituents; $^{14}CO_2$ production from radiolabeled compounds in laboratory-incubated flasks; enumeration of bacteria	Federle and Pastwa, 1988
Shallow aquifer	Phenol, *p*-cresol, naphthalene	Analyses of field site waters for organic and inorganic constituents, computer modeling of contaminant transport, $^{14}CO_2$ production from radiolabeled compounds in laboratory-incubated flasks	Klecka et al., 1990

From Madsen, E. L., *Environ. Sci. Technol.*, 25, 1663, 1991. Reprinted with permission,

Table 2 Examples of Investigations Successfully Proving *In Situ* Biodegradation

Type of site	Contaminants	Evidence	References
		Soil habitat	
Replicated field plots	Pentachlorophenol	Enhanced loss of extractable pentachlorophenol in previously contaminated plots inoculated with the fungi, *Phanerochaete chrysosporium* and *P. sordida*	Lamar and Dietrich, 1990
Replicated field plots	Oil	Enhanced loss of extractable oil components in fertilized plots, corresponding increase in microbial numbers	Westlake et al., 1978
Replicated field plots	Oil	Enhanced loss of extractable oil components in fertilized plots, corresponding increase in numbers of hydrocarbon-degrading bacteria	Raymond et al., 1976
		Aquatic habitat	
Marine samples from Prince William Sound, AK	Oil	Loss of octadecane relative to its less biodegradable branched-chain isomer, phytane; absence of oil from beaches where fertilizer was applied	Pritchard and Costa, 1991
Pond divided into two separate halves	*p*-Nitrophenol	After *p*-nitrophenol (PNP) had been added to and disappeared from one half of the pond, PNP was added to both halves. PNP then disappeared only from the previously exposed, metabolically adapted half of the pond	Spain et al., 1984
		Groundwater habitat	
Shallow unconfined aquifer	Naphthalene and phenanthrene	Biodegradation potential and other microbiological parameters indicated metabolic adaptation in samples from contaminated areas, but not from adjacent, uncontaminated areas. Enhanced	Madsen et al., 1991

Table 2 Continued

Type of site	Contaminants	Evidence	References
		protozoan numbers, only in contaminated area, indicated *in situ* microbial growth resulting from the contaminants.	
Shallow confined aquifer	Chlorinated ethenes	At a highly instrumented confined aquifer site, mass balances were possible. Co-metabolic oxidation of chlorinated ethenes by methanotrophic microorganisms was controlled by pulses of O_2 and CH_4 replicated in time.	Semprini et al., 1990

From Madsen, E. L., *Envir. Sci. Technol.*, 25, 1663, 1991. Reprinted with permission, Copyright 1991, American Chemical Society.

production of wastes that are not readily cycled or biodegraded in order to realize fully the potential of bioremediation.

While bioremediation techniques are often considered to have modern origins, the approach has long been used by man. Microbial sewage treatment and composting systems are well-known examples of how biological processes can be harnessed to societal advantage. New methods of utilizing a growing number of living organisms in waste reduction are continually being developed. Naturally occurring and/or genetically engineered microorganisms may be used in three general types of bioremediation systems, listed here in increasing complexity: land treatment, biofilters, and bioreactors. Each type of system can be adapted for *in situ* treatment of waste, an approach that may offer significant cost, labor, and risk savings over traditional removal and disposal methods.

Waste bioremediation is an emerging field with many basic unknowns and unanswered questions. Determining the exact fates and degradation products of waste in field cases is especially problematic. Further development of waste-tracking techniques is clearly needed. Some risks and limitations must also be considered, and more accurately defined, before bioremediation can be fully exploited, especially with respect to genetically engineered microorganisms. There is a real danger of overselling bioremediation as the answer to all of our waste problems. Meeting the challenge of remediating local and global damage created by the increasing production of toxic and hazardous wastes requires that even the most promising bioremediation methods be applied in a "biorational" manner.

ACKNOWLEDGMENTS

We hereby gratefully acknowledge the support of the National Institutes of Environmental Health Center grant (ES01908), an Office of Naval Research grant (N00014-90-J-1660), and research grants from Sunshine Makers, Inc. and Stroh Brewery Company. In addition, Mr. Bodishbaugh was supported in part by an NRSA grant (ES07031).

REFERENCES

Abramowicz, D. A., Aerobic and anaerobic biodegradation of PCBs: A review, *Biotechnology,* 10, 241, 1990.

Aelion, C. M. and Bradley, P. M., Aerobic biodegradation potential of subsurface microorganisms from a jet fuel-contaminated aquifer, *Appl. Environ. Microbiol.,* 57, 57, 1991.

Ahmed, F. E., Ed., *Seafood Safety,* National Academy Press, Washington, D.C., 1991.

Alexander, M., *Introduction to Soil Microbiology,* 2nd ed., John Wiley & Sons, New York, 1977.

Anon. (Editorial), Environmental pollution: It kills trees, but does it kill people?, *The Lancet,* 340, 821, 1992.

Asplund, G. and Grimvall, A., Organohalogens in nature: More widespread than previously assumed, *Environ. Sci. Technol.,* 25, 1346, 1991.

Atlas, R. M. and Bartha, R., Stimulated biodegradation of oil slicks using oleophilic fertilizers, *Environ. Sci. Technol.,* 7, 538, 1973.

Bachofen, R., Ed., Microorganisms in nuclear waste disposal, Part I, *Experientia,* 46, 777, 1990.

Bachofen, R., Ed., Microorganisms in nuclear waste disposal, Part II, *Experientia,* 47, 507, 1991.

Bailey, G. S., Goeger, D. E., and Hendricks, J. D., Factors influencing experimental carcinogenesis in laboratory fish models, in *Metabolism of Polycyclic Aromatic Hydrocarbons in the Aquatic Environment,* Varanasi, U., Ed., CRC Press, Boca Raton, FL, 1989, 253.

Bedard, D. L., Bacterial transformation of polychlorinated biphenyls, in *Biotechnology and Biodegradation. Advances in Applied Biotechnology Series,* Kamely, D., Chakrabarty, A. and Omenn, G. S., Eds., Portfolio Publishing, The Woodlands, TX, 1990, 369.

Beloin, R. M., Sinclair, J. L., and Ghiorse, W. C., Distribution and activity of microorganisms in subsurface sediments of a pristine study site in Oklahoma, *Microb. Ecol.,* 16, 85, 1988.

Carberry, J. B. and Benzing, T. M., Peroxide pre-oxidation of recalcitrant toxic waste to enhance biodegradation, *Water Sci. Tech.,* 23, 367, 1991.

Cerniglia, C. E. and Heitkamp, M. A., Microbial degradation of polycyclic aromatic hydrocarbons (PAHs) in the aquatic environment, in *Metabolism of Polycyclic Aromatic Hydrocarbons in the Aquatic Environment,* Varanasi, U., Ed., CRC Press, Boca Raton, FL, 1989, 41.

Cleland, G. B., McElroy, P. J., and Sonstegard, R. A., Immunomodulation in C57B1/6 and DBA/2 mice following consumption of halogenated aromatic hydrocarbon-contaminated Great Lakes coho salmon (*Oncorhynchus kisutch*) from Lake Ontario, *J. Toxicol. Environ.,* 24, 477, 1989.

Contie, V. L., Fish as human disease models, *Res. Resources Rep.,* 15, 1, 1991.

Cortese, A. D., Education for an environmentally sustainable future, *Environ. Sci. Technol.,* 26, 1108, 1992.

Criddle, C. S., DeWitt, J. T., Grbic-Galic, D., and McCarty, P. L., Transformation of carbon tetrachloride by *Pseudomonas* sp. strain KC under denitrification conditions, *Appl. Environ. Microbiol.,* 56, 3240, 1990.

Dawe, C. J., Implications of aquatic animal health for human health, *Environ. Health Perspect.,* 86, 245, 1990.

Ember, L., Toxics Release Inventory data show steady drop in emissions, *Chem. Eng. News,* June 15, 1992, p. 13.

EPA (U.S. Environmental Protection Agency), *Assessing Human Health Risks from Chemically Contaminated Fish and Shellfish: A Guidance Manual,* EPA-503/8-89/002, U.S. EPA Office of Marine and Estuarine Protection, Washington, D.C., 1989.

Evans, W. C., Biochemistry of the bacterial catabolism of aromatic compounds in anaerobic environments, *Nature,* 270, 17, 1977.

Federle, T. W. and Pastwa, G. M., Biodegradation of surfactants in saturated subsurface sediments: A field study, *Ground Water,* 26, 761, 1988.

Fiore, B. J., Anderson, H. A., Hanrahan, L. P., Olson, L. J., and Sonzogni, W. C., Sport fish consumption and body burden levels of chlorinated hydrocarbons: A study of Wisconsin anglers, *Arch. Environ. Health,* 44, 82, 1989.

Fridovich, I., Superoxide dismutases, *Ann. Rev. Biochem.,* 44, 147, 1975.

Gibbs, C. F., Methods and interpretation in measurement of oil biodegradation rate, in *Proc. 3rd Int. Biodegradation Symp.,* Sharpley, J. M. and Kaplan, A. M., Eds., Applied Science, London, 1976, p. 127.

Henry, S. M. and Grbic-Galic, D., Influence of endogenous and exogenous electron donors and trichoroethylene oxidation toxicity on trichloroethylene oxidation by methanotrophic cultures from a groundwater aquifer, *Appl. Environ. Microbiol.,* 57, 236, 1991.

Hertzler, D. R., Neurotoxic behavioral effects of Lake Ontario salmon diets in rats, *Neurotoxicol. Teratol.,* 12, 139, 1990.

Hileman, B., Web of interactions makes it difficult to untangle global warming data, *Chem. Eng. News,* April 27, 1992, p. 7.

Hoffmann, G. R., Mutagenicity testing in environmental toxicology, *Environ. Sci. Technol.,* 16, 560A, 1982.

Holloway, M., Soiled shores, *Sci. Am.,* October, 1991, p. 103.

Janssen, D. B., Pentenga, M., Van der Ploeg, J., Pries, F., Van der Waarde, J., Wonink, E., and Van den Wijngaard, A. J., Environmental biotechnology: Transformations of synthetic halogenated compounds by microbial cultures and enzymes, in *BIOS Annual Report.,* Groningen University, Groningen, The Netherlands, 1990, 65.

Keeton, W. T., *Biological Science,* 3rd ed., W. W. Norton, New York, 1980.

Klecka, G. M., Davis, J. W., Gray, D. R., and Madsen, S. S., Natural bioremediation of organic contaminants in ground water: Cliff-Dow Superfund Site, *Ground Water,* 28, 534, 1990.

Kreiss, K., Studies on populations exposed to polychlorinated biphenyls. *Environ. Health Perspect.,* 60, 193, 1985.

Krieger, J., Technique identifies bacteria that degrade wastes, *Chem. Eng. News.* July 27, 1992, p. 36.

Lamar, R. T. and Dietrich, D. M., In situ depletion of pentachlorophenol from contaminated soil by *Phanerochaete* spp., *Appl. Environ. Microbiol.,* 56, 3093, 1990.

Long-de Vallière, C. L., Petrozzi, S., Zürrer, D., Baier, U., and Dunn, I. J., Methods of anaerobic degradation of toxic compounds in chemical and industrial wastes, in *Biological Waste Treatment,* Alan R. Liss, New York, 1989, 35.

Lubchenco, J., Olson, A. M., Brubaker, L. B., Carpenter, S. R., Holland, M. M., Hubbell, S. P., Levin, S. A., MacMahon, J. A., Matson, P. A., Melillo, J. M., Mooney, H. A., Peterson, C. H., Pulliam, H. R., Real, L. A., Regal, P. J., and Risser, P. G., The sustainable biosphere initiative: An ecological research agenda, *Ecology,* 72, 371, 1991.

MacRae, I. C., Microbial metabolism of pesticides and struturally related compounds, *Rev. Environ. Contam. Toxicol.,* 109, 1, 1989.

Madsen, E. L., Determining in situ biodegradation: Facts and challenges, *Environ. Sci. Technol.,* 25, 1663, 1991.

Madsen, E. L., Sinclair, J. L., and Ghiorse, W. C., In situ biodegradation: Microbiological patterns in a contaminated aquifer, *Science,* 252, 830, 1991.

McCabe, A. S. and Fitzgerald, M. R., Media images of environmental biotechnology: What does the public see?, in *Environmental Biotechnology for Waste Treatment,* Sayler, G. S., Fox, R., and Blackburn, J. W., Eds., Plenum Press, New York, 1991, 15.

Metcalf, R. L., Sangha, G. K., and Kapoor, I. P., Model ecosystem for the evaluation of pesticide biodegradability and ecological magnification, *Environ. Sci. Technol.,* 5, 709, 1971.

Murchelano, R. A., Fish health and environmental health, *Environ. Health Perspect.,* 86, 257, 1990.

Neilson, A. H., The biodegradation of halogenated organic compounds, *J. Appl. Bacteriol.,* 69, 445, 1990.

NRC (National Research Council), *Animals as Sentinels of Environmental Health Hazards,* National Academy Press, Washington, D.C., 1991a.

NRC (National Research Council), *Environmental Epidemiology Vol. 1: Public Health and Hazardous Wastes,* National Academy Press, Washington, D.C., 1991b.

NRC (National Research Council), *Environmental Neurotoxicology,* National Academy Press, Washington, D.C., 1991c.

Pelczar, M., Jr., Reid, R. D., and Chan, E. C. S., *Microbiology,* 4th ed., McGraw-Hill, New York, 1977.

Press, F. and Siever, R., *Earth,* 4th ed., W. H. Freeman, New York, 1986.

Pritchard, P. H. and Costa, C. F., EPA Alaska oil spill bioremediation project, *Environ. Sci. Technol.,* 25, 372, 1991.

Rambeloarisoa, E., Rontani, J. F., Giusti, G., Duvnjak, Z., and Bertrand, J. C., Degradation of crude oil by a mixed population of bacteria isolated from sea-surface foams, *Mar. Biol.,* 83, 69, 1984.

Raymond, R. L., Hudson, J. O., and Jamison, V. W., Oil degradation in soil, *Appl. Environ. Microbiol.,* 31, 522, 1976.

Reid, W. V., Conserving life's diversity, *Environ. Sci. Technol.,* 26, 1090, 1992.

Rochkind, M. L., Blackburn, J. W., and Sayler, G. S., *Microbial Decomposition of Chlorinated Aromatic Compounds,* EPA 600/2-86/090, U.S. EPA, Cincinnati, OH, 1986.

Schidlowski, M., The atmosphere, in *The Handbook of Environmental Chemistry, Vol. 1, Part A, The Natural Environment and the Biogeochemical Cycles,* Hutzinger, O., Ed., Springer-Verlag, New York, 1980, 1.

Schlesinger, W. H., *Biogeochemistry,* Academic Press, San Diego, 1991.

Semprini, L., Roberts, P. V., Hopkins, G. D., and McCarty, P. L., A field evaluation of in situ biodegradation of chlorinated ethenes. II. Results of biostimulation and biotransformation experiments, *Ground Water,* 28, 715, 1990.

Shields, J., *Environmental Health: New Directions,* Princeton Scientific, Princeton, NJ, 1990.

Smith, A. E., Breakdown of the herbicide Dicamba and its degradation product 3,6-dichlorosalicylic acid in prairie soils, *J. Agric. Food Chem.,* 22, 601, 1974.

Spain, J. C., Van Veld, P. A., Monti, C. A., Pritchard, P. H., and Cripe, C. R., Comparison of p-nitrophenol biodegradation in field and laboratory test systems, *Appl. Environ. Microbiol.,* 48, 944, 1984.

Stevens, T. O., Crawford, R. L., and Crawford, D. L., Biodegradation of Dinoseb (2-sec-butyl-4,6-dinitrophenol) in several Idaho soils with various Dinoseb exposure histories, *Appl. Environ. Microbiol.,* 56, 133, 1990.

Stroo, H. F., Biotechnology and hazardous waste treatment, *J. Environ. Qual.,* 21, 167, 1992.

Stroo, H. F., Anderson, H. K., and Smith, J. R., How to predict biodegradation rates, *Soils,* April, 1992, p. 20.

Thayer, A. M., Bioremediation: Innovative technology for cleaning up hazardous waste, *Chem. Eng. News,* August 26, 1991, p. 23.

Thorn, P. M., Kreitinger, J. P., and Stroo, H. F., Biodegradation of pesticides in soil, paper presented at 84th Annu. Meet. Air and Waste Manage. Assoc., Vancouver, B.C., Canada, June 16–21, 1991.

Ventullo, R. M. and Larson, R. J., Adaption of aquatic microbial communities to quaternary ammonium compounds, *Appl. Environ. Microbiol.,* 51, 356, 1986.

Walker, J. C. G., *Evolution of the Atmosphere,* Macmillan, New York, 1977.

Westlake, D. W. S., Jobson, A. M., and Cook, F. D., In situ degradation of oil in a soil of the boreal region of the Northwest Territories, *Can. J. Microbiol.,* 24, 254, 1978.

CHAPTER 8

Environmental Bioremediation: Applications and New Horizons

Celia Bonaventura, Joseph Bonaventura, and Donovan F. Bodishbaugh

CONTENTS

1-56670-141-4/96/$0.00+$.50

PRACTICAL BIOREMEDIATION

The appeal of bioremediation, as explained in the previous chapter, is largely that microorganisms do much of the work. In practice, the objectives of waste bioremediation can usually be reduced to bringing desired microorganisms (or biological products) into contact with the waste, and doing whatever is possible to improve conditions for their action. These objectives can be met by a number of engineering approaches and by the use of a variety of organisms and technical means to get them together with the waste. Examples of techniques and systems for various types of remediation projects are described in this chapter. The discussions that follow should in no way be taken as a comprehensive list of potential, or even available, technologies.

For a given remediation project, it is likely that several bioremediation options exist. Assigning rank among feasible bioremediation techniques for a given application is difficult, since many criteria are involved in the selection process, and their relative priorities vary from case to case. Some criteria that are inevitably part of any selection process follow.

COST

The most obvious constraint on any remediation effort is cost, and the business of waste remediation can be very costly (see Chapter 1). Generally, relative cost is a function of the sophistication of the technique (special equipment used, degree of control over bioactivity desired, etc.), and the duration of a remediation project. Bioremediation is still largely a developing field, and many working systems are unique. Costs may be expected to decrease as techniques and equipment become more standardized, although many projects will require a customized remediation system. Unless otherwise specified, cost ranges for various technologies given in the following sections are contractor-supplied values from the EPA Vendor Information System for Innovative Treatment Technologies (VISITT) database (EPA, 1993). The values, which include the price of remediation steps only, were taken from those contractors with field scale technologies in operation. Related costs, such as excavation, transport, and disposal of residual wastes are not included. Costs are highly dependent on specific properties of a given site, and "typical" values may not be meaningfully applied to some specific case-cost projections.

TIME SCALE

As discussed in Chapter 7, bioremediation often occurs via unassisted, natural biodegradation, albeit very slowly. If consideration of time scale were eliminated, then "doing nothing" might be considered a viable bioremediation approach. In fact, doing nothing has been suggested as the best overall remediation option in *some cases* of certain types of contamination, such as marine oil spills (NRC, 1989). Even so, political and legal considerations often effectively eliminate the "do nothing" option. Natural biodegradation of some anthropogenic compounds occurs so slowly as to be practically nonexistent. Obviously, the faster a remediation option works, the better. Consideration of other factors (usually cost) is likely to result in selection of something other than the fastest option.

RISK

Risks to human or environmental health associated with a particular remediation scheme often play a major role in selection of the most desirable technique. Examples include the risk of waste dispersal, as when surface spills threaten groundwater quality, the risk of transportation of toxic wastes through populated or environmentally sensitive areas, or the risk of accidental release of genetically engineered organisms.

Obviously, selection of a biological or any other remediation technique will usually be based on an optimization of all of the above, as well as additional site-specific factors. Rather than attempting to specify which methods should be used in which instances, the following discussion is intended to familiarize the reader with the general nature of established and emerging approaches, and to highlight their advantages and limitations. Suitability of bioremediation techniques to any particular situation must be considered on a case-by-case basis.

BIOREMEDIATION TECHNOLOGIES IN PRACTICE

BIOREMEDIATION OF CONTAMINATED SOIL

A classic waste-contamination scenario is that of organic chemical contamination of soil surrounding a landfill, dumpsite, storage tank, or industrial facility. Although it is one of the most problematic situations to deal with by traditional collection and disposal techniques since it may involve large amounts, frequently many hundreds of tons, of material, it is one of the most promising cases for the application of bioremediation, particularly for *in situ* land treatment (see Chapter 7). It should, therefore, be no surprise that numerous biological approaches for remediation of contaminated soil exist.

The simplest method of soil bioremediation is *in situ* land treatment. Remediative action may be as simple as adjusting nutrient, pH, water, and oxygen levels to encourage growth of indigenous soil microbes by fertilizing, irrigating, and tilling the soil. In some cases, forced aeration of surface soils using compressors or vacuum pumps may be applied. Identification and extensive characterization of the microbial population is useful, but not strictly necessary. Characterization of the fates of waste and waste metabolites produced is necessary, especially since the process is uncontained. Lab flask assays (see Chapter 7) can go a long way toward this end, but the effects of any degradation enhancement measures must also be modeled in the flask assay. Extensive monitoring of metabolite formation during actual remediation is essential to determine if flask assay-based extrapolations hold true. Extensive characterization of soil and waste parameters (Table 1) can be used to determine exactly what enhancement measures are necessary, or if bioremediation is even feasible (reviewed by Middleton et al., 1991; Sims et al., 1990). Direct costs of soil remediation, as reported by private contractors, are quoted per ton or per cubic yard of treated soil. The typical cost range is from about $10 to a few hundred dollars per unit (EPA, 1993).

Inoculation of soil with microbial strains selected for their particular abilities may appear to offer the greatest potential for desired biodegradative action. In

Table 1 Site and Soil Characteristics Identified as Important for *In Situ* Treatment

Site location/topography and slope
Soil type and extent
Soil profile properties
Boundary characteristics
Depth
Texture[a]
Amount and type of coarse fragments
Structure[a]
Color
Degree of mottling
Bulk density[a]
Clay content
Type of clay
Cation exchange capacity[a]
Organic matter content[a]
Acidity (pH)[a]
Redox potential (E_H)[a]
Aeration status[a]
Hydraulic properties and conditions
Soil water characteristic curve
Field capacity/permanent wilting point
Water holding capacity[a]
Permeability[a] (under saturated and a range of unsaturated conditions)
Infiltration rates[a]
Depth to impermeable layer or bedrock
Depth to groundwater,[a] including seasonal variations
Flooding frequency
Runoff potential[a]
Geological and hydrogeological factors
Subsurface geological features
Groundwater flow patterns and characteristics
Meteorological and climatological data
Wind velocity and direction
Temperature
Precipitation
Water budget

[a]Factors that may be managed to enhance soil treatment.

Adapted from Sims et al., *Hazard. Waste Hazard. Mater.*, 7, 117, 1990. With permission.

many cases, such techniques prove ineffective due to incompatibilities between microbes and environmental conditions or technical limitations on inoculant size (Goldstein et al., 1985; Ramadan et al., 1990). Since control over environmental conditions during remediation is minimal, success often depends on selection of the

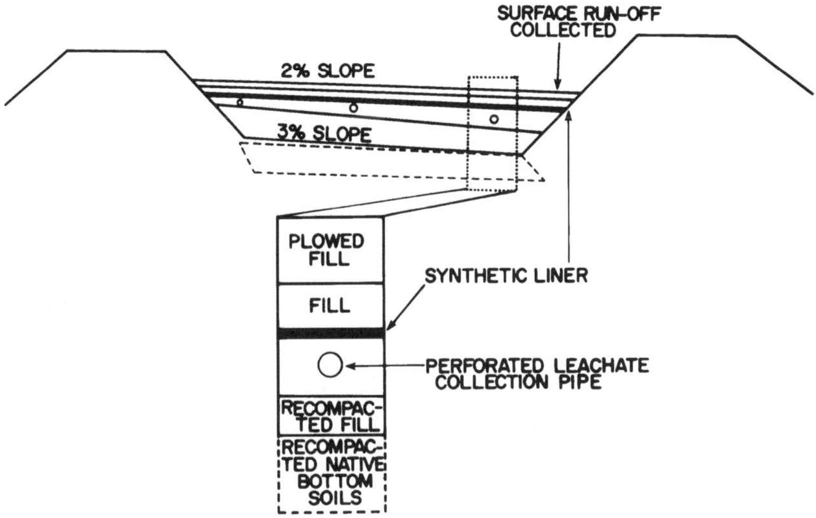

Figure 1 Cross section of a typical prepared bed arrangement, for land farm treatment of contaminated soil. The treated soil is usually returned to its original site following treatment. (Adapted from Ryan et al., *J. Hazard. Mater.,* 28, 159, 1991. With permission.)

proper organisms. It is not enough to demonstrate that a microbial culture can degrade the waste(s) present. For successful remediation, the introduced microbes must be capable of surviving under ambient environmental conditions, and outcompeting the indigenous microbial population.

Land farming, as defined in the previous chapter, is identical to *in situ* land treatment, except that treatment is conducted at a site other than that of contamination. Since land disposal of most wastes is banned in the U.S. under the current terms of the Resources Conservation and Recovery Act (RCRA), land farming can usually only be conducted as an intermediate step in a treatment plan. In practice, this means that treated soil from a land farm must be returned to its original site, or otherwise disposed of, following remediation. Land farm facilities usually take the form of a prepared bed (sometimes called solid phase or prepared bed bioreactors). The typical prepared bed arrangement (Figure 1) consists of synthetic and/or natural low permeability barriers to both vertical and horizontal migration of applied wastes, as well as surface runoff and leachate-collection systems to monitor waste migration and degradation. The design is similar to that of a modern landfill. Monitoring wells located downgradient from the prepared bed (not shown) are used to give early warning of any groundwater contamination. Wastes in the contaminated soil are biodegraded by the same types of soil organisms responsible for *in situ* land treatment, and degradation is facilitated in the same ways. The excavation of soil and redistribution into the surface layer gives greater access and opportunity for various biodegradation enhancement measures (see Table 1). The land farming approach differs from true bioreactors in that the waste is treated in solid phase, it is largely uncontained during treatment, and biodegradation organisms and pathways are generally limited to those that occur *in situ.* Direct costs of prepared-bed land farming are comparable to those quoted above, but the indirect costs of

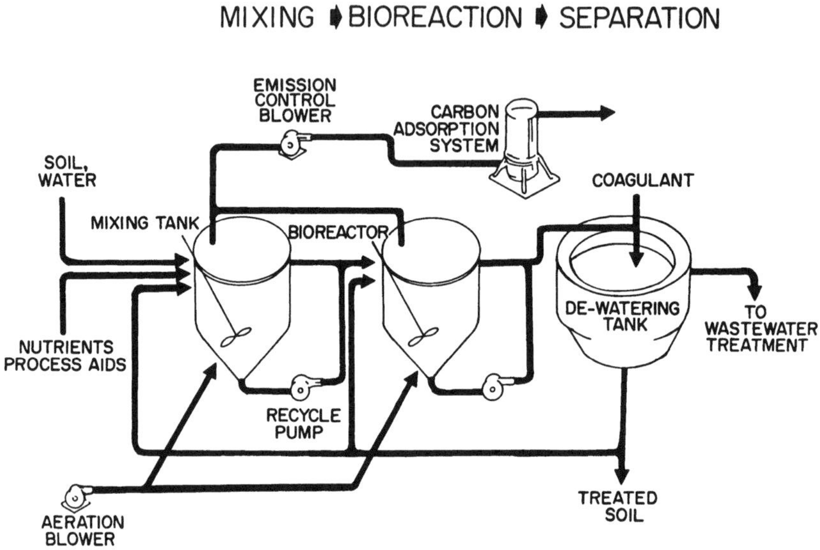

Figure 2 Schematic representation of a sequential bioreactor system, as used for treatment of contaminated soil. Each of the three basic steps may require multiple sequential or parallel reactors. Such a system may be adapted for either batch or continuous operation. (Adapted from Ryan et al., *J. Hazard. Mater.*, 28, 159, 1991. With permission.)

treatment site preparation, transportation, and ultimate disposal are obviously much higher.

Bioreactors for remediation of contaminated soil utilize an aqueous slurry phase system. The soil must be excavated, mixed with water, and introduced into the bioreactor. The primary advantage of the fluid phase is enhanced transport of both nutrients and waste to the degrading microorganisms. In a solid phase (dry soil) system, contact between waste and microbes is largely dependent on proliferation of the microbes into the contaminated soil. In a well-mixed slurry reactor, the waste and microbes can be kept evenly distributed and in continuous contact. Desired concentrations of nutrients and other additives may be easily maintained. Temperature, pH, and other parameters can be more rigidly optimized and kept constant. Furthermore, the reactor can be completely contained, with all outputs regulated, including gaseous emission of volatile wastes and metabolites. The slurry can be kept saturated with oxygen for aerobic degradation, or completely anoxic for anaerobic degradation, as desired. Organisms can thus be utilized that would never thrive or even survive in some soil environments.

In aqueous solution, dispersants that increase the bioavailability of organic wastes to microbes may also be employed. In many cases the degradation process proceeds rapidly at first, then slows over time. This rate change may be explained by the presence of a component or portion of the waste that is tightly adsorbed to the soil particles, and is desorbed only very slowly. Surfactants and other chemicals that accelerate desorption can be used in bioreactor systems.

Bioreactor designs vary greatly in form and complexity. A classic and familiar setup is the sequential reactor system, schematically illustrated in Figure 2. The sealed tank reactors shown represent the highest level of control and containment possible, but similar arrangements of open vats or lined pits are also used. There are three basic steps in this scheme, although each step may require multiple sequential reactor vessels. The process may be adapted to batch or continuous stream processing. The soil-water slurry is first mixed to the desired consistency, and various additives are introduced in the mixing tank(s). The slurry then enters the bioreactor(s) proper, where wastes are biodegraded. Finally, end-products are separated by a combination of mechanical and chemical means, into treatable wastewater and treated soil or sludge. Individual reactor vessels may be equipped with variable feedback recycle lines to fine-tune the quantity and quality of their outputs. Recycling of some activated (partially digested) waste may be used as shown to inoculate new batches. Alternatively, some mechanical arrangement for retaining large amounts of immobilized biomass, such as rotating drums, fluidized beds, or fixed films may be used.

The rate-limiting factors in any bioreactor system are the degree of solubilization of the organic contaminant and the level of active microbial populations maintained (Stroo, 1992). A successful design will seek to maximize both. Table 2 lists some critical factors affecting bioreactor performance, and how they are typically controlled. Bioreactor degradation rates can usually be dramatically enhanced over those possible with land treatment. Slurry phase bioreactors are suitable for remediation of very high concentrations of soluble organic wastes in soil or sludges. They are recognized as capable of handling up to at least 250 g/kg levels of organic wastes (EPA, 1990). Recalcitrant wastes, such as organohalogens, may require pretreatment; and some wastes, such as organic corrosives and most inorganics, may not be degradable by bioreactor systems. As with solid-phase soil treatment, contractor-supplied cost ranges for slurry phase bioreactor remediation vary widely. A comprehensive range would be on the order of 30 to $600 per ton or cubic yard of contaminated soil (EPA, 1993).

BIOREMEDIATION OF GROUNDWATER

Another common case of waste contamination is that of a contaminated subsurface aquifer. While this scenario both shares the causes of and frequently accompanies soil contamination, groundwater remediation presents unique problems, risks, and considerations. Among the most obvious of these are that groundwater is mobile, whereas soil is generally stationary, and that people and livestock frequently use untreated groundwater for drinking. Thus, there will often be an additional urgency factor associated with groundwater cleanup that may justify more drastic and expensive measures.

The usual approach in remediation of contaminated aquifers is groundwater pumping, accompanied by surface treatment to eliminate the water-soluble wastes, as shown schematically in Figure 3. The treated water is then "recharged" into the aquifer, via one or more injection wells, at some point upgradient to the contaminated zone. Pump-and-treat operations can incorporate bioremediation in at least two ways. The most obvious is to use biological (bioreactor) surface treatment, but, like any pump-and-treat approach, this method is only able to degrade wastes in the mobile, aqueous phase. It is important to recognize that most organic wastes have low water solubilities, and aquifer-associated soils will often contain larger volumes

Table 2 Characteristics Limiting the Slurry Phase Bioreactor Process

Limiting characteristic	Reasons for impact	Action to minimize
Variable waste composition	Inconsistent biodegradation caused by variation in biological activity	Dilution of waste stream; increase mixing
Nonuniform particle size	Inconsistent microbes, solids, organic distribution	Physical separation
Biodegradability	Low rate of destruction inhibits process	Addition of microbial culture capable of degrading particularly difficult compounds; increase residence time
Water solubility	Contaminants with low solubility are harder to biodegrade	Addition of surfactants or other emulsifiers
Temperature outside 15–35°C range	Less microbial activity outside this range	Temperature monitoring and adjustment
Nutrient deficiency	Lack of adequate nutrients for biological activity	Monitoring and adjustment of carbon, nitrogen, phosphorus levels and ratio
Oxygen deficiency	Lack of oxygen is rate limiting	Oxygen monitoring and adjustment
Insufficient mixing	Minimizes contact between waste, nutrients, and microorganisms	Optimize mixing characteristics
pH outside 4.5–8.8 range	Inhibition of most biological activity	Sludge pH monitoring; addition of pH buffering compounds
Microbial population density	Low density results in low biodegradation rate	Culture monitoring; addition of cultured strains
Water and air emissions	Potential environmental and/or health impact	Posttreatment measures (air scrubbing, carbon filtration, etc.)
Presence of elevated levels of dissolves: -Heavy metals -Highly chlorinated organics -Some pesticides and herbicides -Inorganic salts	Can be highly toxic to microorganisms, resulting in inhibition of degradation	Pretreatment measures to reduce concentration of toxins to noninhibiting range

Adapted from EPA, *Slurry Biodegradation,* EPA/540/2-90/016, U.S. Environmental Protection Agency, Washington, D.C., 1990. With permission.

of organic wastes than the water itself. The conviction that pump-and-treat measures can be effective has led to an appreciable effort in this direction, at numerous hazardous waste sites (HWS). However, the goal of remediating aquifers to drinking water standards by such techniques may be unrealistic in many, if not most, cases. Curtis C. Travis, director of the Center for Risk Management and head of the Risk Analysis Section of the Health and Safety Research Division at Oak Ridge National Laboratory, has recently addressed the question of whether it is technically possible to restore contaminated groundwater to an environmentally sound condition (Travis, 1992). He argues, convincingly, that pump-and-treat technologies cannot do the job. Contamination levels at remediation sites are

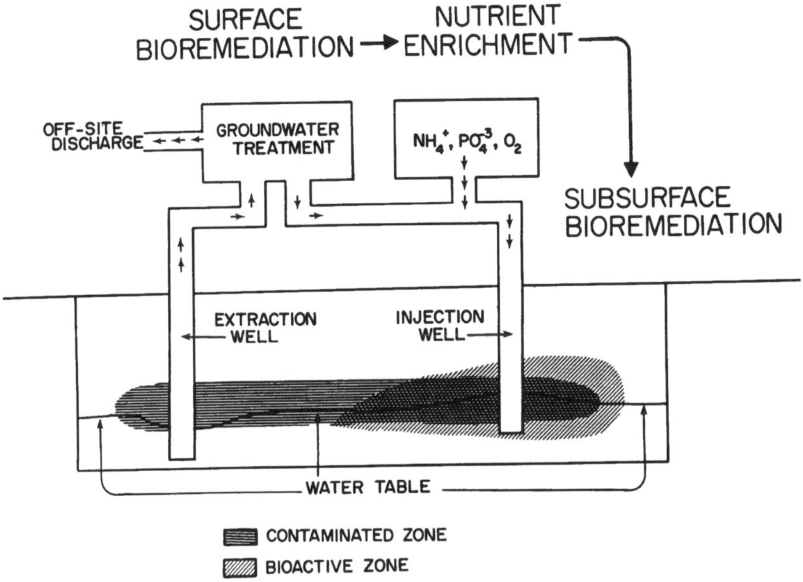

Figure 3 Schematic representation of a groundwater pumping system for *in situ* treatment of contaminated groundwater. In addition to surface treatment implementation of bioremediation (bioreactor), the reinjected groundwater can be charged with oxygen and nutrients to stimulate biodegradation of soilbound wastes by indigenous microorganisms as shown (the bioactive zone). In some versions, living cultures of introduced microbes may be injected as well. The extent of the bioactive zone is dependent on soil permeability and pumping parameters. Optional surface discharge contains nonbiodegradable wastes, and may require further treatment or disposal. (Adapted from Ryan et al., *J Hazard. Mater.,* 28, 159, 1991. With permission.)

typically two to three orders of magnitude above allowable drinking water limits. Based on practical experience gained in pumping and treating contaminated aquifers over the past 10 years, Travis argues that treatment typically brings about a drop in pollutant concentrations by a factor of 2 to 10, followed by leveling out with no further decline. Cessation of pumping is often followed by a rebound in aquifer waste concentrations. Once again, the problem is largely that sites are typically contaminated with organic wastes that do not readily dissolve in the aqueous phase. They either remain adsorbed to the soil matrix, float on the top, or sink to the bottom of the water table. Wastes therefore only slowly seep into the groundwater, at a rate that is diffusion limited, and cannot be significantly changed by groundwater pumping. Pump-and-treat measures may dramatically reduce pollutant concentration in the aqueous phase of the aquifer, but when the pumps are switched off, pollutants gradually leach out of the soil, and the aqueous concentration rises again. Many leading hydrologists have concluded that literally hundreds to thousands of years of pumping could be required to purge some contaminated aquifers of their organic waste contaminants (Hall, 1988). Travis draws the unfortunate conclusion that, "Not a single aquifer in the United States has been confirmed to be successfully restored through pumping and treating." (Travis, 1992). The implication is that, while pump-and-treat measures may be useful to limit dispersal of a waste plume into the water table, massive excavation of soil is usually required to remove the source of the problem.

A second, more recently developed bioremediation approach is subsurface *in situ* remediation. The treated water can be nutrient and oxygen enriched prior to recharge, to stimulate aerobic biodegradation of soilbound, water-insoluble wastes by indigenous soil microorganisms (see Figure 3). Actual oxygen (O_2) content of the water can be boosted by air pumps, or alternative oxygen sources such as hydrogen peroxide (H_2O_2) may be added. Surfactants and other organic waste desorbing chemicals can also be added to increase waste bioavailability. If a surface bioreactor is used, some portion of the active microbial biomass can be recharged with the water, providing continuous inoculation of the contaminated aquifer and soil. While stimulation of aerobic metabolism is the objective of most systems, the reinjected groundwater can be enriched with nitrate to stimulate growth and enhance the biodegradative action of anaerobic denitrifying microbes. This approach has recently been shown to be effective in degrading aromatic constituents of gasoline, with toxicity reductions comparable to those seen in aerobic degradation (Carroquino et al., 1992). As with *in situ* soil treatment, the success of subsurface aquifer bioremediation is largely determined by waste and soil characteristics. Soil permeability is especially important to the success of nutrient enrichment and inoculation efforts.

Considering the limits of conventional groundwater remediation methods, *in situ* bioremediation may offer greater potential than any other technique for restoration of contaminated aquifers. A number of vendors make impressive claims about the *in situ* degradation efficiencies of their systems. In considering such claims, it is advisable to look for field-proven systems, with long-term monitoring results that indicate no rebound of aquifer waste concentrations following project closure. The development of such systems is a fairly recent occurrence, and their use is not yet standardized or even fully accepted. Some states restrict or prohibit the introduction of microorganisms into groundwater, for example. Considering how widespread aquifer contamination is, due to surface spills, leaking underground storage tanks, and landfills, this would have to be highlighted as a priority area for future research. Clearly, recycling and waste minimization efforts are critical elements in a sound groundwater management strategy.

Contractor-supplied cost ranges for groundwater remediation are difficult to compare or even interpret. Some vendors quote costs in the same units used for soil treatment (cost per pound, ton, or cubic yard of treated material). In such cases, dollar figures usually cover the same range given above for soil remediation. Others quote cost per gallon of groundwater pumped, with a price range of about 1 to 10¢ per gallon (EPA, 1993). Since the pump-and-treat approach is a closed loop, either measure is meaningless without knowing the waste extraction and/or degradation efficiency.

BIOREMEDIATION OF SURFACE WATERS

Over the years, a number of lakes and rivers have become seriously contaminated with various industrial wastes in many parts of the U.S. In some cases, the sources of the pollution have been reduced or even eliminated; and public demand for remediation to a condition safe for fishing and other recreational uses is growing. Unfortunately, technologies for surface water remediation are not nearly as well developed as those for soil or even groundwater. Part of the problem lies in the sheer size of many water bodies. It is technically and environmentally impractical to divert a large, flowing water body from its course for treatment. Also, as with underground aquifers, the water contamination problem is largely a sediment contamination problem. Many persistent wastes become tightly bound to bottom sediments from which they slowly leach out, and cannot readily be removed by water treatment in any case. The conventional treatment has typically been to dredge up and remove bottom sediment in the most polluted areas, but such measures can themselves be environmentally devastating, and risk remobilization of toxicants accumulated over many years. Furthermore, the excavated sediment must still be treated and/or disposed of as toxic waste. Workers are turning to *in situ* bioremediation almost as a last resort.

Surface water bioremediation technologies are largely being developed in place. An ongoing example is General Electric's (G.E.) Fort Edward, New York, site on the upper Hudson River. For many years G.E. legally released polychlorinated biphenyls (PCBs) into the Hudson from a plant that manufactured capacitors. When PCBs became priority environmental pollutants in the early 1980s, at least 20 miles of the river bottom were found to be contaminated seriously downstream of the plant. G.E. began looking for remediation options at that time. In 1991, G.E. conducted an extensive field research program to characterize natural degradation of PCBs at this site (General Electric, 1992). They discovered that the indigenous consortia of microorganisms was exceptionally good at degrading PCBs. Presumably, since the PCBs have been present in this site for a significant time period (at least 35 years), microorganisms there have adapted to utilize the material as a food source. Both anaerobic and aerobic biodegradation have been identified as part of the natural process of remediation. As noted elsewhere, the natural process can be exceedingly slow. Field tests in cylindrical caissons sunk into the river sediment at this site have been carried out to identify the variables that can be manipulated to enhance *in situ* biodegradation of PCBs. It was found that addition of inorganic nutrients, the organic co-metabolite biphenyl (see Chapter 7), and oxygen significantly increased PCB degradation rates. Addition of selected PCB-degrading bacterial cultures did not dramatically improve biodegradative efficiency. No more than 60% of the PCBs were degraded in any laboratory or field experiments, a finding attributed to tight sediment adsorption of the least water-soluble PCB compounds (Harkness et al., 1993). More information on degradation rates, products, and variability under natural conditions is required for a realistic evaluation of the role that bioremediation may play in this and other surface-water sites contaminated by organic waste.

BIOREMEDIATION OF MARINE OIL SPILLS

The volume-dilution power of the sea has long been exploited by man to dispose of unwanted wastes. While concern about accumulation of wastes in marine environments is increasing, especially for coastal waters, marine remediation

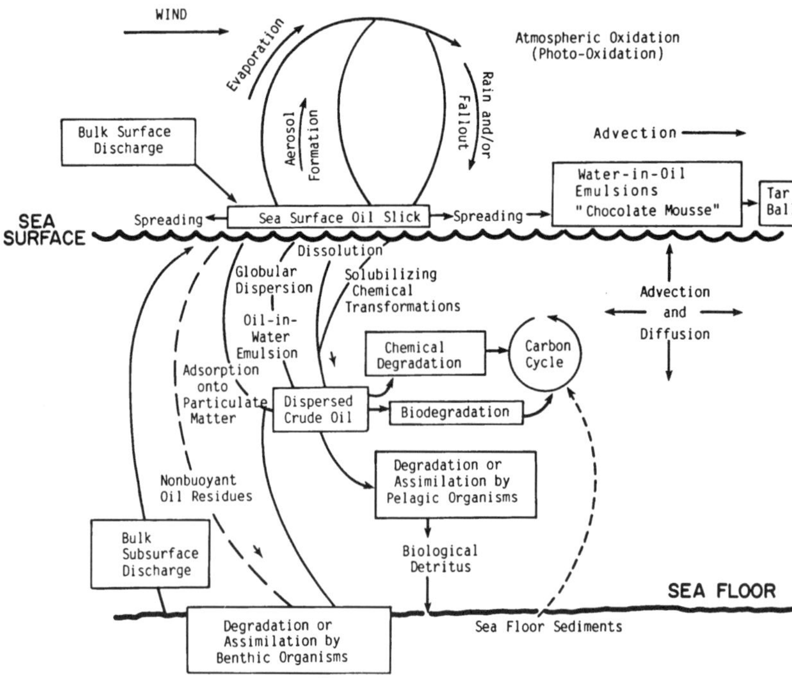

Figure 4 Diagram of the numerous dispersion components, fates, and degradative pathways of spilled oil on the surface of the sea. Since crude oil is a mixture of thousands of individual compounds, and the relative importance of each pathway will vary among compounds, the complexity of this diagram is even greater than it appears. (From NRC, *Oil in the Sea: Inputs, Fates, and Effects,* National Academy Press, Washington, D.C., 1985. With permission.)

efforts remain nearly nonexistent. The notable exception to this rule is the case of crude-oil and refined-petroleum-product spills. Tanker spills account for only 13% of the estimated 3.2 million metric tons of annual marine petroleum hydrocarbon inputs (NRC, 1985). Yet, they have remained the focus of research efforts related to remediation of marine oil contamination. The potential for truly massive spills from modern supertankers and the readily visible direct impact on affected areas have captured the attention of the public and sensitized regulatory and industry groups to the local destructive potential of such accidents. Petroleum is a complex mixture of thousands of individual compounds, and the degradation pathways of spilled oil are numerous and complex (Figure 4). Biodegradation, especially by microbes, is believed to be one of the primary mechanisms of ultimate removal of petroleum hydrocarbons from marine and shore environments (reviewed by Atlas, 1988; NRC, 1985). Acceleration of this natural process is the objective of bioremediation efforts. Bioremediation has yet to become an established spill-response technology, but some attempts to implement it have been encouraging. The inability of established nonbiological techniques to cope with recent large spills has led to increased interest in bioremediation. Special problems associated with marine oil spills include the uncontained nature of the waste, potential huge size of the contaminated area, and moderate to extreme difficulty in access for remediative and monitoring activities.

As with other forms of *in situ* bioremediation, natural biodegradation of marine oil spills may be enhanced by inducing changes in either the microbial population or the availability of microbial nutrients. Most researchers have concluded that nutrient availability is the chief limitation, and most research has been directed toward enhancing this end (OTA, 1991). Marine oil spill cleanups represent some of the largest *in situ* remediation projects ever attempted. The March 1989 spill of 11 million gallons of crude oil from the supertanker *Exxon Valdez* into Prince William Sound, Alaska, provided a testing ground for a large number of nutrient enrichment technologies. EPA and Exxon spent about \$8 million on a joint program to test and apply such measures (Thayer, 1991). Open-ocean nutrient enrichment is problematic. Physical dispersion of oil and nutrients makes both application and assessment difficult. The impossibility of maintaining discrete treated and control areas of a floating slick makes any evaluation of success suspect. In Prince William Sound, therefore, beaches were primarily targeted for bioremediation. The microorganisms present in the beach sediment were considered to be wide enough in diversity and degradative capability to carry out the oil biodegradation desired. Following appraisal of small-scale tests, large-scale applications of nutrients to Alaskan beaches were carried out. Eventually, about 110 miles of beach were treated with many tons of nutrient materials. Both water-soluble and oleophilic fertilizers were found to be effective, with visible clearing of oiled beaches evident in many cases. The results obtained indicate that, for the conditions encountered, the bioremediative action of indigenous bacteria can safely be accelerated two- to fourfold over control beaches by a single addition of nutrients. A second application 3 to 5 weeks later boosted this figure to as high as five- to tenfold (OTA, 1991). Precise analysis of the process and steps involved in the bioremediation carried out and final assessment are still under way. Assessments of the utility of inoculation or "seeding" oil spills with selected microorganisms has, thus far, been inconclusive (Pritchard and Costa, 1991). Once again, lack of long-term background or control biodegradation-rate data makes evaluation of success difficult. Some nonliving biological products were tested on the beaches of Prince William Sound, also with inconclusive results (OTA, 1991).

The greatest theoretical problem with bioremediation as a first response to oil spills is the time course of action. Movement of oil slicks toward sensitive areas (i.e., coastal wetlands, beaches, shellfish beds, etc.) may necessitate a rapid nonbiological response, such as chemical dispersal or burning to prevent contamination. In such cases, bioremediation may prove to be most useful as a secondary treatment option, or when paired with nonbiological methods of degradation enhancement. A novel proposed first response to oil spills utilizes titanium dioxide (TiO_2) coated, floating glass microbeads to catalyze photooxidation of the oil (Jackson et al., 1991; Rosenberg et al., 1992). When these beads are applied to a floating slick, a microfilm of oil coats the beads; and the semiconducting compound TiO_2 absorbs sunlight energy, catalyzing the oxidation of the organic microfilm (reviewed by Gerischer and Heller, 1992). The resulting breakdown products are more water soluble and more bioavailable, resulting in the more rapid onset of biodegradation.

BIOREMEDIATION OF HEAVY METAL CONTAMINATION

Bioremediation technologies are usually associated with degradation of organic wastes, as well as a few biodegradable inorganics, such as nitrates and cyanides. As mentioned in the previous chapter, alternative applications of biological principles exist for remediation of sites contaminated with nonbiodegradable inorganic wastes. Among such wastes, heavy metals are of primary concern, due to their widespread use and release, toxicity, persistence, and, in some cases, ability to biomagnify up the food chain. Concentration of metals in waste by conventional methods (dewatering, controlled emission incineration, chemical extraction, etc.) is usually energy intensive and expensive. Special abilities of some microorganisms and higher plants are just beginning to be exploited for remediation of sites contaminated with heavy metals. Microbes degrade organic wastes by utilizing them as food sources (see Chapter 7). Their ability to concentrate metals is similarly based on normal biological function. This may take one of several forms.

Perhaps the best-known example of microbial metal metabolism is the "mining" microbe *Thiobacillus ferrooxidans*. This bacterium and other related species derive energy by breaking down metallic sulfides. Many heavy-metal ores, such as cadmium, copper, lead, mercury, uranium, and zinc, often occur as complexes with sulfur. Microbial metabolism of the sulfide releases the free metal. Sulfuric acid is produced as a by-product, which contributes to further metal leaching. Technologies utilizing *T. ferrooxidans* as an inexpensive, nonpolluting alternative to high-temperature ore smelting have seen limited use since the late 1970s, and development of larger-scale applications continues (reviewed by Merson, 1992). Proposed bioremediation uses of sulfur-oxidizing bacteria include removal of sulfur from coal and fuel oil, and use in biofilters to scrub sulfur dioxide (SO_2) from various industrial emissions. While heavy metals are not actually accumulated by such bacteria, the potential exists for bioreactor treatment of metal-contaminated solid wastes, resulting in extraction of valuable metals for reuse, especially when sulfides are also present in high concentration. Bioreactor designs that utilize sulfide-oxidizing microbes to release heavy metals from contaminated soils and sludges are now being explored, with promising early results (TNO, 1991). Other bacteria make their living catalyzing the reverse reaction: anaerobic reduction of sulfates to sulfides (see biogeochemical cycling discussion in Chapter 7). Sulfides produced in this manner readily complex with heavy metals and precipitate. Indeed, many commercial ore deposits are believed to have been produced by biogenesis (reviewed by Zajic, 1969). Sulfate-reducing bacteria may prove to be valuable for precipitation of dissolved metals in waste slurries, using anaerobic bioreactor technology.

All organisms require a number of specific metals for their growth and function. Some of these essential minerals either have very thin geochemical distributions, or are only found in forms unavailable for biological use. Heterotrophic organisms (those that metabolize existing organic material) may obtain sufficient essential minerals from their diet. Autotrophs (those that synthesize their organic material from simple, inorganic constituents) must have some means of obtaining them directly from their environment. Plants and autotrophic bacteria have evolved various means of extracting essential nutrients, including metals, from their environment. In some cases, these organisms can concentrate metals several thousand-fold over ambient concentrations (reviewed by Zajic, 1969). Many bacteria can leach metals from solid ores, as described above. Others rely on contact with dissolved metals. In some cases, high-affinity, metal-binding compounds called siderophores are secreted. The siderophores bind specific chemical forms of metals, and the metal-siderophore complex is then absorbed back into the organism for utilization. Another mechanism of metal accumulation is bioadsorption. The surfaces of many bacteria and algae have a slimy cell envelope or capsule that is capable of passively adsorbing certain dissolved metals, usually via a charge-mediated attraction. Other bacteria absorb and utilize mineral complexes, and have evolved special ways of packaging and handling the residual heavy-metal-containing metabolites as a protection against metal toxicity. Often this culminates in secretion of heavy metal precipitates or insoluble complexes. Manipulation of culture nutrient levels and environmental conditions can stimulate such activity, leading to remarkable concentration efficiencies for metals such as arsenic, cadmium, chromium, cobalt, copper, nickel, mercury, tellurium, and zinc (reviewed by Mergeay, 1991). Practicable bioremediation technologies exploiting these abilities have yet to be perfected, but the potential for metal-concentrating bioreactors has been recognized, and is the object of current research.

Again, it should be stressed that microbes are not capable of rendering heavy metals permanently harmless. For actual removal of metals from treated waste or contaminated biomass, some final extraction step is required. When microbial action is used to liberate bound metals from soil or sludge, this step will probably be chemical or physical (i.e., precipitation, evaporation). In the case of bioaccumulation, a fixed biomass arrangement, or some method of harvesting metal-laden microbes from the slurry (i.e., filtration, centrifugation, etc.) would have to be used for final recovery of the metal-rich fraction.

A special class of proteins called metallothioneins merits special mention with regard to potential applicability in environmental remediation. Metallothioneins are low-molecular-weight proteins with a very high sulfur content. They are produced by many microorganisms, as well as higher animals, and have a high affinity to form complexes with certain heavy metals, notably cadmium, copper, and zinc (Brady, 1982; Higham et al., 1984; Mehra et al., 1988). The primary purpose of metallothioneins appears to be protection against nonessential heavy metals. Heavy metals are toxic due to their interference with sensitive biochemical processes and/or adverse oxidative events they initiate. Binding to metallothionein renders them much less reactive. The gene for metallothionein production is typically inducible, so that an organism exposed to a potentially toxic metal rapidly turns on metallothionein synthesis as a protective mechanism (Ecker et al., 1986; Lin and Kosman, 1990). Other possible biological roles include transport of metals for insertion into metalloproteins.

Metallothionein-producing microbes have been suggested as a biological metal-sequestering system for wastewater treatment (Romeyer et al., 1988). Microbes could be selected for high production of metallothioneins to sequester unwanted heavy-metal contaminants in bioreactors or possibly used for *in situ* inoculation. Production of metal-binding proteins by cultured microorganisms for subsequent extraction is also of potential utility for treatment of metal-contaminated wastes and environments. Immobilized metallothioneins could be used in biofilters and bioreactors to remove dissolved metals from waste streams.

Appreciation of the potential of natural systems to regulate levels of aquatic toxicants has led to the development of so-called "constructed wetlands" for bioremediation of complex wastes. Wetlands have long been observed to have a buffering ability on surface waters, with respect to circulating nutrient and pollutant levels. In other words, wetlands have the capacity to store excess nutrients or wastes, and to release stored excesses under the right environmental conditions (for a collection of reviews on this subject, see Hammer, 1989). A constructed wetland is an artificial habitat, most visibly made up of vascular plants and colonial algae, which also provides a structural and nutritional support for an associated, highly heterogeneous microbial community. One of the most promising applications of constructed wetlands is for *in situ* bioremediation of metal contamination. It is not always known to what extent the observed metal removal in natural wetlands is due to bacterial action, and what is due to higher plant or algal activity. In any case, many of these organisms exist in a symbiotic arrangement; and multitrophic cultured systems are increasingly being viewed as an alternative to monocultures, or even heterogeneous bacterial cultures. Field tests on acid mine-drainage effluent have indicated that such systems are capable of removing metals via multiple-pathway biological action (Batal et al., 1989). The use of both natural and constructed wetlands for heavy metal abatement is of great potential value, but questions remain about the eventual fates of the metals. Some means of extraction such as removal of plant or sediment material is necessary to prevent remobilization of metals from dead organic material, or trophic transfer to grazing animals.

BIOFILTRATION OF AIRBORNE POLLUTANTS

Another emerging application of bioremediation, with potential yet to be fully realized, is biodegradation and/or removal of environmentally undesirable compounds from air via biofilter technology. At first glance, this idea may seem almost laughable. To some, it may appear analogous to the tongue-in-cheek proposal to remove smog from the Los Angeles basin by mounting huge fans on the San Gabriel Mountains, and blowing it out over the Pacific Ocean! However, this technology has been shown, primarily by groups working in Europe, to be suitable for large-scale biodegradation of volatile organic compounds (VOCs).

A typical, simple biofilter arrangement is shown schematically in Figure 5. The idea is, quite simply, to direct the flow of a VOC-laden gas stream through a bed containing mixed cultures of microorganisms that mineralize the VOCs. The actual filter bed is a wet, biologically active layer of compost, peat, or soil that provides both a structural support matrix and a source of inorganic nutrients to the microbial population. Inert structural additives, such as plastic beads, may be included to decrease flow resistance and reduce structural changes due to aging. Compost mixtures made from municipal wastes, wood chips, bark, or leaves have usually been the media used in European gas biofilters. U.S. designs have typically utilized soil beds, which suffer from lower biodegradation capacity, larger space requirements, and higher gas flow resistance (Leson and Winer, 1991). Given sufficient retention time in the filter, the organic contaminants diffuse into the aqueous biolayer (or biofilm) surrounding the filter bed particles, where they are degraded by microbial metabolism. The final products are microbial biomass, resulting from growth of the microorganisms, and mineral end-products. Volatile end-products will be carried out in the gas effluent stream. Nonvolatiles will accumulate in the filter media, which eventually has to be changed. Naturally occurring microorganisms are usually present in quantities adequate to handle easily biodegradable

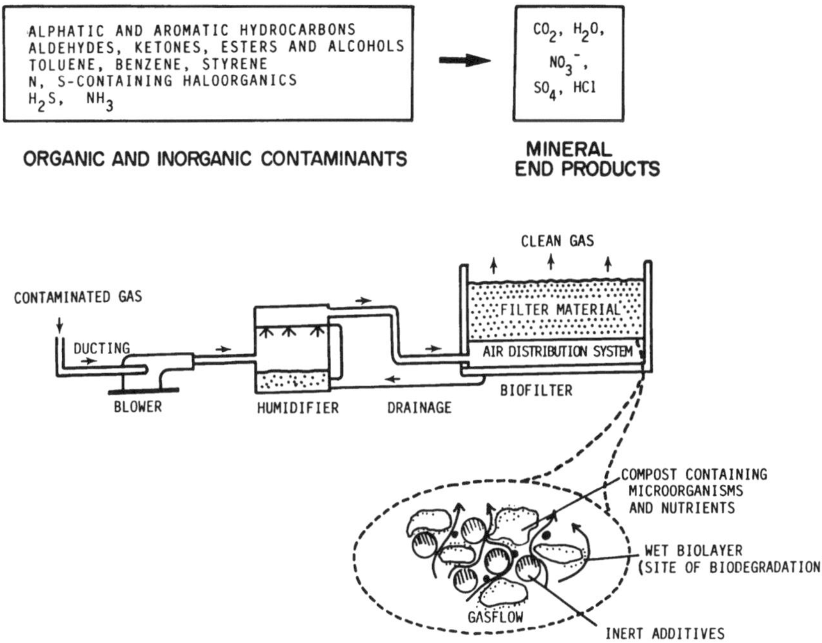

Figure 5 Schematic representation of a typical biofilter used to remediate contaminated gas streams. Ideally, organic and some inorganic compounds are converted to mineral end-products by microorganisms in the filter bed. Volatile end-products will be flushed out with the "clean" gas effluent, while nonvolatiles will be retained in the filter bed. Humidification is important to prevent dessication of the filter bed.

compounds like alcohols, ethers, simple aromatics, etc. More degradation resistant chemicals, such as nitrogen- and sulfur-containing organics, and especially chlorinated organics and aliphatics, may require inoculation with selected strains of microbes to achieve desired degradation efficiencies.

Biofilter technology has arisen primarily from odor abatement efforts related to sewage treatment and agricultural operations (reviewed by Leson and Winer, 1991). In the last decade, research has increasingly been directed at development of applications for airborne toxins and regulated pollutants. Design of biofilters is complex, being highly dependent on gas stream chemistry and flow rate, yet is governed by well-established kinetic and mass balance principles used in general stream reactor design (reviewed by Ottengraf, 1986). While every application must be evaluated individually, biofilter technology represents a VOC abatement option that is competitive on both an efficiency and cost basis in many cases. The total operating and maintenance cost range has been estimated at 0.60 to $1.50 per 100,000 standard cubic feet of off-gas treated (Leson and Winer, 1991). Biodegradation efficiencies in excess of 90% have been achieved in a number of industrial installations (see review by Dragt and Ottengraf, 1989).

As previously mentioned, most exploitation of aerial biofilter technology has taken place in Europe. In a recent, comprehensive review of aerial biofiltration, the authors assert that inadequate research and development support, lack of government regulations, and a shortage of English language descriptions of existing technology have led to a lack of appreciation and application in the U.S. (Leson and Winer, 1991). As many as 500 aerial biofilters may be functioning in Germany and the Netherlands, ranging from small, simple systems associated with farms and food processing plants to large-scale facilities at chemical plants, foundries, print shops, and paint shops.

Aerial biofiltration is a developing technology that has yet to see widespread application in the U.S. It has received extensive attention in this review because it is the only proven biological system that can degrade airborne pollutants. At present, it is an emission control technology, applicable only to pollutant point sources. While no remediation technique has yet been established to remove ambient levels of organic pollutants from air, extensions of biofilter technology may yield such techniques in the future. Potential technical problems include maintenance of degradation efficiency with variable off-gas flow rates, accumulation of damaging metabolic by-products in the filter bed (inorganic acids from sulfur, nitrogen, and chloride-containing wastes, for example), and the possibility of release to the atmosphere of undesirable amounts of microbial or fungal organisms or spores.

FUNGI AND BIOREMEDIATION

Throughout this chapter we have generally referred to organisms involved in bioremediation as microbes. It is common to read the terms microbial and bacterial as synonymous, and most bioremediation systems in use utilize bacteria for biodegradation of wastes. A rapidly emerging area of bioremediation science utilizes fungi or fungal products. Fungal enzymes are effective at degrading some wastes that prove resistant to most bacterial action. Encouraging results from studies of fungal metabolism suggest that many more remediation problems are approachable with fungi than was previously thought. Notable successes have been made in soil and wastewater treatment with wood-rotting fungi, most notably the white-rot fungus, *Phanerachaete chrysosporium* (reviewed by Wainwright, 1990). Few organisms are capable of degrading wood on their own, primarily because of the biodegradation resistance of lignin, the aromatic carbohydrate polymer that largely makes up vascular plant cell walls and binds adjacent cells together. Among biopolymers, lignin is second only to cellulose in abundance, yet is degraded by a narrower range of microbes than any other (Kirk and Farrell, 1987). The wood-rot fungi earn their name by producing a host of oxidative enzymes, especially non-specific ligninases and cellulases. Fortuitously, these ligninases can also degrade a wide variety of environmental contaminants, notably organohalogens. It has been shown that white-rot fungus ligninases greatly facilitate the degradation of PCBs, polycyclic aromatic hydrocarbons (PAHs), and chlorinated dibenzodioxins (reviewed by Wainwright, 1990). The biodegradative activity of fungi is, in general, accomplished by materials excreted into the environment, acting outside the living organism in contrast to bacterial action where, in general, the degradative action is carried out by internal metabolic processes. It is therefore less important that the substrate be completely solubilized in order to be biodegraded. Simply being in aqueous contact gives secreted enzymes access to the wastes. Since extraction of secreted enzymes from cultured organisms is more straightforward than internalized ones, fungal products may prove to be especially adaptable for use in nonliving bioreactors, or as immobilized enzymes in biofilters. Various bioremediation techniques can be adapted to utilize wood-rotting fungi, but special limitations may apply. Successful use of white-rot fungus as an *in situ* soil-remediating organism has taken place. They are normally cultured on wood chips, and inoculated into the soil by plowing the chips into the ground. Unfortunately, the fungi tend to stay associated with the wood chips, since they are not soil organisms. This problem can be partially overcome by using a large inoculum or by strain selection. Bioreactors that use intact, living fungi may make use of drip, spray, or rotating drum arrangements to keep beds of wood chip cultures in contact with (but not continuously submerged in) an aqueous waste slurry.

FUTURE PROSPECTS FOR ENVIRONMENTAL BIOREMEDIATION

ENHANCED BIOREMEDIATION VIA GENE MANIPULATION

Modern techniques of molecular biology have ushered in a new era for the prospects of biotechnology. Genetically engineered organisms, primarily plants, are used extensively today in agriculture (reviewed by Olson, 1986). Genetically engineered microorganisms (GEMs) have long been used for biomedical purposes, such as the production of peptide hormones for human therapy. The earliest and perhaps most successful such example is bacterial production of human insulin (reviewed by OTA, 1984). The genes coding for insulin have been inserted or "transfected" into the common bacterium *Escherichia coli* (for details see Goeddel et al., 1979). The "transgenic" bacteria are then cultured into large colonies, which produce functional insulin but have no use for it themselves. Portions of the culture can be harvested and the human protein extracted from the bacterial biomass. Microbes have been engineered to produce insulin and other pharmaceutical chemicals, with a product yield of up to 40% of the total cell protein content (Keeler, 1991). Potential future roles of GEMs, often cited by biotechnology proponents, include the production of totally new, highly specific enzymes. The ability to custom-build genetic sequences (designer genes) that yield microbial protein products with novel activities is, at present, beyond the state of the art. Current technology is restricted to transfecting the genes of one microbe into another or to modification of regulatory genetic sequences, leading to altered production levels of normal gene products.

Molecular biology is among the most jargon-prone of all scientific fields. A concise theoretical and technical summary is difficult, and is largely beyond the scope of this discussion (see Watson et al., 1987 for general reference). Some understanding of basic concepts is required, however, to understand the potential utility and problems associated with the use of GEMs in bioremediation. The following is a simplified overview of the genetic engineering process.

Microbial DNA sequences that code for the proteins responsible for desired bioactivities must be identified before they can be used. This is a science in itself. The gene or genes are then replicated or "cloned" by extracting all of the DNA from the original hosts, cutting out and isolating copies of the desired genes, and then

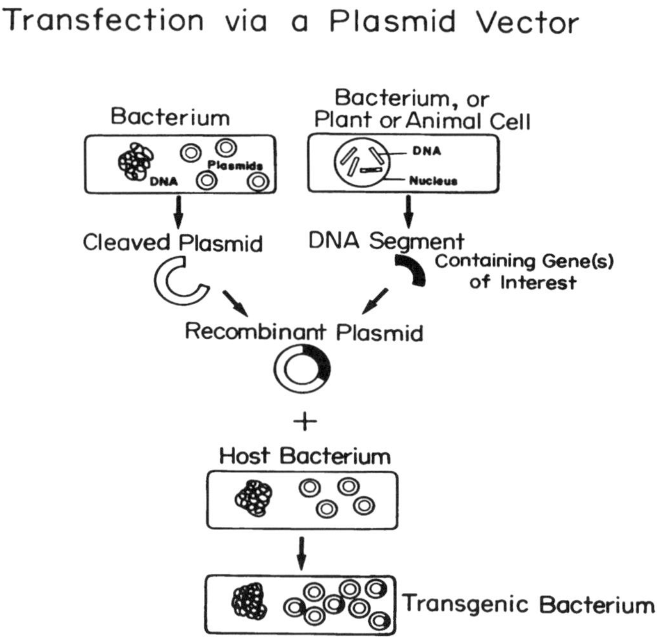

Figure 6 Typical procedure for introduction of recombinant DNA into bacteria via a genetically engineered plasmid vector. In this illustration, a plant or animal gene is inserted into an indigenous bacterial plasmid, which is then transfected into the host in multiple copies. The progeny of transfected bacteria will also be transgenic. (Adapted from Olson, S., *Biotechnology: An Industry Comes of Age,* National Academy Press, Washington, D.C., 1986. With permission.)

using recombinant techniques to introduce this genetic information into new host cells. A number of technical advances in recent years, including improved separation techniques for DNA fragments and methods of rapidly replicating DNA outside of living organisms, have made this tedious process more feasible.

All cells carry most of their DNA in ordered chains called chromosomes although the number, size, and form of chromosomes vary widely. Bacteria differ from animals in that they commonly carry part of their DNA in the form of plasmids: small, independently replicating loops of DNA. Plasmids are one of the most commonly used carriers or "vectors" for introducing recombinant DNA (DNA-containing manipulated gene sequences) into bacteria, as illustrated in Figure 6. When the GEM undergoes division, it passes on copies of the engineered plasmid to each daughter cell, leading to rapid amplification of the engineered culture. In some cases, multiple copies of plasmids can be introduced into bacteria, leading to higher expression levels of the plasmid genes and their protein products. Recombinant DNA can also be transfected into the chromosomal DNA of living cells by viral vectors, as well as by a number of chemical and physical techniques that soften cell and nuclear membranes to allow passage of DNA fragments.

A number of scenarios whereby genetic engineering might be used to attain bioremediation-associated objectives may be proposed. One would be to modify waste-degrading bacteria by transfecting them with genes that improve their survival and competitiveness in particular environments. An example is the work of several researchers on bacterial hemoglobin expression (reviewed by Bailey, 1991). It is very unusual for any single-celled organism to produce hemoglobin, which is most commonly found in the circulatory systems of higher animals. *Vitreoseilla* is an aerobic bacterium that lives in poorly aerated environments. It appears that production of hemoglobin by *Vitreoseilla* under low-oxygen conditions improves the efficiency of oxygen absorption. Thus, the evolution of an oxygen supply-regulated hemoglobin expression system allows this microorganism to thrive in environments that are unfavorable for other aerobic microbes (Boerman and Webster, 1982). The gene for *Vitreoseilla* hemoglobin has been cloned and transfected into *E. coli* chromosomal DNA, resulting in increased metabolism at low oxygen conditions by 1.5- to 3.3-fold (Khosla et al., 1990). A similar strategy has been used to increase commercial production of antibiotics by *Streptomyces* spp. without external aeration. In this case, a multicopy plasmid vector increased culture cell densities up to 54% (Magnolo et al., 1991).

Other approaches to improving microbial degradation performance suggest themselves. Biodegradative genes can often be isolated from bacterial populations at contaminated waste sites. These genes are frequently found on plasmids (reviewed by Lindow et al., 1989) and could be transfected into other bacteria by the technique described above. Microbes that normally thrive in a given environment could be transfected with degradative enzyme genes from bacterial species that are good biodegraders, but poorly adapted to the environment of interest. Alternatively, genetic sequences that regulate production of adjacent gene products, called "operons," could be inserted to give remediation workers greater control over production of either normal or transfected degradative enzymes. There exist operons that are induced (turned on) by chemical exposure, by changes in environmental parameters, and by starvation conditions. Deletion of genes that produce enzymes catalyzing undesirable degradation pathways is also theoretically possible. Once created, GEM cultures might be used in the same types of systems described previously for naturally occurring microbes, with some potential improvements. By combination of different genetic sequences, microorganisms that degrade pollutants via the most desirable pathway, with the exact specificity and activity levels desired can theoretically be designed (reviewed by Bailey, 1991). In cases where genetic mechanisms of toxicity resistance have been identified, such as heavy metals (reviewed by Mergeay, 1991), microbes that can function in the presence of waste mixtures that are too toxic for natural cultures to tolerate might be engineered. If multiple steps in pollutant degradation are required, a single engineered organism might carry the appropriate multiple-step metabolic machinery, making GEMs especially attractive for remediation of recalcitrant wastes. Of course, GEM cultures might also be used as living chemical factories that overproduce biodegradative enzymes for subsequent extraction and use in nonliving bioreactors in a manner analogous to the insulin example cited above (reviewed by Stephanopoulos and Vallino, 1991). The possibilities are virtually endless, making it easy to understand the appeal this technology has for many in the remediation field.

To date, most bioremediation-related, transgenic research has focused on combining bacterial biodegradation capabilities. In the early 1970s, Ananda Chakrabarty, then with G.E., began working to create an oil-degrading "superbug" by inserting plasmids from bacterial species with different biodegradative capabilities into one organism. The resulting microbes contained no recombinant DNA, only nonindigenous, naturally occurring plasmids. This distinction has become important in the debate over regulatory treatment of GEMs. While we use the term GEM to refer to any genetically manipulated microbe, many workers now distinguish between organisms that are genetically engineered and those that are "genetically altered" (contain no recombinant DNA). Chakrabarty's early work has become a landmark case in many ways with respect to both legal and regulatory issues surrounding genetic engineering.

The former began with Chakrabarty's attempts to patent his novel bacterium, an action without precedent. The ensuing litigation culminated in the 1980 U.S. Supreme Court decision *Diamond v. Chakrabarty,* where the patenting of GEMs was narrowly ruled legal. This decision was of tremendous importance to the fledgling biotechnology industry in the U.S., since there would be little incentive to spend millions of dollars on transgenic research if the results could not be patented. The issue remains contentious. Some feel it inappropriate, even immoral, to hold a patent on any form of life. The issue of protection of intellectual property, with regard to genetically engineered organisms and products, is of major international concern today. Many countries and collective trade bargaining groups, including the European Community, refuse to recognize certain patents on organisms. It has been reported that this issue lies at the heart of the U.S. refusal to become a signatory to the biodiversity convention at the June 1992 Rio de Janeiro Earth Summit (Hileman, 1992).

If anything, the regulatory issues are even stickier. Although they possess high potential for a number of bioremediation applications, GEMs cannot be freely applied or even field-tested. Current regulation of "genetically manipulated organisms" by EPA is under the Toxic Substances Control Act (TSCA), and restricts their utilization to highly controlled environments. Chakrabarty's proposal to release his bacteria to test their ability to degrade a real oil spill stirred intense debate, and was eventually rejected because of the risk of uncontrolled and unexpected reactions. There is widespread concern that efforts to increase growth rate or bioactivity by genetic manipulation may remove natural limitations on distribution or numbers of organic chemical-degrading microbes. A readily visualized scenario is one in which a microbe has been engineered specifically to mineralize spilled gasoline. If uncontrolled in its growth and distribution, such an organism could theoretically wreak havoc on stored fuel supplies, even to the point of infecting the gas tanks of automobiles! Although this example may seem extreme, the concern and its basis are quite real. To date, GEMs have not been licensed for use in bioremediation systems. Uncertainty about the acceptance of GEMs by regulatory agencies and the public have discouraged most bioremediation companies from development of systems beyond the laboratory bench level. In fact, EPA has yet to review a commercial application for the use of GEMs in bioremediation (Thayer, 1991). Many in the field feel it unlikely that this will change in the near future, unless some genetically programmed limitations on growth can be developed and proven to work. Examples might be cell death after a fixed number of divisions, or inclusion of genes for lethal products that are induced by changes in conditions or availability of food supply when biodegradation goals are completed. Even given such controls, current and proposed federal regulations appear to make it unlikely that such strains will ever be used for uncontained *in situ* remediation. Contained bioreactors are probably the first place where GEMs will find application (see following section). Industry advocates of engineered-organism use, such as the Applied Biotreatment Association of Washington, D.C., are actively lobbying for minimal regulation of naturally occurring and genetically altered (but not engineered) microorganisms in bioremediation. Their argument is that naturally occurring genetic sequences, even when transfected into another organism, pose far less threat than actual novel, recombinant sequences. For the time being, even most bioremediation companies remain cautious about recombinant-engineered organisms. Many in the research and environmental communities are highly suspicious of the use of GEMs, and oppose any regulatory relaxation. Regulation under TSCA of all introduction or enhancement of microbial populations, even naturally occurring ones, has even been proposed (see discussion in Alpert, 1989). No consensus has been reached on this issue. Like many technologies under government regulation, evaluation of genetic engineering suffers from frequent conflicts of interest. Few administrators in the regulatory or legislative communities have the training or experience necessary for an informed opinion. Yet many, perhaps most, of those who are best qualified to advise them have vested interests in the future of the biotechnology industry. Much of the GEM-related research and development being conducted today is in the private laboratories of biotechnology companies. Results of this work often remain proprietary secrets, making an exact assessment of the state of the art difficult. The bioremediation value of released GEMs remains an unknown quantity.

PRODUCTS FOR BIOREMEDIATION FROM ENGINEERED MICROORGANISMS

The genetic-engineering aspect of bioremediation has come to resemble the pharmaceutical industry, in that the legal barriers, costs, and risks associated with new-product development can be as great as the technical ones. For the short term, many companies and academic investigators are focusing on less controversial applications of GEMs, such as biological materials production. While production of biochemicals by captive microorganisms is frequently mentioned as an application of genetic engineering technique with tremendous potential for bioremediation, full-scale applications have yet to be realized. Unlike pharmaceutical GEM applications, utilization of immobilized enzymes for bioremediation largely remains a hypothetical technology. The difference is at least partially due to the relatively recent emergence of bioremediation as a serious industry. We are just beginning to apply living biological systems to waste remediation. Nonliving biological systems are another step beyond. In any case, engineering microorganisms to produce high quantities of bioactive chemicals is a proven technology. The organisms containing these enzymes need not be in a viable or a replicating state in order for this bioactivity to be expressed. Bioreactors, for example, could readily make use of powdered or freeze-dried organisms, or a partially purified preparation from them that contains the waste-degrading enzymes. Immobilized-enzyme technology has advanced to the point that such materials can be kept in a stationary phase and would remain readily separable from liquid or gaseous streams after the remediative treatment is employed. Thus, the potential exists for nonliving, enzyme-based, bioaccumulating reactors, as well as biodegrading ones. Furthermore, biological products may see use in *in situ* treatment of waste as well.

As an example of ongoing research, we can examine the recent work of Dr. Chakrabarty, now at the University of Illinois at Chicago, who has turned his oil-degrading microbe research toward biosurfactant production (Keeler, 1991). Oil-eating bacteria, such as *Pseudomonas aeruginosa,* must solubilize oil in order to ingest and degrade it. They do so by secreting oil-emulsifying surfactant chemicals. These biosurfactants are much more degradable and less toxic than synthetic ones, and should therefore do less environmental damage if used as oil-slick dispersants. GEMs that overproduce biosurfactants could be safely cultured in a controlled facility, and the chemicals harvested for formulation into oil dispersants.

Pioneering breakthroughs in practical application of new technology are frequently made by those with an edge in cost of operation or access to materials. Large-scale production of microorganisms requires large-scale investment in culture facilities. One interesting source of biomass already in existence is the "spent" brewing yeast of the beer brewing industry. In the mid-1980's, Delta Biotechnologies, a British biotechnology company, began work on the so-called "Delta concept." The process began with the insertion of genes for desirable biochemical products into brewing yeast. Appropriate suppressor and activator gene sequences were included so that the genes would remain silent during normal beer fermentation. Following recovery of the finished beer, the added genes could be turned on by chemical induction, for production of a valued second product. The edge of course, is that a huge quantity of living yeast is cultured at effectively zero cost since it is a normal by-product of the beer-making process. The program is still in the research and development phase. Although this concept has not yet been elaborated in regard to culture of organisms or biological products of potential value for environmental remediation (Delta has thus far focused on pharmaceutical chemicals), it illustrates how new bioremediation technologies may develop.

ENVIRONMENTAL EVALUATION BY GENE-PROBING TECHNIQUES

There are two methods traditionally used to monitor populations of microorganisms in bioremediation. The most common approach has been to monitor pollutant and metabolite concentrations as an indirect measure of the extent of biological activity that has occurred. The problems associated with this objective have already been discussed. This approach reflects the tendency of remediation workers to focus on the waste, rather than on the biology. From a microbiologist's standpoint, the more rigorous approach is to monitor directly the bacterial population itself. Yet classical culture-and-identify methods are tedious and time consuming, and require extensive expertise in culturing technique and microbial taxonomy. An assay that detected bioactivity potential, yet directly measured actual properties of the microbes, would be much better suited to the needs of bioremediation workers than either of the traditional approaches. Furthermore, for bioremediation purposes, the exact identity of the microorganisms present is of less interest than their degradative capabilities. New molecular biology techniques, termed "gene probes." make it possible to test a small, mixed microbial population for specific degradative enzyme genes (reviewed by Olson, 1991). Gene probing can also give an indication of the natural abundance of organisms having the potential to degrade specific pollutants at a given site.

Gene probes may be of two types: DNA or RNA probes. DNA probes consist of complementary DNA strands (cDNA) that bind with extremely high affinity to the DNA sequence that makes up the gene of interest. The probe is chemically labeled or radiolabeled for later detection. When incubated with a DNA extract from the sample to be tested, the probe will bind genetic sequences that are complementary to the probe. Excess, unbound probe is removed, and the amount of bound probe (and therefore of target DNA) is measured. Development of a gene probe obviously requires knowledge of the nucleic-acid sequence of the target gene; but unknown, related genes may be detectable as well. Genes that code for enzymes catalyzing specific degradations vary in sequence from one organism to another, but often have regions where sequences are highly conserved among even totally unrelated species. Knowledge of such conserved regions enables probe designers to incorporate them into probes, making detection of unknown biodegradative species possible.

The presence of degradative enzyme genes indicates the potential for biodegradation, but not necessarily that the activity exists. DNA probes cannot distinguish between genes that are actively being expressed and those that are "silent." The presence of messenger RNA (mRNA) indicates that the gene products are actually being synthesized, and for this reason molecular biologists have developed RNA probes. Since mRNA binds its specific DNA sequence with high affinity, the procedure is similar to that of the DNA probe. A sample mRNA extract is probed with a labeled DNA (not cDNA) probe sequence. While RNA probes yield more useful information, they require larger samples and more expertise, since mRNA may be present in very small amounts and is notoriously unstable.

These techniques may be used as part of the initial site characterization process or as a means of monitoring changes in populations and activities of microorganisms in contaminated sites during the time course of remediative action. They may also be used to detect previously unknown, actively biodegrading microbial species. Like any sensitive assay, they require care and skill to succeed, but development of new methods continues to improve performance. Now available are a number of gene probes for bioremediation, including TCE, PCB, and various aromatic organic degrading enzymes, as well as binding proteins for arsenic, cadmium, chromium, and zinc (Olson, 1991).

RESEARCH NEEDED TO ADVANCE BIOREMEDIATION TECHNOLOGIES

An often-stated priority for bioremediation research is development of novel treatment technologies. When generally applied, this objective is obvious to the point of being ludicrous; but there are specific areas where extant technologies appear especially inadequate for current and anticipated needs. One is the treatment of subsurface aquifers and associated sediments that are contaminated by organic chemicals. Reliable methods for treatment and remediation assessment that do not require aqueous extraction of the contaminants are needed. Another area is treatment of contaminated sediments in surface water systems. Methods that do not require total disruption of flowing water bodies and that do not jeopardize downstream ecosystems are needed here.

Several workshops and congresses have been held to enumerate bioremediation research needs. The following recommendations have been abstracted in large part from the summary report of a bioremediation research needs workshop conducted by EPA in April 1991 (EPA, 1991).

SITE EVALUATION

There is a clear need for improved methods of site evaluation. Continued use and development of new methods described previously, including lab bench assays and gene probe technology, may provide some of the desired information. Site evaluation procedures, with regard to bioremediation success, can be likened to evaluating a field with regard to what crop it is most suited to produce. Agricultural specialists in the area of microbiological growth and expression of bioremediative activity are, however, few and far between. Current systems have largely been developed by trial and error. Bioremediative approaches need to be applied with the understanding that each local environment requires individual attention and detailed site evaluation. Feedback from field tests to the research community is definitely needed, and the information circuit between the research scientist and the environmental manager is not well developed. Comparative evaluations of bioremediative technologies by cleanup planners generally require not only that the initial pollutant be known, but also that the identities and concentrations of the products of bioremediation be determined as well. In a contaminated area that is subject to a bioremediation effort, performance feedback to researchers with regard to the transport, fate, and possible toxicity of the metabolites produced is of tremendous value for refinement of methods. Moreover, site evaluation processes must incorporate expertise from those knowledgeable in other remediation technologies, as well as bioremediation experts. Coupled and integrated methods of containment, destruction, and biodegradation of pollutants are certain to yield more cost-effective cleanup solutions than procedures advocated by experts in a single remediation technology. An ultracompetitive promotion of any technique at the expense of all others will be counterproductive in the long run.

BIOAVAILABILITY OF WASTES

The primary limitation to the widespread use of many bioremediation approaches is often judged to be the extent to which the pollutant is available to the microbial population. Development of methods that make pollutants more bioavailable to current systems is clearly needed. This process begins with an expansion of knowledge of the environmental chemistry and natural degradation pathways of many pollutants. Environmental conditions and the physical or chemical state of the waste are often limiting factors in field degradation rates.

Recognized areas for research advancement include how sorption and desorption kinetics can be altered at contaminated sites to enhance the rate of bioremediation. Many organic pollutants do not readily enter the bioactive, aqueous phase of soil and sediment environments. Their bioavailability to the microbial population might be appreciably increased by the use of appropriate surfactants, dispersants, chelators, or emulsifiers. Technologies that enhance remediation of sites where the wastes are in nonaqueous phase compartments and the microbial population is present primarily in the aqueous phase need to be developed.

A related problem concerns the physical matrix in which pollutants are found (i.e., soil- or clay-particle size and makeup, soil porosity and permeability, presence of variable composition pockets, etc.). This matrix largely determines the rate at which the pollutants become bioavailable. As a minimal first step, the physical matrix in which waste is complexed needs to be defined and then analyzed with respect to waste bioavailability and technologies that can potentially improve the situation.

The bioavailability of many chemicals diminishes with time as a result of weathering and/or aging phenomena. The extent to which this can be reversed and the time window in which appropriate bioremediation technologies can be employed requires further definition.

Finally, beyond their ability to degrade materials, microorganisms also have the capacity in some cases to mobilize, immobilize, or fix compounds or ions in such a way that they are rendered susceptible to further treatment. The first stage of immobilization or solubilization by microorganisms may not in itself constitute bioremediation but may require coupling with the action of a biodegrading or bioaccumulating organism. Development of practical means of exploiting such bioactivities under field conditions is a definite research need. A scenario where one organism solubilizes a sediment-bound waste, perhaps through secretion of biosurfactants, and another organism subsequently biodegrades the waste is an example of the applications that might arise from future research.

PROCESS DESIGN

Some technologies based on the activity of biological systems or organisms appear to be feasible in small-scale laboratory tests but are not readily expandable to field use due to engineering limitations. The technology of scaling up lab-bench experiments to make them practicable in bioreactors or *in situ* remediation efforts requires considerable development.

As discussed in Chapter 7, the evaluation of *in situ* biodegradation is itself a difficult project. Practical means of evaluating the degree of pollutant biodegradation in complicated environmental settings need to be further developed. Although complete mass-balance analyses are desirable, they are seldom obtainable. In light of this deficit, the durability and stability of bioremediative technologies will require other indicators. New and better models for the processes involved may be required. If adequate models are developed, their use would greatly facilitate the design of cost-effective, safe, and practical bioremediative technologies to be applied in large-scale operations. As has been explained previously, environmental remediation may require multistage processes, such as anaerobic followed by aerobic biodegradation, desorption followed by biodegradation, or chemical followed by biological treatment. At present, modeling of such systems is largely by trial and error variation of process design, with monitoring of initial waste- and end-product concentrations only. A true multiparameter model would require computer-assisted evaluation and more definitive progressive measures of bioremediation than are typically used today. Formation and decay rates of intermediate degradation products, rates of nutrient utilization, and growth rate/attrition rate of cultures are examples of such parameters.

An often-overlooked point with regard to biodegradation processes is that systems that are effective at high pollutant concentrations may not work or may not be efficient at low levels of pollutants. The converse is also true. To complicate matters further, various components of a pollutant mixture are likely to be degraded at different rates. The utilization of living organisms requires an understanding that rates of biological processes are controlled by concentration and bioavailability factors and that variations in these factors will produce variations in effective pollutant decontamination. This is particularly the case where the biological system is relying on a process of co-metabolism to degrade the primary environmental pollutant. Knowledge of transport rates and mass balances of nutrients, electron acceptors, and electron donors in the immediate area where bioremediation is to be effected is desirable but seldom available. Again, feedback indicators of effectiveness may allow for a developing "art" of bioremediation in spite of the lack of controlled environmental conditions in which remediation is required.

A final note on the process aspects of bioremediation and associated research needs involves the potential of microbial activity to be induced, i.e., to increase after a time of exposure to a given pollutant. For some microorganisms, a better awareness of the process of induction and factors that govern it may allow for better methods of introducing microorganisms into an environment or bioreactor system where they are able to adapt and respond. Preinducing or activating cultures or inoculants prior to use may prove to accelerate greatly the onset of biodegradative action.

WASTE MIXTURES

A major recommendation of the recent EPA workshop (EPA, 1991) was that innovative and novel processes be explored for dealing with complex waste mixtures. In some cases, some components of complex wastes may be toxic to microorganisms that are needed to degrade other components. Alternatively, waste components may act antagonistically or synergistically on degradative efficiency in unpredictable ways. As stated above, it must also be anticipated that relative compositions of mixed wastes will change with biodegradation. Exploration of complex waste mixtures with regard to the capacity of microorganisms to convert them into environmentally innocuous compounds or to reduce their toxicities to acceptable levels must be a research priority.

SUMMARY AND CONCLUSIONS

In the previous chapter, general principles of and strategies for bioremediation of waste were discussed, without reference to specific organisms or specific technical systems utilized. While not intended to be a comprehensive list, this chapter uses examples of both existing and experimental systems to illustrate the application of biological techniques to actual waste problems. Common scenarios requiring remediation include contaminated soil, groundwater, and surface waters. Bioremediation technologies exist for all of these.

Nondegradable wastes, notably heavy metals, can be extracted from mixed-waste streams by biochemical action and may be concentrated by the method of bioaccumulation, whereby wastes are absorbed and stored in the tissues of the organisms utilized.

Biofilters are an emerging area of bioremediation that utilize intact organisms or immobilized enzymes to degrade or accumulate waste components of flowing, gaseous streams. Biofilter systems for VOCs from point sources exist and are in wide use in many European countries.

Bioremediation technologies utilizing fungi and fungal products are an emerging area with tremendous promise for degradation of recalcitrant pollutants. The wood-rotting fungi, especially white-rot fungus *(Phanerachaete chrysosporium),* produce a variety of oxidizing enzymes, including nonspecific ligninases and cellulases, capable of biodegrading a variety of wastes resistant to bacterial degradation, including PCBs and PAHs.

Future directions and emerging technologies of bioremediation include GEMs. Genetic engineering combines desirable genetic characteristics from two or more species into a single microbe to create a combination of characteristics not found in nature. While the potential consequences of releasing GEMs into the environment are the subject of intense debate and controversy, research continues on potential bioremediation applications. The potential exists for the use of GEMs in sealed bioreactors that would minimize the risk of release into the environment. Alternatively, GEMs could be cultured for production of large amounts of biological products (i.e., waste-degrading enzymes) that could be extracted for use in nonliving bioremediation systems (biofilters, biochemical reactors, etc.).

Another application of molecular biology under development is the use of "gene probes" for identification and screening of microorganisms for use in bioremediation. In this technique, genetic sequences that code for the desired gene product (a particular oxidative enzyme, for example) are determined, and samples of microbial populations are tested for the presence of the corresponding DNA or RNA sequences. This process may represent a much more rapid and sensitive screening technique than traditional culture, separation, and identification of intact microbes, and may identify unknown species that possess the desired abilities.

Bioremediation is an emerging field, the full potential of which is as yet unknown. There is a tremendous need for further basic research and development, especially in the areas of site and waste characterization, waste-technology matching, and integration of multiple remediation techniques. Furthermore, most current applications of bioremediation are for cases of single waste or well-defined mixture contamination. Many sites in need of remediation are contaminated by complex waste mixtures. The potential of biological techniques for remediating mixed-waste sites and the myriad of biological, chemical, and physical interactions possible must be more thoroughly investigated and understood to exploit fully the biological techniques for waste remediation.

ACKNOWLEDGMENTS

We hereby gratefully acknowledge the support of the National Institutes of Environmental Health Center grant (ES01908), an Office of Naval Research grant (N00014-90-J-1660), and research grants from Sunshine Makers, Inc. and Stroh

Brewery Company. In addition, Mr. Bodishbaugh was supported in part by an NRSA grant (ES07031).

REFERENCES

Alpert, S., The Application of Molecular Genetics Techniques to Bioremediation Processes, EPRI GS-6591, Project 2655–28, Electric Power Research Institute, Palo Alto, CA, 1989.

Atlas, R. M., Biodegradation of hydrocarbons in the environment, in *Environmental Biotechnology,* Omenn, G. S., Ed., Plenum Press, New York, 1988.

Bailey, J. E., Toward a science of metabolic engineering, *Science,* 252, 1668, 1991.

Batal, W., Laudon, L. S., and Wildeman, T. R., Bacteriological tests from the constructed wetland of the Big Five Tunnel, Idaho Springs, Colorado, in *Constructed Wetlands for Wastewater Treatment: Municipal, Industrial and Agricultural,* Hammer, D. A., Ed., Lewis Publishers, Chelsea, MI, 1989, 550.

Boerman, S. J. and Webster, D. A., *J. Gen. Appl. Microbiol.,* 28, 35, 1982.

Brady, F. O., The physiological function of metallothionein, *Trends Biochem. Sci.,* April, 1982, p. 143.

Carroquino, M. J., Gersberg, R. M., Dawsey, W. J., and Bradley, M. D., Toxicity reduction associated with bioremediation of gasoline-contaminated groundwaters, *Bull. Environ. Contam. Toxicol.,* 49, 224, 1992.

Dragt, A. J. and Ottengraf, S. P. P., Fundamental aspects of biofiltration and some practical experiences, in *Milieubiotechnologie, 1989,* Postacademisch Onderwijs Natuurwetenschappen, Eindhoven, The Netherlands, 1989.

Ecker, D. J., Butt, J. R., Stemberg, E. J., Neeper, M. P., Debonck, C. D., Gorman, J. A., and Crooker, S. J., Yeast metallothionein function in metal detoxification, *J. Biol. Chem.,* 261, 16895, 1986.

EPA (U.S. Environmental Protection Agency), *Slurry Biodegradation,* EPA/540/2-90/016, EPA, Washington, D.C., 1990.

EPA (U.S. Environmental Protection Agency), *Summary Report: High-Priority Research Needs Workshop,* Report of EPA sponsored workshop held April 15–16, Washington, D.C., 1991.

EPA (U.S. Environmental Protection Agency), *Vendor Information System for Innovative Treatment Technologies (VISITT),* Computer Database, Version No. 2, EPA, Washington, D.C., 1993.

G.E. (General Electric Company), *1991 In Situ Hudson River Research Study: A Field Study on Biodegradation of PCBs in Hudson River Sediments. Final Report,* General Electric, Schenectady, NY, 1992.

Gerischer, H. and Heller, A., Photocatalytic oxidation of organic molecules on TiO_2 particles by sunlight in aerated water, *J. Electrochem. Soc.,* 139, 113, 1992.

Goeddel, D. V., Kleid, D. G., Bolivar, F., Heyneker, H. L., Yansura, D. G., Crea, R., Hirose, T., and Kraszewski, A., Expression in *Escherichia coli* of chemically synthesized genes for human insulin, *Proc. Natl. Acad. Sci. U.S.A.,* 76, 106, 1979.

Goldstein, R. M., Mallory, L. M., and Alexander, M., Reasons for possible failure of inoculation to enhance biodegradation, *Appl. Environ. Microbiol.,* 50, 977, 1985.

Goldstein, R. and Porcella, D., Applying genetic ecology to bioremediation, *Exploratory Res. Lett.* (Electric Power Research Institute, Palo Alto, CA), 9, 3, 1991.

Hall, C. W., Practical limits to pump and treat technology for aquifer remediation, in *Groundwater Quality Protection Preconference Workshop Proceedings,* Water Pollution Control Federation, Dallas, 1988, p. 7.

Hammer, D. A., Ed., *Constructed Wetlands for Wastewater Treatment: Municipal, Industrial and Agricultural,* Lewis Publishers, Chelsea, MI, 1989.

Harkness, M. R., McDermott, J. B., Abramowicz, D. A., Salvo, J. J., Flanagan, W. P., Stephens, M. L., Mondello, F. J., May, R. J., Lobos, J. H., Carroll, K. M., Brennan, M. J., Bracco, A. A., Fish, K. M., Warner, G. L., Wilson, P. R., Dietrich, D. K., Lin, D. T., Morgan, C. B., and Gately, W. L., In situ stimulation of aerobic PCB biodegradation in Hudson River sediments, *Science,* 259, 503, 1993.

Higham, D. P., Sadler, P. J., and Scawen, M. D., Cadmium resistant *Pseudomonas putides* synthesizes novel cadmium proteins, *Science,* 225, 1043, 1984.

Hileman, B., Earth Summit concludes with agenda for action, but little funding, *Chem. Eng. News,* July 6, 1992, p. 7.

Jackson, N. B., Wang, C. M., Luo, Z., Schwitzgebel, J., Ekerdt, J. G., Brock, J. R., and Heller, A., Attachment of TiO_2 powders to hollow glass microbeads: Activity of the TiO_2-coated beads in the photoassisted oxidation of ethanol to acetaldehyde, *J. Electrochem. Soc.,* 138, 3660, 1991.

Keeler, R., Bioremediation: Healing the environment naturally, *R&D Mag.,* July, 1991, p. 34.

Khosla, C., Curtis, J. E., DeModena, J., Rinas, U., and Bailey, J. E., Expression of intracellular hemoglobin improves protein synthesis in oxygen-limited *Escherichia coli, Biotechnology,* 8, 849, 1990.

Kirk, T. K. and Farrell, R. L., Enzymatic "combustion": The microbial degradation of lignin, *Annu. Rev. Microbiol.,* 41, 465, 1987.

Leson, G. and Winer, A. M., Biofiltration: An innovative air pollution control technology for VOC emissions, *J. Air Waste Manage. Assoc.,* 41, 1045, 1991.

Lin, C. M. and Kosman, D. J., Copper uptake in wild type and copper-deficient *Saccharomyces cerevisiae, J. Biol. Chem.,* 265, 9194, 1990.

Lindow, S. E., Panopoulos, N. J., and McFarland, B. L., Genetic engineering of bacteria from managed and natural habitats, *Science,* 244, 1300, 1989.

Magnolo, S. K., Leenutaphong, D. L., DeModena, J. A., Curtis, J. E., Bailey, J. E., Galazzo, J. L., and Hughes, D. E., Actinorhodin production by *Streptomyces coelicolor* and growth of *Streptomyces lividans* are improved by the expression of a bacterial hemoglobin, *Biotechnology,* 9, 473, 1991.

Mehra, P. K., Tarbett, E. B., Gray, W. R., and Wingo, D. R., Metal-specific synthesis of two metallothioneins and gamma-glutamyl peptides in *Candida grabrata, Proc. Natl. Acad. Sci. U.S.A.,* 85, 8815, 1988.

Mergeay, M., Towards an understanding of the genetics of bacterial metal resistance, *Tibtech,* 9, 17, 1991.

Merson, J., Mining with microbes, *New Scientist,* January 4, 1992, p. 17.

Middleton, A. C., Nakles, D. V., and Linz, D.G., The influence of soil composition on bioremediation of PAH-contaminated soils. *Remediation,* Autumn, 1991, p. 391.

NRC (National Research Council), *Oil in the Sea: Inputs, Fates, and Effects,* National Academy Press, Washington, D.C., 1985.

NRC (National Research Council), *Using Oil Spill Dispersants on the Sea,* National Academy Press, Washington, D.C., 1989.

Olson, B. H., Tracking and using genes in the environment, *Environ. Sci. Technol.,* 25, 604, 1991.

Olson, S., *Biotechnology: An Industry Comes of Age,* National Academy Press, Washington, D.C., 1986.
OTA (Office of Technology Assessment, U. S. Congress), *Commercial Biotechnology: An International Analysis,* OTA-BA-218, U.S. Government Printing Office, Washington, D.C., 1984.
OTA (Office of Technology Assessment, U.S. Congress), *Bioremediation for Marine Oil Spills — Background Paper,* OTA-BP-O-70, U.S. Government Printing Office, Washington, D.C., 1991.
Ottengraf, S. P. P., Exhaust gas purification, in *Biotechnology, Vol. 8,* Rehm, H. J. and Reed, G., Eds., VCH Verlagsgesellschaft, Weinheim, 1986.
Pritchard, P. H. and Costa, C. F., EPA Alaska oil spill bioremediation project, *Environ. Sci. Technol.,* 25, 372, 1991.
Ramadan, M. A., El-Tayeb, O. M., and Alexander, M., Inoculum size as a factor limiting success of inoculation for biodegradation, *Appl. Environ. Microbiol.,* 56, 1392, 1990.
Romeyer, F. M., Jacobs, F. A., Masson, L., Hanna, Z., and Brousseau, R., Bioaccumulation of heavy metals in *Escherichia coli* expressing an inducible synthetic human metallothionein gene, *J. Biotechnol.,* 8, 207, 1988.
Rosenberg, I., Brock, J. R., and Heller, A., Collection optics of TiO_2 photocatalyst on hollow glass microbeads floating on oil slicks, *J. Phys. Chem.,* 96, 3423, 1992.
Ryan, J. R., Loehr, R. C., and Rucker, E., Bioremediation of organic contaminated soils, *J. Hazard. Mater.,* 28, 159, 1991.
Sims, J. L., Sims, R. C., and Matthews, J. E., Approach to bioremediation of contaminated soil, *Hazard. Waste Hazard. Mater.,* 7, 117, 1990.
Stephanopoulos, G. and Vallino, J. J., Network rigidity and metabolic engineering in metabolite overproduction, *Science,* 252, 1675, 1991.
Stroo, H. F., Biotechnology and hazardous waste treatment, *J. Environ. Qual.,* 21, 167, 1992.
Thayer, A. M., Bioremediation: Innovative technology for cleaning up hazardous waste, *Chem. Eng. News,* August 26, 1991, p. 23.
TNO (Netherlands Organization for Applied Scientific Research), Getting bugs to remove heavy metals from bulk wastes, *Applied Research,* TNO Newsletter, No. 38, August/September, 1991, p. 1.
Travis, C. C., Toxic waste in groundwater: Can it be removed?, *J. NIH Res.,* 4, 49, 1992.
Wainwright, M., Novel uses for fungi in biotechnology, *Chem. Ind.,* 2, 31, 1990.
Watson, J. D., Hopkins, N. H., and Roberts, J. W., *Molecular Biology of the Gene,* 4th ed., Benjamin/Cummings, Redwood City, CA, 1987.
Zajic, J. E., *Microbial Biogeochemistry,* Academic Press, New York, 1969.

CHAPTER 9

Quantitative Modeling and Analysis in Environmental Studies

Donald P. Gaver

CONTENTS

INTRODUCTION

Most biological systems are complex, being made up of many subtly purposeful, interacting parts. Whenever such complex systems interact with each other or the environment it is useful, or even essential, to introduce a simplified way of thinking about and expressing the possible outcomes. To do so is to employ a *model.* In this chapter we discuss and illustrate some of the types of mathematical

1-56670-141-4/96/$0.00+$.50

models that have been developed and found informative by those who study and attempt to control biological and physical environmental systems. In particular, we examine mathematical models for the interaction, eventual disposition or fate, and effect, of biological and chemical agents that have been released into the physical environment by mankind, have dispersed and accumulated, are potentially harmful to the natural environment, including mankind, and hence are candidates for remediation.

Biological scientists are more accustomed to work with and think in terms of physical biological models, e.g., laboratory animals, than with the quantitative mathematical models that are our topic. However, the use of mathematical models in biology has a long and honorable history: early examples include population growth models by Volterra and Lotka, and the genetic models of Mendel, Fisher, and others. An excellent single reference is the book by Murray (1980). Statistical models that represent individual physical variations, such as in height, weight, blood pressure, pulse rate, and many other physiological characteristics, are in routine use, as are models that describe human response to doses of drugs, medicines, or exposure to toxic agents in various forms and concentrations. These latter will be reviewed in this chapter.

Environmental problems encountered by mankind typically involve control of the production and emission, dispersion, interactions, location, fates, and effects of numerous biological, chemical, and physical (such as ionizing radiation) components within the environment. Understanding and control of the many interrelated processes involved is well beyond the scope of simple field experimentation alone, just as it transcends the boundaries of the single traditional scientific disciplines and technologies. An overall logical framework is needed to assemble the various system components so that options can be expeditiously compared on the bases of costs and effectiveness. Mathematical, often computer-activated, models are a useful, and increasingly utilized, tool for economically exploring the cost and effectiveness of different options for environmental usage and control. This exploration makes use of scientific information and available data pertinent to the particular situations of concern, for example at those sites that are candidates for remediation. In addition, modeling efforts now guide the collection of suitable meaningful data, and help to organize and focus their summary and analysis. Such focusing is accomplished by initially combining data and theoretical understanding from the various relevant scientific disciplines and technologies in an attempt to address meaningful questions that are generally, if simplistically, of the form: "If option X is used for remediating hazardous waste site (HWS) Y, what will be the cost, how long will be required, and what will be the change in the habitability, or risk, associated with the site?" It is increasingly recognized that there is considerable uncertainty attached to quantitative responses of the types described. Thorough and careful appraisal and communication of that uncertainty to policy makers and the public is the responsibility of the growing fields of *risk analysis* and *risk communication.*

This chapter considers several of the generic modeling areas encountered in environmental studies. No attempt is made to be detailed or encyclopedic. We attempt to present the flavor of the area and to put the reader into contact with relevant literature and references thereto. The aim is to provide an overview of models for certain of the many major components of a *risk characterization* analysis or study.

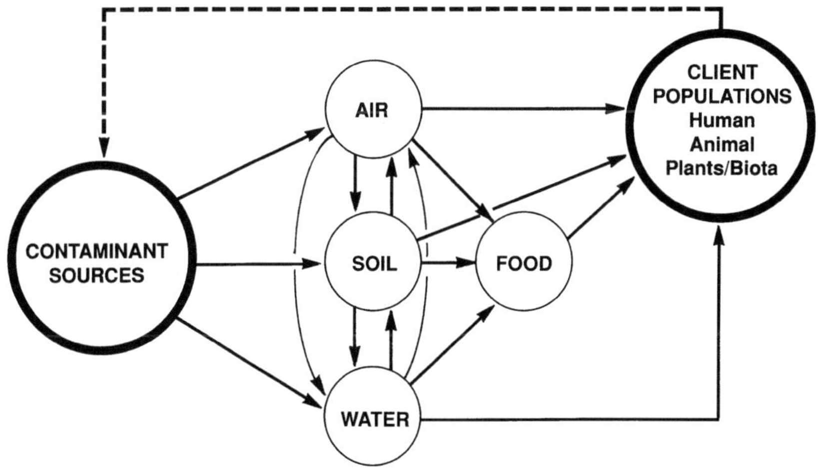

Figure 1 Influence diagram for flow of contaminants (external to clients).

A graphical flow chart or *influence diagram* appears as Figure 1. The links between the round nodes show the direction of influence of contaminants that enter the environmental system and are transported to such client populations as humans, animals, and plants. We use a dotted arrow to denote the influence of client populations, such as humans, upon the contaminant sources. Needless to say, models for the behavior of all sources and links will not be exhaustively discussed in this chapter.

The next section outlines modeling issues connected with *groundwater,* a vital resource, and also a primary medium of transport of pollutants. The history of models in this area is long, and the subject is highly technical, requiring the intellectual tools of physics and engineering, applied mathematics, numerical analysis and statistics, as well as chemistry and biology. Groundwater is one of the many media by which human beings, plants, and animals come into contact with chemical and biological pollutants and toxicants. Maintenance of its quality is a matter of primary concern to the U.S. Environmental Protection Agency (EPA).

In the third section we consider *compartmental* and *physiologically based pharmacokinetic models* for the description of the transport patterns of concentration of toxic agents, and also medicines, between and within bodily organs. Thus, given exposure to various chemicals via drinking, cooking, and washing water (but also through air, food, etc.) it is natural to ask about their ultimate concentrations in the blood that enters such organs as the liver or kidney. Pharmacokinetic models describe the time-dependent relationship of that concentration, or dosage, in terms of exposure routes to the host organism, e.g., a human being, and the scheme by which blood flows, and its contents modified, in that organism on the way to target organs. Intake from groundwater provides one of the many exposure routes referred to above.

The fourth section provides an introductory account of the active new field of *exposure monitoring and modeling.* Reflection suggests that careful effort is needed to relate the origin and subsequent transport of pollutants, e.g., through groundwater,

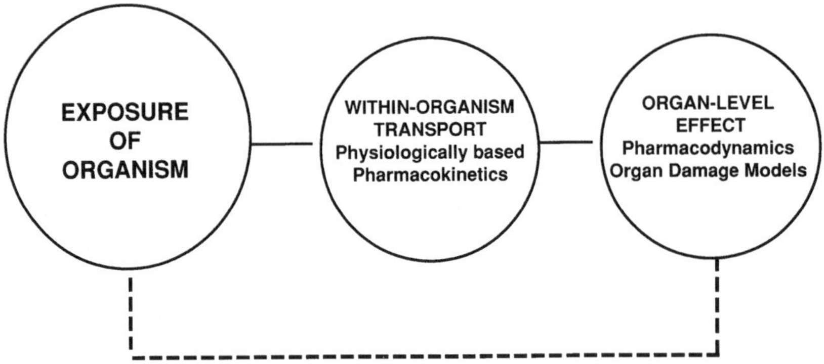

Figure 2 Influence diagram for flow of contaminants (internal to human and animal clients).

to the exposure and dosage of humans and other endpoints, such as the DNA adducts discussed in another chapter.

The topic of the fifth section is that of *dose-response models.* Such models connect dosage levels, i.e., concentrations of toxic chemicals or biological agents, to response at the organ, or host organism, level; response may be in terms such as illness or death of the organism or modification of an organ, i.e., the proliferation of cells or the conversion of healthy cells to precancerous or cancerous. Such events are called *endpoints*; current attention appears to be focused on increasingly more biologically basic, and subtle, endpoints. The mathematical models that have been proposed reflect biological phenomena to as great a degree as possible, but that degree is limited in practice by the complexity and imperfect knowledge of the variety of biological processes by which alien substances actually alter organs. Figure 2 describes the interactions envisioned, and modeled, in the categories alluded to above.

The sixth section discusses some of the issues associated with model-based *risk assessment.* The challenge is to represent multistage organ alteration processes credibly by models that are simple enough to be estimated from available and relevant data. Major interest focuses on the effects of chemical or biological agents on human beings. Often dosages of such agents are low, but prolonged, although response to sudden impulse dosage can be of concern, e.g., that caused by accidental spill. Direct experimental human response information is seldom available, so indication of effect is often sought in animal experimentation; *animal* includes rodents, but also fish, frog embryo, and many other *biological models.* For reasons of economy and time, such experimentation is often done at relatively high doses. It is current practice to extrapolate such animal data to indicate the animal responses at low doses, and across species to a corresponding response in man. The extrapolations are typically made in terms of mathematical models. Such assessments are uncertain and vulnerable for various reasons, and the results have been questioned. The techniques that have been proposed to circumvent the criticisms seem currently to be twofold: to increase the size of the animal experiments, thus allowing more confident determination of low-dose effects, and to improve upon the biological credibility of the basic models. However, very large animal studies are costly, and hence rare. References are given later to the current discussion on extrapolation problems.

The seventh section outlines, in two epidemiologic case studies, issues that were confronted when quantitative approaches, including the use of mathematical and statistical models, were employed to analyze risk to humans exposed to chemically polluted drinking water. Techniques are discussed, as are problems associated with interpretation of results.

The various sections described above ultimately pertain to *risk assessment, characterization,* and *analysis.* The task of risk assessment, characterization, analysis, and management, is to estimate the effect on environmental clients, e.g., humankind but not exclusively so, of admission of certain chemicals and biological agents into the environmental system. Such admission may be appealing in many ways, for instance economically, but the appeal must be balanced against possible adverse effects. Risk analysis attempts to quantify such effects in an atmosphere of considerable uncertainty. In each of the sections we point to efforts made to quantify the uncertainties inherent in the modeling efforts and their applications. Numerous references are presented to supplement the discussion. There remains much to be done to reduce and characterize such uncertainty.

MODELING GROUNDWATER: A SPECIFIC SYSTEM RELEVANT TO REMEDIATION ISSUES

Water is essential for sustaining life on Earth. It is also the medium that transports many of the pollutants that are introduced into the natural environment. *Groundwater,* which is the earth surface's source of most of the water that humans come into contact with, is an element of the *hydrosphere,* which itself acts as a system. The latter encompasses all waters above, on, and below the surface of the earth. Water moves cyclically through this closed system, being exchanged from state to state by evaporation, precipitation, plant transpiration, and other processes including human usage and consumption: i.e., it enters the groundwater subsystem in recharge zones and leaves in discharge areas. While doing so it is consumed by humans, plants, and animals, contributes to disposing of their waste products, and is faced with increasingly many demands as a medium for disposing of agricultural and technological by-products.

Groundwater modeling refers to the description, in quantitative numerical terms, of the flow, and residence or storage, of water into and out of regions near the surface of the earth. Although such flow is of natural origin it is affected by man's activities such as pumping out of and into those regions. See van der Heijde et al. (1988), Schwartz et al. (1990), Anderson and Woessner (1992), and numerous articles in the journal *Water Resources Research* and elsewhere. The objectives of modeling are both to build a scientific basis for understanding the many interacting processes involved and to provide specific information for managing an essential, scarce resource. Public concern is with the adequate supply of water for human consumption and usage, and increasingly with the quality of that supply. Models help to inform policy-makers of the effects of regulations placed on water usage for direct human consumption and waste disposal, and also on the impacts of usage for waste disposal by the agricultural, manufacturing, electric power, transportation and defense subsystems of the economy. They provide information concerning the status and cost effectiveness of remediation efforts.

Models for such purposes use geological, chemical, and biological science and principles of fluid mechanics to describe the time and spatially varying and interacting flows of relevant fluids: freshwater in aquifer containment, but also, where relevant, the extent of intruding saltwater, petroleum, and solutions of chemicals such as fertilizers and pesticides, radionuclides, and other items. Specialized multicomponent models are used to assess the chemical-biological content of groundwater: important classes are *solute transport models* and *pollutant transformation and degradation models.* In these, the processes of change of transported or resident pollutants are described and predicted, using appropriate chemical and biological science. An important objective is to track the availability for consumption by mankind of various substances that enter the water cycle at various remote points and reach aquifers that yield drinking water.

Since groundwater carries contaminants, its motion is of primary importance. That motion is best understood and mathematically modeled in fully saturated regions inhabited by porous materials such as rock or soil. Aquifers are such regions: they store water accessed by wells that furnish water for consumption. Widely accepted mathematical models for aquifer flow are based on physical conservation laws that govern fluid flow; these models are called the Navier-Stokes equations. For use in the porous media groundwater environment, these in turn may be simplified by use of the semiempirical Darcy law; see Schwartz et al. (1990), Chapter 3 for a discussion of the extent of this law's applicability. The result is a linear second-order partial differential equation for *pressure head* that involves a hydraulic conductivity "parameter" (actually a function of space as that reflects regional variation), a storage term, and a source-sink term. The latter represents natural flow into and out of the aquifer via, say, rainfall, plus the influence of pumping for direct usage plus recycling remediation, i.e., the recharge and discharge areas previously mentioned. This setup is called the *porous media* model. Given these data requirements, plus boundary and initial conditions, partial differential equations can be solved numerically to describe pressure head as a function of space and time throughout the aquifer. From head information, flow velocities can be derived, and these allow prediction of the concentration of a dissolved chemical solute at a specified place and time in an aquifer, since the chemical solution is regarded as being largely transported by the process of *advection,* i.e., flow along with the general groundwater. In addition, the solute *disperses;* the dispersion flux is sometimes viewed as depending on the current concentration gradient (Fick's law). Further models are needed, and to some extent available, for describing chemical reactions, abiotic transformation, and biological processes between many possible inorganic and organic solute types; see Schwartz et al. (1990), Chapter 4. Basic spatial and temporal change in solute concentration is also modeled by second-order partial differential equations, coefficients of which depend on the flow rates, obtained as sketched earlier, and on local dispersivities.

Despite the availability of the fundamental fluid-mechanical science on which modeling of flows in saturated media can be based, various approximations are required in practice. Hydraulic conductivity "parameters" may not be conveniently nearly constant, much less well-known, over the entire region of interest, so that in practice regions are divided into *hydrostratigraphic* units (regions of similar hydrogeological properties) within each of which nearly the same parameter values are assumed to prevail; see Anderson and Woessner (1992), Chapter 3, for practical details. It has been noted that good prediction of dispersive effects depends on accurate calculation of the spatially varying velocity field to a high degree of resolution, which is difficult because of lack of information concerning influential local variations in hydraulic conductivity. It is thus difficult to make highly accurate predictions of contaminant concentration at various locations in an aquifer, e.g., near a well location.

Contaminants typically enter the saturated regions addressed earlier through unsaturated zones, e.g., through soil, possibly from near-surface disposal, or spills, of contaminants. The type of liquid saturation in these zones is water or nonaqueous-phase liquids, plus gases. Although Darcy's law remains approximately applicable in unsaturated flow the hydraulic conductivity parameter, *K,* is now a function of the head, so the partial differential equation becomes nonlinear, increasing difficulties with numerical solutions. Several approximate analytical forms have been proposed to describe $K = K(h)$; see Schwartz et al. (1990), Chapter 3. Furthermore, unsteady and transient flows in the unsaturated zones are the rule, whereas the flow in saturated regions tends to be steady. These features further complicate model parameter specification and equation solution. Transport of contaminants in unsaturated regions also is subject to a greater number of physical and biological processes that are interactive, complex, and less well understood than are those in the saturated regions.

Still more serious complications arise when modeling the flow of water and dissolved contaminants through rocky media that is extensively fractured. Such fractures form haphazardly placed channels with apertures of varying sizes, and hence flow capacities. The hydraulic conductivities of fractured media may actually change with applied stress, so that pumping from a well in such regions can alter the effective porosity of the region, changing its flows. Various modeling options have been employed to handle the fractured-media flow problem. A first approach is to make a continuum approximation: one represents fractured-media flow as that in an equivalent porous media, defining a hydraulic conductivity parameter to be incorporated into the aforementioned partial differential equation; this is now called an equivalent porous media model. This approach is inaccurate if the region is connected but very sparsely fractured; i.e., if a small local region is being considered. A second approach assumes dual porosity, meaning that the porosities of the rock matrix and of the fractures are separately identified, as are flows within and between rock and fractures; the model is now two coupled partial differential equations and is referred to as a dual porous media model. Finally, if fracture flow predominates (rock is relatively impermeable), the flow is modeled as occurring through a network; this is the discrete fracture model. Practical uncertainties exist as to an actual fracture configuration, which add to the uncertainty of prediction of flows or water plus transported pollutants.

Various methods of ascertaining properties of underground regions are in use so that roughly appropriate choices of flow and transport models can be made. These include evidence from bore holes, the application of *kriging* (interpolation between spatially separated observations; see Cressie, 1991), use of tracer information, inverse problem solution (inference of concealed hydraulic parameters from head and flow observation), and others; see Anderson and Woessner (1992), Chapter 8, on the *calibration process,* wherein a model is adjusted to fit available data and the success of the fit is examined.

UNCERTAINTY, VARIABILITY, AND STOCHASTIC MODELS

Sizable heterogeneity and irregularity of media through which water—and contaminants—pass and are stored has prompted the development of *probabilistic* or *stochastic* (= chance = random) models to supplement the earlier deterministic versions. For example, Gelhar (1986) treats the logarithm of the hydraulic conductivity, K, as a spatially correlated Gaussian random process, which he then uses to characterize the induced head statistics, e.g., its variance. He utilizes mathematical techniques to deduce that "the large-scale transmissivity (hydraulic conductivity) of an aquifer is obtained by averaging the logarithms of local transmissivities that are measured." It is stated that the same approach can be used to evaluate an effective large-scale dispersion coefficient (the "macrodispersivity") relevant to solute transport. More recent work by Glimm and Sharp (1991) and by Zhang (1992) expands on the characterizations of heterogeneous porous media—the habitat of groundwater—as general random fields. The assumption of the log-Gaussian (or log normal) distribution for hydraulic conductivity goes back at least to Freeze (1975). It is legitimate to ask about the sensitivity of results based on this approach to its assumptions, i.e., that of Gaussian distributions, for in other areas there occur extensively tail-dispersed and otherwise non-Gaussian distributions.

Note that randomness may enter the groundwater picture not only by way of media characterization, as above, but through representation of the recharge process, which is influenced by the irregular occurrence of rain, snow, and heat. Chemical and biological processes occurring in the subsurface environment are also so affected. In summary, significant fluctuation and variability (sometimes called *structural randomness)* may well occur in media, recharge, and discharge, and in the properties of the items transported therein. Early models ignored such effects: more modern models include such realisms.

An additional component of uncertainty or randomness results from errors inherent in the measurement process for determining properties of the media and inputs and outputs at particular sites from observations; this component may be called *measurement error* or *statistical randomness.* In the literature (see Anderson and Woessner, 1992), measurement errors are also frequently characterized by a Gaussian or normal distribution: the familiar "bell-shaped curve." However, an extensive statistical literature on robustness suggests caution; see Tukey (1984, p. 614). In particular, care should be taken in the application of ordinary least-squares regression techniques: these tend to be undesirably responsive to isolated outlying observations. Such comments potentially apply to kriging, a technique for smoothing and interpolating noisy spatial observations that is increasingly applied in groundwater studies. Cressie (1991) considers *robust kriging,* a technique that down-weights isolated maverick observations that are the result of aberrant measurements.

Since both structural randomness and random measurement errors seem plausible and have been invoked, their blending in a *Bayesian approach* suggests itself (see Berger, 1985). Such an approach is briefly mentioned by Anderson and Woessner (1992), who refer to work by Freeze et al. (1990). Briefly, Bayesian theory (named for the Reverend Thomas Bayes, 1763—see Berger, 1985) represents uncertainty in a physical or biological parameter by calculating its probability distribution, the so-called *posterior distribution.* The components of this distribution are a *prior distribution,* which incorporates general information about the unknown value obtained from other situations and expert judgment, as well as a likelihood function that represents the information given by measurements on the specific situation of concern. For the mechanics, see Berger (1985). The Bayesian methodology is capable of formally incorporating information from other, similar sites into the estimation of properties of a site of current concern, as well as measurements taken at the latter. The classical statistical approach omits formal treatment of information other than that obtained at a particular site. There apparently exist computer codes for carrying out such processes; several references are given by Anderson and Woessner (1992), Chapter 8. A Bayes version of kriging exists; for an account, again see Cressie (1991).

Another approach to the uncertainty problems in the groundwater arena is that of *expert systems,* or more generally *artificial intelligence* (AI). Several publications and expert system tools such as knowledge bases and inference engines are referenced in Schwartz et al. (1990), Chapter 7. For a way into AI and uncertainty ideas, see Pearl (1988).

Remediation of contaminant content of groundwater is often attempted by pumping fresh (low or uncontaminated) water into and out of aquifers in an attempt to dilute and flush out existing contamination. The effectiveness of this methodology depends upon the nature of the contaminant; if the contaminants are multiphasic nonsoluble (immiscible) and have been trapped in pores of rocks or soils, their removal by pumping technology is extremely slow; see Travis (1992).

COMPARTMENT AND PHARMACOKINETIC MODELS

Models that envision either the physical environment or an animal or human body as a collection of interlinked but homogeneous *compartments* are called *compartment* or *pharmacokinetic models.* They behave as follows: a substance enters one or more compartments according to a specified time pattern; once in the compartment it resides there for a characteristic time before it changes form, disappears, or passes to another compartment, where the process continues; cycling back and forth between compartments may occur. The substance may in fact be a combination of more elementary substances. The agent that carries the specified substance can be blood or other bodily fluids, so the amount present is naturally expressed as a concentration. In the case of animal or human body compartments, the modeler's goal may be to predict the dosage and transformation of a medicine, drug, anesthesia, or possibly toxin, at a particular organ, such as the liver. The varying rate at which dosage is eventually administered to the target organ depends upon the rate of substance entry into the blood stream, e.g., from the lung or stomach, and thereafter on the rate of transfer within and between organs that precede the target organ in natural order, and also on the blood flow rate. The term *pharmacokinetics* is used to describe this process when the context is generally biological. Classical pharmacokinetic models are expressed in terms of systems of linear ordinary differential equations with constant coefficients that reflect the rate at which concentration changes in the various compartments. Physiologically based pharmacokinetic models (PB,PK for short) utilize physiological/biological interpretations of mechanism to specify the equation coefficients, which may not be constant in concentration, and may in fact be nonconstant. PB,PK models are, theoretically, capable of validly representing intraspecies biological responses, and to give promise of useful interspecies extrapolations. See Bischoff (1987) for more details. Quite recent work by Bois et al. (1994) has applied modern Bayesian statistics to fit a pharmacokinetic-physiological toxicokinetic model to assess fraction of tetracholoroethylene metabolized at a given dose level, taking account of individual variability plus estimation uncertainty.

Analogous models can be used to describe the flow or transfer of dissolved chemicals, pesticides, or waste material from the earth's surface through soil and rock to groundwater in aquifers. Presumably the predicted concentrations of such items in, say drinking or washing water, which may be understood as the output of an appropriate *environmental component (compartment) model,* provide inputs to a *pharmacokinetic (compartment) model* eventually describing the dosage of (some transformation of) the dissolved and transported chemical to a particular target organ within a human body or another organism. This dosage could, in turn, provide input to a *pharmacodynamic model,* e.g., a multi-hit or clonal expansion model (see Moolgavkar and Luebeck, 1990), or Portier (1989) to predict the occurrence of carcinogenetic material in an animal or human organism. A nice overview of PB,PK compartment models is given by Andersen (1987). A list of open problems awaiting solution to provide improved risk analysis tools is in Rhomberg (1988). More detail and further references occur in the Dose-Response Models section below.

PB,PK models have been tested empirically by several groups of investigators. For example, see Andersen et al. (1987) and Reitz et al. (1987), who studied methylene chloride, and Travis et al. (1990), who examined benzene. In both of these latter cases data from rats, mice, and humans were modeled with the objective of explaining or relating concentrations of items in question (methylene chloride and benzene, respectively) and their metabolites in target organs. These studies have as an objective the reconciliation of various empirical investigations on the basis of plausible biological mechanism.

The strategy currently followed in applications of PB,PK thinking is, first, to identify physiologically meaningful compartments; in the methylene chloride case four were used: liver (the primary metabolizer), fat, organs such as brain, heart, kidney, and other viscera, and muscle; in the benzene case bone marrow was added, for it and liver are the primary organs that metabolize benzene. The differential equation coefficient values are then specified from previous related experimental studies. Explicit recognition is given to nonlinear process-saturation terms such as the Michaelis-Menton expression used in modeling metabolism and also to gastric absorption rate as a function of dose level. In Travis et al. (1990) the Michaelis-Menton parameters, V_{max} and K_m were actually determined by fitting the model to existing data. After the model is fully specified, it is asked to explain data on concentrations of the dosed material or its metabolites at target organs. In both Reitz et al. (1987) and Travis et al. (1990), objectives are predictions of responses, i.e., concentration in an important organ, or other endpoints, as a function of time. In general, agreement appears qualitatively reasonable to good. However, some such published comparisons appear to depend on parameter values obtained from the data being described, so the term "prediction" is perhaps a bit generous. To test true predictive capability of a model it would be necessary to apply it to an independent data set: apparently this has been done in some cases with reasonable success.

A good modern introduction to pharmacokinetic modeling is Gillette and Jollow (1987). There the reader will find several references to validation of models; although details are not given, they can presumably be obtained from the authors.

UNCERTAINTY, VARIABILITY, AND STOCHASTIC MODELS

It is clear that the response of even quite similar organisms to controlled doses of a medicine or possible toxin will vary somewhat inexplicably, suggesting the need for models that enhance the original, deterministic, ordinary-differential-equation setups. An initial step has been to provide or adapt models to express concentrations as randomly varying around a mean. Thus the models represent the *number* of elementary particles of a foreign element, e.g., a chemical, in a compartment as fluctuating randomly, and often independently, governed by fixed (and known) transition probabilities. This phenomenon can be called *outcome variability*. Later approaches capitalized on the large numbers (or particles) involved by using continuous (diffusion) approximations (see Lehoczky and Gaver, 1977). The latter approach has the advantage of close contact with the initial deterministic models (the latter gives precisely the mean of the later diffusion approximation) and of furnishing a description of random fluctuation around that mean that is the familiar Gaussian. However, the scale of random fluctuation is on the order of the square root of the mean, which may be smaller than that observed in practice. Outcome variability can be negligible in practice.

A second, and possibly dominant, source of variability is in the parameters of the differential equation themselves: it is plausible that these vary in time within individual organisms, and, possibly more importantly, between organisms. This variability is entirely analogous to the structural variability mentioned in connection to groundwater in models in the previous section. Various researchers have incorporated the between-organism component into analyses: Sheiner and Beal (1992) have provided NONMEM (Nonlinear Mixed Effects Model), which computes an approximate joint Bayesian posterior distribution of the various parameters in a given compartment model (this is treated as log-multivariate Gauss/normal) and thence to compute the (posterior-based) estimate of, say, a concentration at a target organ; PREDPP is a package used by NONMEM for this purpose. The latter calculation is accomplished either by Monte Carlo simulation ("bootstrapping") from the above parameter posterior or by an approximate numerical calculation. Other investigators, e.g., Farrar et al. (1989), have taken the bootstrapping route as well. A further step, taken by some investigators, is to invoke Bayesian formalism to infer, predict, and *control* (in the case of a drug or anesthesia) the concentration of a substance at a target organ in a particular individual, in the face of uncertainty as to how that individual is processing the substance input, and allowing for measurement error; see Bois et al. (1994).

The above approaches all treat model coefficients as constants that are picked from a population described by a distribution but thereafter held fixed. To date it appears that little successful work has been done to characterize the effect of random temporal parameter variability on target concentration variation in an individual organ, although such a model might well be more realistic than those discussed. Models exist in which parameters change because of the presence of pharmaceuticals or toxins; see Jackson (1993) and Gaver et al. (1994).

Although a considerable amount of effort has been devoted to modeling uncertainty in pharmacokinetic compartment model parameters, very few of the results seem to have been adapted for the use of compartment modelers in the environmental sciences. See MacKay and Peterson (1991), who are concerned with modeling the environmental fate of organic chemicals (pesticides, PCBs, wood preservatives, incineration by-products, etc.) that, purposefully or inadvertently, enter the environment and proceed through soil, water and air to expose humans, plants, and animals. The effect of uncertainties in the various parameters (transport and partition) can be addressed by the resampling or bootstrapping approach of Efron and Tibshirani (1986), assuming existence of information concerning their distribution. The same is true in compartment-model studies of persistent contaminants that concentrate in food chains; see Moriarty (1984), who fits a three-compartment model for dieldrin absorption in tissue; stochastic models are mentioned in passing but are not actually applied.

HUMAN EXPOSURE MODELING AND ANALYSIS

Environmental regulation is intended to protect human public health and welfare from adverse effects of environmental pollution. Environmental remediation involves reduction of environmental pollution levels to or below some tolerable level, and the maintenance of that condition. Thus, formal regulatory demands and common sense suggest that the level of pollutant to which actual human beings are exposed when in a neighborhood of one or more polluting sources is of direct relevance. Consequences of experiencing various levels of different pollutants via various exposure routes (air, water, food, soil) must often be assessed by models: candidates are PB,PK models that predict the within-organism transfer of polluting chemicals by way of intake from external sources to organs; thereafter, one or more *dose-response,* or *pharmacodynamic* models (see next section) are invoked to convert organ-level dosage to biological responses or endpoints such as the occurrence of carcinogenic material or other disease forms.

The importance of quantifying the link between the presence of pollutants in a medium and resulting human exposure has stimulated the development and use of a number of computer models. In 1987 the EPA established the Center for Exposure Assessment Modeling in Athens, GA. According to Ambrose and Barnwell (1989), that office supplies "predictive exposure assessment techniques for aquatic, atmospheric, terrestrial, and multimedia pathways for organic chemicals and metals"; the techniques are in the form of computer modeling packages that are made available to users on diskettes. A general description of a number of the available models is contained in Ambrose and Barnwell (1989). Such models are generally descriptive of pollutant concentrations in the physical environment but do not appear to make a quantitative connection between such concentrations and the actual uptake of such pollutants by humans, or other inhabitants of the surrounding ecological system.

Efforts to quantitatively monitor individual human exposure to pollutants, e.g., within particular closed spaces or near hazardous sites, are described by Ott (1990). The approach has been called Total Human Exposure (THE). Briefly, there are two versions of THE. In the *direct approach* a probability sample of individuals is selected that is representative of a particular exposed population. With the aid of personal exposure meters carried by those sampled, supplemented by their activity diaries, the attempt is made to relate individual exposure to sources of pollution in a great many media. In addition, the body burden of various pollutants is measured. The results from the probability sample can be applied to make statistical inferences concerning population exposure or dose. The latter data can be used as input to dose-response models (see next section), so as to infer that population's risk. The direct exposure monitoring approach has been called the Total Exposure Assessment Methodology (TEAM); apparently a number of TEAM field studies have been conducted in various cities. Among the reported findings is that the number of indoor sources of toxic agents exceeds the number of outdoor sources; the dosage levels indoors can also be much higher than those outdoors owing to exposure to cleaning fluids, paints, and CO_2 in enclosed conditions such as homes and offices.

The *indirect approach* to exposure assessment is more truly *exposure modeling*. Field data serve as input to characterize pollutant concentrations in various microenvironments (locations of homogeneous pollutant concentration) and the randomly varying times spent in these by individuals (see Duan, 1991). Such models can possibly be used to augment direct exposure assessments to predict human exposure under changed conditions. For more references and details, see Ott (1990).

DOSE-RESPONSE MODELS

Suppose a dose of a single, or specified combination, of chemicals is imbibed by a human subject, or a wild or laboratory animal. It is of interest to relate the *response* of such subjects to the type and level or concentration of the dose received. This activity is called dose-response modeling, and is an important part of a quantitative risk assessment.

It is usual to assess individual response in binary terms: either a subject has reached some specified endpoint, e.g., exhibited tumors of a certain type in a target organ at time of observation (in animal cases, sacrifice), or it has not. Less frequently, more complicated responses are considered; e.g., Zhu et al. (1992), in a developmental toxicology context, record and analyze multiple responses of female rats and mice (and their offspring) exposed to the toxin, 2,4,5-T. We concentrate here on binary responses, although information may well be lost in some cases by doing so.

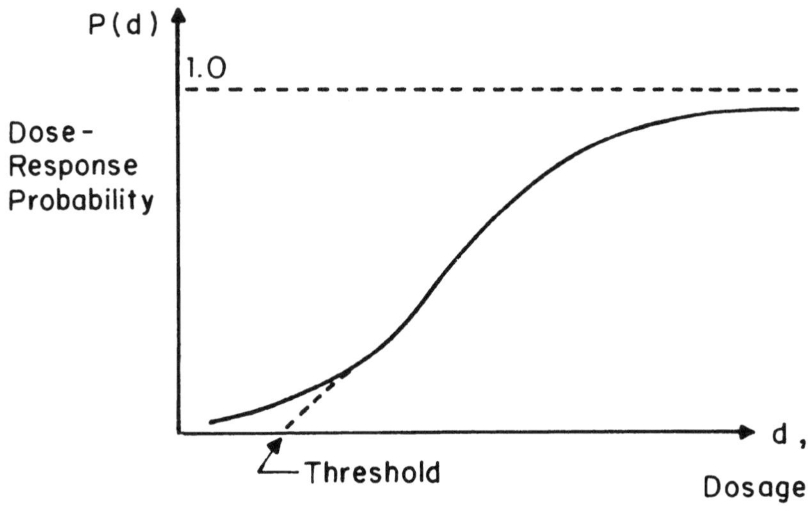

Figure 3 Dose response curve.

If individuals subjected to a specified dosage level are representative of a given population some will, and the remainder will not, exhibit a positive response. It is useful to think of the fraction that do respond as an estimate of the *probability* that a random number of the population will respond. The relationship between dosage level or concentration, *d*, and the probability of a particular response as a function of *d*, i.e., *P(d)* will typically appear as shown in Figure 3. One anticipates that the probability of a positive response, such as exhibition of one or more tumors in the liver after sacrifice, will be monotonically increasing as depicted, at least in large enough samples so that random fluctuations are small. However, reversals do occur and may have biological significance, i.e., are not individual sampling variations.

Mathematical representations of dose-response relationships are of two general types. The *simple statistical* type selects a standard statistical model such as the Gaussian/normal "bell-shaped curve," or, alternatively, the so-called Weibull distribution first introduced as a descriptor of mechanical system failure time, and converts that model into a dose-response model. A good introduction to such setups is provided by Kalbfleisch and Prentice (1980), who also discuss the fitting of such models to data, accounting for complications that occur when data are missing (test animals have died before the planned time of sacrifice), and additional data are available to adjust for differences in individual age, gender, or environmental conditions such as temperature. A particularly effective and popular model of this latter type is the so-called *Cox regression model;* see Cox (1972, 1984).

A conceptually appealing alternative and supplement to the simple statistical models are called *pharmacodynamic models*; these attempt to mathematically represent some of the detailed biological mechanisms that influence organ response. Excellent examples are the *multistage cell-proliferation clonal expansion* models described by Moolgavkar and Venzon (1979), Moolgavkar and Luebeck (1990), and Kopp-Schneider et al. (1991). The importance of cell proliferation in the cancer-development process has been noted early; a modern account is by Cohen et al. (1991), and a deterministic dynamic simulation model has been presented by Ellwein and Cohen (1992). There are many papers in this general area; see also Hart et al. (1986, p. 217). All of the above models explicitly consider the following multistage process that is currently thought to lead to cancerous tumors: first, an *initiator event* occurs randomly at single-cell level and causes permanent genetic damage. After such an event, cell division yields an increasing number of precancerous clones. The clones so generated may independently die and replicate. *Promoter events,* usually considered to be a second gene-damaging event, may lead to the initiated cells becoming a cancerous lesion or tumor, a typical dose-response endpoint.

The above types of models have been converted to dose-response models by various authors. The procedure has generally been to express initiation, proliferation and completion rates as linear functions of dosages, somehow expressed (any, but often nondecreasing *positive,* function of time-dependent concentration is presumably minimally acceptable); see Crump and Howe (1984), Murdoch and Krewski (1988), and Kodell et al. (1991), for example. All of these approaches resemble each other in that they *postulate* a simple mathematical relationship between dose or exposure and biologically based model parameters such as the aforementioned rates. While the above step is natural, it seems possible that models that more faithfully reflect the actual mechanistic interactions between potential toxic agents and cells at a molecular level could have increased credibility over wider ranges of dosage. In particular, the low-dose and across-species response is of interest in risk assessment; see next section and Portier (1989, pp. 256–259). One approach in this general direction has been reported by Freedman and Shukla (1991), wherein references to other related work are given.

An alternative class of biologically and physiologically based models are those that strive to represent the behavior of some distinguished organ, particularly when the latter is subject to toxic insult. A well-studied class concerns the liver; a good overview of the "distributed parallel tube" liver model (and its predecessors and competitors) is given by Robinson (1992). Such models view the liver as a collection of enzyme-lined tubes through which blood flows carrying a substrate to be removed by enzyme action. Natural functional heterogeneities in the liver require formal mathematical recognition: this can be accomplished by the addition of one model parameter in a partially randomized setup.

RISK ASSESSMENT

An important application for the models discussed here is to quantitative *risk assessment.* The general objective of risk assessment is to estimate the type and incidence of detrimental biological effects associated with the introduction of various levels of pollutants into the environment. Of major but not exclusive concern are the biological effects upon human beings; the term pollutants refers not only to chemical and biological substances but also to ionizing and electromagnetic radiation and other physical agents, and to any combinations thereof. The term "biological effects" has often referred to cancer, but should also include genetic and developmental defects and other detrimental outcomes including psychological and behavioral abnormalities. Of course, all such outcomes can in principle occur in various combinations and severities, depending upon the exposure and the nature and current status of the recipient of that exposure.

The *risk* attributable to a particular agent or substance can be thought of as a combination of the *hazard* or detrimental effect, i.e., toxicity, of that agent, *given* a level of exposure, and the extent and pattern of *exposure* to the agent. The task of risk assessment is to identify agents whose toxicity is a health threat to specified subpopulations, given exposure or dosage at specified levels, and to estimate the likely extent of exposure (dosage) of those subpopulations. Quantitative risk assessment then often presents its conclusions in simple numerical form. The stark form of the numerical statements typically quoted in the news media provokes concern by the general public, and skepticism by scientists who are aware of the various inferential difficulties encountered in obtaining those numbers; (see Davis, 1992; Freedman and Zeisel, 1988). Nevertheless, attempts to quantify suitably defined risks will be intensively sought in a continuing attempt to rationalize and communicate, and in particular to assess the effects of both modifying pollutant introduction into the environment, and of environmental remediation. Problems of cost-effective risk reduction and of wise resource allocation are of great interest and concern; see *The Economist* (1992) and Keunreuther (1991) for more on the economics of waste and pollution management, and the cost effectiveness of regulations as opposed to taxes on polluting technologies.

Efforts to improve the quality and credibility of risk assessment results have taken several forms. One is to clearly express the current state of scientific knowledge concerning cancer initiation mechanisms; see Hart et al. (1986), for example. The Hart committee's summary is in the form of 31 principles that address mechanisms of carcinogenesis, tests of cancer induction, epidemiology, exposure assessment, and risk assessment. The general multistage nature of the cancer development process has been recognized as the first principle derived from the mechanisms of carcinogenesis in the above review, and has been incorporated into the biologically based pharmacodynamic models described in the previous section.

Assessment of human response to low doses of possibly toxic agents, alone or in combination, is an important regulatory issue. Realistic estimates of exposures to hazardous agents experienced by human beings in real-world conditions are often comparatively low, if protracted. Furthermore, exposure to chemical toxicants in addition to normal background exposures may well be at a comparatively low level. However, available experimental results with laboratory animals are typically at a relatively high dose level, i.e., near to a maximum tolerable dose for the species or somewhat below. The problem of extrapolation from such laboratory results to the low values of risk-assessment interest poses difficult questions that have not been totally satisfactorily addressed in many cases.

For example, interest has focused on the possible existence of a *threshold* dose, a concentration below which a particular toxic agent would have zero response. The position of the Hart committee is expressed in its principle 3: "At the present state of knowledge, mechanistic considerations such as DNA repair and other biological responses, in general, do not prove the existence of, the absence of, or the location of a threshold for carcinogenesis."

Low-dose threshold phenomena must be experimentally investigated by actually submitting sets of experimental animals to a sequence of lower and lower doses, observing the numbers that respond, and interpreting the dose-response trend. Recent experimental work investigates the dose-response relationship of

N-nitrosodiethylamine and N-nitrosodimethylamine and the development of esophageal and liver cancer, and other types as well. In this study (Peto et al., 1991a, b), a large number, 4080, of Colworth/Wistar rats were given varying, but in particular low, drinking-water doses of the above agents, and the incidence of neoplasms, e.g., in the liver, was noted. Conclusion: at low doses the fraction of animals exhibiting lifetime neoplasms was nearly proportional to dose, "with no indication of any 'threshold'."

In another very recent study, Portier et al. (1992) investigated the response to dioxin, 2,3,7,8-TCDD, administered regularly (biweekly) after DEN initiation, of female Sprague-Dawley rats. The response was taken to be concentration of a dioxin-mediated protein in the liver. This concentration was predicted using two biologically plausible mathematical models, an "additive" and an "independent" version. The additive model was interpreted to fit study data better than the "independent" model, but both fit adequately, and in neither case was there strong evidence of a threshold or strongly sigmoidal nonlinear response. From a policy viewpoint this finding is interpreted to mean that safe exposure levels are lower than would be the case if a threshold were demonstrable.

The Portier et al. (1992) study assesses the *uncertainty* associated with its inferences by use of the resampling or "bootstrapping" methodology referred to earlier; see Efron and Tibshirani (1986). In Peto et al. (1991b) a formal two-parameter Weibull model is fitted but no uncertainty statements seem to be made about parameter estimates or extrapolated low-dose responses; perhaps this omission merely reflects the authors' skepticism concerning the parametric models' validity at truly low doses.

Conventional statistical attention to random sampling uncertainty, e.g., as addressed by bootstrapping, largely ignores the effect of model uncertainty. However, concern with interest and attention to this feature of much quantitative risk assessment is naturally evident. Note that an ultimate objective often is to provide credible and defensible cross-species "mouse to man" extrapolations of the effect of suspected toxins or carcinogens. It seems to be generally agreed that it is reasonable to regard "chemicals for which there is sufficient evidence of carcinogenicity in animals as if they presented a carcinogenic risk to humans," see Hart et al. (1986), principle 8. However, this statement is vague and there are difficulties with its quantitative and qualitative interpretation; see Freedman and Zeisel (1988) and accompanying (discordant) discussion. See also Crouch and Wilson (1979) and Crouch et al. (1987), who have tried to show a statistical association between mouse and rat response; the reality of that association has been questioned by D. Freedman (Stanford statistics colloquium, 1992). More recently Talcott (1992) has pointed out the many places in an environmental risk assessment that are vulnerable to uncertainties, and recommends systematic attention to these so as to limit the use of arbitrary conservative assumptions and safety factors.

In summary, quantitative risk analysis based on models used for extrapolation of animal experiment dose responses to other species, notably humans, is demonstrably still somewhat inexact. The recognition of this fact has stimulated further research on the fundamental biological phenomena, and consequently more attention to development of mathematical models faithful to that of phenomena. Such work demands, and stimulates, fruitful interplay between representatives of different scientific disciplines. Issues of international, national, and local economics stimulate the acquisition of sound information on the risks potentially associated with exposures to substances produced by our technologies, so that cost-effective choices can be knowledgeably made.

EPIDEMIOLOGIC CASE STUDIES

In this concluding section we review approaches taken to assessing potential human risk from contaminated water in two different areas of the United States: Woburn, MA, and Battle Creek, MI. The discussion points up the difficulties of such epidemiological studies. Uncertainties include imperfect knowledge concerning exposure, and dose-response relationships when doses are chronic and to several toxic compounds when also many routes of entry and responses are possible. Nevertheless, serious attempts at quantitative assessments are valuable in that they focus energy on specific issues and questions, and on the revealed deficiencies in data and theory that are candidates for improvement. Recognition of sources of uncertainty in assessments also contributes to better understanding of the value of these assessments, and stimulates efforts to reduce and quantify uncertainties.

CASE 1: THE WOBURN WELL WATER CASE

We summarize and discuss the model-based statistical analysis of a water-associated ecotoxicity situation (Lagakos et al., 1986). To summarize: in 1979 it was discovered that two out of eight water wells servicing Woburn, MA, were contaminated with various organics: trichloroethylene, tetrachloroethylene, chloroform, tricholorotrifluoroethane, and dichloroethylene. Groundwater tests under eastern Woburn, where the two implicated wells (designated *G* and *H*) were located, revealed EPA-priority pollutants. The wells were closed in 1979. Subsequent studies by the Massachusetts Department of Public Health revealed a cancer mortality rate for Woburn that was significantly higher than that for the state and the six adjacent communities. Woburn's childhood leukemia rate appeared to be elevated for the 1969–1979 period: 12 cases were diagnosed when 5.3 were expected (p-value = 0.008).

Lagakos et al. obtained data and performed statistical analyses to assess possible association between access to water from wells *G* and *H* and incidence rate of childhood leukemia. They also attempted to relate such water consumption to adverse pregnancy outcomes and childhood disorders; their logic was that the latter health effects are of shorter latency than leukemia and thus may be more sensitive indicators than is leukemia incidence.

The data that were analyzed statistically consisted first of 20 cases of childhood leukemia diagnosed in Woburn over roughly the *G* and *H* pumps' active period (1964–1983). Exposure to *G* and *H* water was scored by year and cumulatively, according to the residence history. A telephone sample survey was conducted in 1982 in order to obtain information on incidence of APO/CD during the period 1960–1982, along with mother's residence history. Care was taken in the survey, but since it managed to contact something like 50% of Woburn residents with listed phones, and since the many types of adverse responses were grouped into categories, the use of the survey data was criticized by some discussants of the paper.

Statistical analysis of the 20 leukemia cases was conducted with the aid of a statistical model, the so-called *Cox failure time regression model* (Cox, 1972). That model relates the age-dependent rate of early occurrence of leukemia to exposure. Strength of association is measured by the degree of positivity of a regression parameter, which turned out to be positive with moderate statistical significance. A logistic regression approach when applied to data on pregnancy outcomes and childhood disorders also indicated an association of some of these responses and access to *G, H* water.

Lagakos et al. carefully attempted to check for survey biases, such as could be caused by overreporting among those exposed to *G, H,* and underreporting among those unexposed; they concluded that these biases were negligible. (Note: this effort did not satisfy at least one discussant of the paper.) The effect of inexact exposure estimation was also assessed by redoing calculations based on a coarser partitioning of *G, H* exposure levels than that first used. Such a step resembles classical errors-in-variables strategies, e.g., the Wald grouping method (see Fuller, 1987). The procedure resulted in the same significant associations as previously detected.

In summary, inference concerning risk of leukemia and CD/APO was conducted using data on exposure to Woburn wells *G* and *H* and mathematical models deemed appropriate for such investigation. The investigations concluded that their analysis, while showing positive association, did not explain the entirety of Woburn's leukemia excess. Little evidence was found of increase in spontaneous abortion or low birth weight, with increased *G, H* exposure, but perinatal/stillbirth rate was up, as were (strongly) eye/ear and chromosomal/oral cleft anomalies. Other positive associations were found as well.

Six discussants commented upon the reported study. All were free in pointing out deficiencies, many of which were acknowledged by the authors. Prominent among the deficiencies noted were (apparently unavoidable) difficulties with exposure assessment and the survey data, and the possibility of overinterpretation of positive indications of association because of the "multiple testing" phenomenon; also, some doubt was cast on the accuracy of approximations used to calculate *p*-values, and to the sensitivity of the latter values to deletion/addition of single observations. All such valid comments contribute to a better understanding of the difficulties of conducting convincing studies of environmental risk; their recognition presumably will lead to improvements in future studies.

CASE 2: BATTLE CREEK HEALTH STUDY

In 1981 an aquifer servicing the Battle Creek, MI, area was found to be contaminated with various volatile organic chemicals (VOCs). The wells were subsequently closed. Groundwater contamination with the same VOCs was later detected in Dowagiac and in Springfield, adjacent to Battle Creek. The Michigan Department of Public Health proceeded to conduct a comprehensive epidemiologic study of the potential health effect of the contaminated drinking water (see Freni and Bloomer, 1988); this is called the Battle Creek Health Study.

An initial literature review indicated that adverse effects of chronic exposure to particular VOCs had been observed only at levels much higher than those discovered in the Battle Creek drinking water. However, this information did not extend to situations involving multiroute exposure to multiple VOCs. Consequently, a *retrospective cohort study* was designed: a cohort of exposed people was compared to a reference or control cohort of unexposed people with respect to incidence of diseases or other health parameters during a follow-up period. The reference cohort was from neighborhoods comparable to the contaminated areas with respect to age, size, and value of dwellings; the individuals selected for both cohorts were comparable with respect to age and sex. There were some refusals to participate (about 20% in exposed cohorts and 30% in reference cohorts), a common occurrence in such studies that is potentially biasing.

The quality of raw data on extent of exposure of the exposed cohort was "extremely poor," according to Freni and Bloomer (1988). It was necessary to construct a mathematical model to infer from the results of available well-monitoring data the time of start of residential well contamination and the total accumulated exposure of individuals; these latter estimates were supplemented by interview data, and when possible converted to inferred dosage in drinking water; estimated dosages varied considerably across individuals. As was true in the Woburn Study, dosage levels were indirect, and hence uncertain, although the Battle Creek Study devoted much thought and energy to quantifying individual exposure by drinking water but also from showering and bathing.

Health data were obtained from several sources: interviews, in which subjects were asked to recall diseases experienced in the past, and their date of diagnosis; medical records; clinical examination for subjects at least five years old; mortality and hospital discharge rates. Efforts to obtain a variety of possibly informative data seemed greater than those made in the Woburn Study.

It was judged likely that an interview bias existed: those in exposed areas, who became aware of possible exposure through the media, by word of mouth, or when interviewed, may recall more disease than others. However, analysis of the uncorrected 1976–1980 mortality data for Battle Creek is reported to have shown that the city had "significantly higher" rates for eight of the state's ten leading causes of death; these include heart disease and cancer. The magnitudes of the effects are not reported.

Statistical models were brought into play to analyze the above data. The simplest and most traditional of these is the *odds ratio,* defined in terms of these numbers: for specified risk factor, *R,* and specified disease (or indicator thereof), *D,* denote

n_{DR} = Number exposed to risk factor *R* that exhibit disease *D*;
$n_{\not{D}R}$ = Number exposed to risk factor *R* that do not exhibit disease *D;*
$n_{D\not{R}}$ = Number *not* exposed to risk factor *R* that exhibit disease *D;*
$n_{\not{D}\not{R}}$ = Number *not* exposed to risk factor *R* that do not exhibit disease *D.*

Then the odds ratio (for *D,* given *R)* is computed as

$$\mathrm{OR}(D; R) = \frac{\text{Odds for } D\text{, given } R}{\text{Odds for } D\text{, given not-}R} = \frac{(n_{\not{D}R}/n_{\not{D}R})}{(n_{D\not{R}}/n_{\not{D}\not{R}})}$$

For example: Suppose 250 individuals (= study participants) consumed water with high VOC concentration (risk factor *R);* of these 50 = n_{DR} exhibited liver disease. A reference or control group of 500 individuals consumed water with low or normal VOC concentration; of these $10 = n_{DR}$ exhibited liver disease. Then the odds ratio for liver disease, *D*, given the high-VOC risk factor, *R*, is

$$\mathrm{OR} = \frac{50/200}{10/490} = 12.3$$

an attention-getting number that strongly suggests an association of liver disease, *D*, with high-VOC risk factor, *R*. (No such levels were found in the Battle Creek Study.) A numerical value close to unity indicates neutrality so far as effect of the risk factor goes. A numerical factor less than unity suggests that absence of the specified risk factor is associated with greater disease prevalence in the sample analyzed. This effect may be caused by the action of an unsuspected additional risk factor in the reference group. The odds ratio statistic is very frequently employed in the Battle Creek Study, often adjusted for estimated exposure or dosage by stratification. In such cases the numbers in the various categories are small, so the sampling variability is large. As a result, only a few of the calculated odds ratios reach statistical significance at the (modest) level $p = 0.10$.

Another frequently utilized statistical measure of effect was the *rate* ratio or relative risk; in words,

$$\mathrm{RR}(D;\, R) = \frac{\text{Number exhibiting } D \text{ per person-month exposed to } R}{\text{Number exhibiting } D \text{ per person-month of not-}R}$$

This index controls broadly for exposure.

Statistical modeling of realistically individually variable exposure or dosage responses were conducted by *multivariable regression analysis*, particularly the *proportional hazard* or *Cox model* (see Cox, 1972), and the *logistic regression model* (Cox and Oakes, 1984). These models were also utilized in the Woburn Study; both may now be fitted to data using standard package computer programs, as may a variety of other relevant *generalized linear models* (see McCullagh and Nelder, 1983); the latter techniques were available in 1986, and their application to the Battle Creek data could be of interest.

The statistical procedures carried out on the data involved the above models when viewed as appropriate, plus others. [The simple general conclusion was that no positive or significant (below the $p = 0.1$ level) relationship between exposure to VOCs and adverse effects on health.] In fact, a weak reverse effect was noted: the data suggest an excess of abnormal response values in the reference or control cohort. This effect has not been explained.

DISCUSSION

Despite careful attempts to control the relevance and quality of the data used, the associated health effects with contamination levels in both Woburn and Battle Creek appeared small. Furthermore, they often did not reach conventional levels of statistical significance.

Such outcomes may well have several explanations. One is that, despite considerable care, the data obtainable retrospectively on individual dosages and responses were simply inadequate to allow the detection of rather weak and individual-specific effects. Small effects are difficult to detect when exposed populations are homogeneous, but the samples studied may not have been homogeneous enough in response to the "contaminant treatment" level under investigation, to reveal the latter's influence over and above that of the prevailing background. Furthermore, the more sophisticated regression models used do not take account of the fact that data on dosage and health-effect responses are surrogate, in that they represent the variation of causal variables indirectly, in broad error-afflicted summary, and they are not designed to reflect the variation of individual responses to contaminants and background.

To make further progress in quantitative epidemiologic studies, it appears necessary to obtain increasingly pertinent and inclusive dose and response data, where guidance for what data are needed will derive from improved understanding of biological response to chemical intruders. That same understanding will permit the construction of models that can better be relied upon to predict health effects in a trustworthy manner.

SUMMARY AND CONCLUSIONS

This chapter furnishes an overview of a selection of the many mathematical models used in representing the flow, transformation, and fate of toxic substances in the environment, and eventually within organisms such as the human body. The presentation is in no way encyclopedic; for example, there is no discussion of air pollution generation, transport, and human exposure, a significant omission. References are provided to repositories of models of transport and exposure. The functionality of dose-response models, particularly for cancer, is described in some literary detail, without mathematical elaboration; references will allow interested readers access to details. Such mechanistic models are not now widely available at a fundamental level for many other health effects, a deficiency which appears to offer multifold research opportunities.

The users of models must face the task of *linking* or *integration* of *models* of various stages, from pollution generation through transport, dispersion, and transformation, to eventual exposure and dosage of an organism to produce various health effects. The literature on methodology for this important linkage problem, with its attendant uncertainties, appears to be minuscule; see, however, Smith in Chapter 5 of Bloom and Poskitt (1988) who examine DNA adduct formation caused by ethylene oxide. This work emphasizes mechanistic differential equation tools at the expense of attempts to characterize interindividual and other variability by stochastic models. It provides a start on the road toward credible interspecies extrapolation.

Through the consideration of two case studies concerning health effects or risks associated with contaminated drinking water, we appreciate the difficulty of establishing a causal link between presence of an elevated contaminant level and strong evidence of corresponding health effects. The credible reconstruction of historical exposure from available data is seen to be essential, but difficult and fraught with often unquantified uncertainty. Additionally, the variable susceptibility of those exposed can only add to the difficulty of linkage. This area appears to be a prime candidate for future research. For example, data on the response to toxic chemical agents in drinking water of individuals who are suffering from various forms of disease might well reveal magnified health effects. To learn this, the appropriate data must be available and be analyzed. Eventually, after the physical transport and biological mechanisms are well enough understood, the impact of multiple results may be usefully anticipated and appropriate steps taken. More research appears necessary before this is reliably possible.

In various branches of science and technology it has been found useful to consider together, and possibly to judiciously combine, data on similar situations so as to achieve stronger quantitative estimates of effects; see Gaver et al. (1992). In medical and social science circles in the U.S. this activity is frequently called *meta analysis;* the term *overviews* is preferred by the prominent British biostatistician R. Peto. It appears likely that the techniques of this field, notably but not exclusively hierarchical Bayesian analysis, will find a place in the toolkits of those who analyze environmental transport and fate, exposure, and dose-response relations and data. The basic notion is that at least certain classes of situations, such as certain Superfund sites eligible for cleanup, may have features in common that can be invoked to strengthen individual assessments of cleanup status assessment, for example. This would involve suitably combining (*not* uncritical pooling) of data from several similar sites with that of a specific site of interest so as to improve the estimate of its condition after a particular effort has been made. Likewise, assessments of health effects of certain pollutants that have been conducted at different times or places might be profitably brought together in an overall analysis. Such actions have precedent, but do not now appear to be commonly employed. For a status report on meta analysis in various fields, see Gaver et al. (1992).

As was stated initially, those who make quantitative studies of environmental risk are essentially never able to conduct planned or designed experiments on human subjects. Consequently they are confronted with *observational data* that potentially contain biasing and confounding factors. A good account of effective statistical analyses of such difficult data is given by Cochran (1983) (edited and completed by L. Moses and F. Mosteller). This material should not be ignored by those health-effects analysts attempting to draw conclusions from environmental and observed response data. In particular, the careful use of modern nonlinear regression techniques in company with appropriately biologically motivated mechanistic models is to be encouraged in order to adjust for measured covariates such as age, gender, weight, or others while also properly modeling the biological phenomena. Recent work by Piegorsch and Casella (1992) is in this direction; they consider mouse genotoxicity data in the context of a hierarchical logistic or binomial model. See also the references in the previous three sections.

The aim of this chapter has been to describe some of the ways in which quantitative modeling and analytical techniques have been applied in the environmental areas of concern in this book. It is hoped that the result will be stimulating and helpful, especially to those new to this approach.

ACKNOWLEDGMENTS

The following individuals assisted in obtaining background for the above presentation: C. Travis, D. Hoel, S. Freni, S. Lagakos, B. Johnson, H. Gardner, S.

Beal, P. J. Robinson, R. L. Carpenter, and the other members of the Conte Institute Monograph Writing Group. (The selection of information to present and its organization, accuracy, and emphasis are the author's responsibility alone.)

REFERENCES

Ambrose, R. B. and Barnwell, T. O., Environmental software at the U.S. Environmental Protection Agency's Center for Exposure Assessment Modeling, *Environ. Software,* 4, No. 2, 1989.

Andersen, M. E., Tissue dosimetry in risk assessment, or what's the problem here anyway?, in *Pharmacokinetics in Risk Assessment, Drinking Water and Health,* Vol. 8, National Academy Press, Washington, D.C., 1987.

Andersen, M. E., Clewell III, H. J., Gargas, M. L., Smith, E. A., and Reitz, T. H., Physiologically based pharmacokinetics and the risk assessment process for methylene chloride, *Toxicol. Appl. Pharmacol.,* 87, 185, 1987.

Anderson, M. P. and Woessner, W. W., *Applied Groundwater Modeling,* Academic Press, San Diego, 1992.

Beck, M. B., Water quality modeling: A review of the analysis of uncertainty, *Water Resour. Res.,* 23, 1393, 1987.

Berger, J. O., *Statistical Decision Theory and Bayesian Analysis,* 2nd ed., Springer-Verlag, New York, 1985.

Bischoff, K. B., Physiologically based pharmacokinetic modeling, in *Pharmacokinetics in Risk Assessment, Drinking Water and Health,* Gillette, J. R. and Jollow, O. J., Eds., National Academy Press, Washington, D.C., 1987.

Bois, F. Y., Gelman, A., Jiang, J., and Maszle, D. R., A toxicokinetic analysis of tetrachloroethylene metabolism in humans, California EPA Office of Environmental Health Hazard Assessment Technical Report, August, 1994.

Bloom, A. D. and Poskitt, P. K. F., Genetic Risk Assessment, Monogr. No. 1, Environmental Health Institute, Pittsfield, MA, 1988.

Breslow, N., Biostatistics and Bayes, *Stat. Sci.,* 5, 269, 1990.

Cochran, W. G., *Planning and Analysis of Observational Studies,* Moses, L. E. and Mosteller, F., Eds., John Wiley & Sons, New York, 1983.

Cohen, S. M., Portier, D. T., and Ellwein, L. B., Pivotal role of increased cell proliferation in human carcinogenesis, *Mod. Pathol.,* 4, 371, 1991.

Cox, D. R., Regression models and life-tables (with discussion), *J. R. Stat. Soc. B,* 34, 187, 1972.

Cox, D. R. and Oakes, D., *Analysis of Survival Data,* Chapman & Hall, London, 1984.

Cressie, N. A. C., *Statistics for Spatial Data,* Wiley-Interscience, New York, 1991.

Crouch, E. and Wilson, R., Interspecies comparison of carcinogenic potency, *J. Toxicol. Environ. Health,* 5, 1095, 1979.

Crouch, E., Wilson, R., and Zeise, L., Tautology or not tautology?, *J. Toxicol. Environ. Health,* 20, 1, 1987.

Crump, K. S. and Howe, R. B., The multistage model with a time-dependent dose pattern: Applications to carcinogenic risk assessment, *Risk Anal.,* 4, 163, 1984.

Davis, B., What Price Safety? Risk Analysis Measures Need for Regulation, But It's No Science, *Wall Street Journal,* p. 1, August 6, 1992.

Doty, C. B. and Travis, C. C., The effectiveness of groundwater pumping as a restoration technology, Report ORNL/TM-11866, Oak Ridge National Laboratory, Oak Ridge, TN, 1991.

Duan, N., Stochastic microenvironment models for air pollution exposure, *J. Exposure Anal. Environ. Epidemiol.,* 1, 235, 1991.

Duan, N., Sauls, H., and Holland, D., Micro-environment versus personal monitoring: Estimation of exposure to carbon monoxide, in *Methods for Assessing Exposure of Human and Non-Human Biota,* Tardiff, R. G. and Goldstein, B., Eds., John Wiley, New York, 1991.

The Economist, Environmentalism runs riot, p. 11, August 8–14, 1992.

Efron, B. and Tibshirani, R., Boostrap methods for standard errors, confidence intervals, and other measures of statistical accuracy, *Statistical Science*, 1, 54, 1986.

Ellwein, L. B. and Cohen, S. M., Simulation modeling of carcinogenesis, *Toxicology and Applied Pharmacology*, 113, 98, 1992.

Farrar, D., Allen, B., Crump, K., and Shipp, A., Evaluation of uncertainty in input parameters to pharmacokinetic models and the resulting uncertainty in output, *Toxicol. Lett.* 49, 371, 1989.

Freedman, D. and Zeisel, H., Cancer risk assessment: From mouse to man (with discussion), *Stat. Sci.,* 3, 3, 1988.

Freedman, H. I. and Shukla, J. R., Models for the effect of toxicant in single-species and predator-prey systems, *J. Math. Biol.* 20, 15, 1991.

Freeze, R. A., A stochastic-conceptual analysis of one-dimensional groundwater flow in nonuniform homogeneous media, *Water Resour. Res.,* 11, 725, 1975.

Freeze, R. A., Massman, J., Smith, L., Sperling, T., and James, B., Hydrogeological decision analysis: 1. A framework, *Ground Water,* 28, 738, 1990.

Freni, S. C. and Bloomer, A. W., *Report on the Battle Creek Health Study,* Michigan Department of Public Health, 1988.

Fuller, W. A., *Measurement Error Models,* John Wiley & Sons, New York, 1987.

Gaver, D. P., Draper, D., Goel, P. K., Greenhouse, J. B., Hedges, L. V., Morris, C. N., and Waternany, C., *Combining Information. Statistical Issues and Opportunities for Research,* National Academy Press, Washington, D.C., 1992.

Gaver, D. P., Jacobs, P. A., and Carpenter, R. L., Stochastic models for cell signalling and toxic effects on cells. Technical report, Department of Operations Research, Naval Postgraduate School, Monterey, CA, 1994.

Gelhar, L. W., Stochastic subsurface hydrology from theory to applications, *Water Resour. Res.*, 22, 135S, 1986.

Germann, P. F., Smith, M. S., and Thomas, G. W., Kinematic wave approximation to the transport of *Escherichia coli* in the vadose zone, *Water Resour. Res.,* 23, 1281, 1987.

Gillette, J. R. and Jollow, D. J. (Chairs), *Pharmacokinetics in Risk Assessment, Drinking Water and Health,* National Academy Press, Washington, D.C., 1987.

Glimm, J. and Sharp, D. H., A random field model, *J. Stat. Phys.,* 62, 1/2, 415, 1991.

Hart, R. W., et al., Chemical carcinogens: A review of the science and its associated principles (Report of the U.S. Interagency Staff Group on Carcinogens), *Environ. Health Perspect.,* 67, 201, 1986.

Harte, J., *Consider a Spherical Cow: A Course in Environmental Problem Solving*, University Science Books, Mill Valley, CA, 1988.

Hull, L. C., Miller, J. D., and Clemo, T. M., Laboratory and simulation studies of solute transport in fracture networks, *Water Resour. Res.,* 23, 1505, 1987.

Jackson, R. C., *The Theoretical Foundations of Cancer Chemotherapy Introduced by Computer Models,* Academic Press, New York, 1993.

Jensen, K. H. and Mantoglou, A., Application of stochastic unsaturated flow theory, numerical simulations, and comparisons to field observations, *Water Resour. Res.,* 28, 269, 1992.

Kalbfleisch, J. D. and Prentice, R. L., *The Statistical Analysis of Failure Time Data,* John Wiley & Sons, New York, 1980.

Keunreuther, H., Managing hazardous waste: Past, present, and future. From a workshop on hazardous waste, *Risk Anal.,* 11, 19, 1991.

Kodell, R. L., Krewski, D., and Zielinski, J. M., Additive and multiplicative relative risk in the two-stage clonal expansion model of carcinogenesis, *Risk Anal.,* 11, 483, 1991.

Kopp-Schneider, A., Portier, C. J., and Rippman, F., The application of a multistage model that incorporates DNA damage and repair to the analysis of initiation/ promotion experiments, *Math. Biosci.,* 105, 139, 1991.

Lagakos, S. W., Wessen, B. J., and Zelen, M., An analysis of contaminated well water and health effects in Woburn, Massachusetts, *J. Am. Stat. Assoc.,* 81, 583, 1986.

Lehoczky, J. P. and Gaver, D. P., A diffusion-approximation analysis of a general n-compartment system, *Math. Biosci.,* 36, 127, 1977.

Mackay, D. and Peterson, S., Evaluating the multimedia fate of organic chemicals: A level III fugacity model, *Environ. Sci. Technol.,* 25, 427, 1991.

McCullagh, P. and Nelder, J. A., *Generalized Linear Models,* Chapman & Hall, London, 1983.

Miller, J. G., *Living Systems,* McGraw-Hill, New York, 1978.

Moolgavkar, S. H. and Luebeck, E. G., Two-event models for carcinogenesis: biological, mathematical, and statistical considerations, *Risk Anal.,* 10, 323, 1990.

Moolgavkar, S. H. and Venzon, D. J., Two event model for carcinogenesis: Incidence curves for childhood and adult cancer, *Math. Biosci.,* 47, 55, 1979.

Moriarty, F., Persistent contaminants, compartmental models and concentration along food chains, *Ecol. Bull.,* 36, 35, 1984.

Murdoch, D. J. and Krewski, D., Carcinogenic risk assessment with time-dependent exposure patterns, *Risk Anal.,* 8, 521, 1988.

Murray, J. D., *Mathematical Biology,* Springer-Verlag, New York, 1980.

Nielsen, D. R., van Genuchten, M. Th., and Biggar, J. W., Water flow and solute transport processes in the unsaturated zone, *Water Resour. Res.,* 22, 895, 1986.

NONMEM, *User's Guide,* Beal, S. L. and Sheiner, L. B., Eds., NONMEM Project Group, University of California, San Francisco, 1990.

Ott, W. R., Total Human Exposure: basic concepts, EPA field studies, and future research needs, *J. Air Waste Manage. Assoc.,* 40, 966, 1990.

Pearl, J., *Probabilistic Reasoning in Intelligent Systems,* Morgan Kaufman, San Mateo, CA, 1988.

Peto, R., Gray, R., and Grasso, P., Effects on 4080 note of chronic ingestion of N-nitrosodiethylamine or N-nitrosodimethylamine: A detailed dose-response study, *Cancer Res.,* 51, 6415, 1991a.

Peto, R., Gray, R., Brantom, P., and Grasso, P., Dose and time relationships for tumor induction in the liver and esophagus of 4080 inbred rate by chronic ingestion of N-nitrosodiethylamine or N-nitrosodimethylamine, *Cancer Res.,* 51, 6452, 1991b.

Piegorsch, W. W. and Casella, G., Hierarchical Generalized Linear Models, NIEHS technical paper, 1992.

Portier, C. J., Utilizing biologically based models to estimate carcinogenic risk, in *Scientific Issues in Quantitative Cancer Risk Assessment,* Moolgavkar, S. H., Ed., Birkhauser, Boston, 1989.

Portier, C. J., Tritschler, A., Kohn, M., Sewall, C., Clark, G., Edler, L., Hoel, D., and Lucier, G., Ligand/receptor binding for 2,3,7,8–TCDD: Implications for risk assessment, *Fundam. Appl. Toxicol.,* 20, 48, 1993.

Purdue, P., Stochastic theory of compartments: An open, two-compartment, reversible system with independent Poisson arrivals, *Bull. Math. Biology,* 37, 269, 1975.

Reitz, R. H., Nolan, R. J., and Schumann, A. M., Development of multispecies, multiroute pharmacokinetic models for methylene chloride and 1,1,1-trichloroethane (methyl chloroform), in *Pharmacokinetics in Risk Assessment, Drinking Water and Health,* Vol. 8, National Academy Press, Washington, D.C., 1987.

Rhomberg, L., *Prospects for Interagency Cooperation in Using Pharmacokinetics in Risk Assessment,* Report of the Interagency Pharmacokinetic Group (EPA, U.S. Consumer Product Safety Commission, FDA), Washington, D.C., 1988.

Robinson, P. J., Physiologically based liver modeling and risk assessment, *Risk Anal.,* 17, No. 1, 1992.

Rogers, P. P. and Fiering, M. B., Use of systems analysis in water management, *Water Resour. Res.,* 22, 146S, 1986.

Schwartz, E. W., et al., *Groundwater Models. Scientific and Regulatory Applications,* National Academy Press, Washington, D.C., 1990.

Smith, T. J., Extrapolation of laboratory findings to risks from environmental exposures: Male reproductive effects of ethylene oxide, *Genetic Risk Assessment,* Chap. 5; Monogr. No. 1, Bloom, A. D. and Poskitt, P. K. F., Eds., Environmental Health Institute, Pittsfield, MA, 1988.

Talcott, F. W., How certain is that environmental risk estimate? *Resources,* 107, 10, 1992.

Travis, C. C., Toxic waste in groundwater: Can it be removed? Research commentary, *J. NIH Res.,* 4, 49, 1992.

Travis, C. C., Quillen, J. L., and Arms, A. D., Pharmacokinetics of benzene, *Toxicol. Appl. Pharmacol.,* 102, 400, 1990.

Tukey, J. W., *The Collected Works of John W. Tukey,* Vol. 4, Jones, L. V., Ed., Wadsworth Publishing, Belmont, CA, 1984.

van der Heijde, P. K. M., El-Kadi, A. I., and Williams, S. A., *Ground-water Modeling: An Overview and Status Report,* Report GWMI88-10, International Ground Water Modeling Center, Institute for Ground-water Research and Education, Colorado School of Mines, Golden, CO, 1988.

Zhang, Q., A multi-length-scale theory of the anomalous mixing-length growth for tracer flow in heterogeneous porous media, *J. Stat. Phys.,* 66, 1/2, 485, 1992.

Zhu, Y., Krewski, D., and Ross, W. R., Multinomial dose-response models for developmental toxicity experiments, *Appl. Stat.,* 43, 583, 1992.

CHAPTER 10

Legal and International Considerations: One World

Barry L. Johnson and Deborah Weimer

CONTENTS

1-56670-141-4/96/$0.00+$.50

"No knowledge can be more satisfactory to a man than that of his own frame, its parts, their functions and actions."

Thomas Jefferson, 1814

INTRODUCTION

This advice from Thomas Jefferson refers to the importance of knowing the workings of our bodies. Because we live in an increasingly complex world and environment, Jefferson's advice on the importance of knowing how things work is relevant here to the workings of laws, regulations, and agreements that pertain to hazardous wastes. For instance, global political systems have changed remarkably in recent years due to the restructuring of many national governments and a move toward increased regional collaboration in trade (e.g., the North American Free Trade Agreement) and greater international trade competition. It is therefore important to know what these changes portend for governments' environmental and economic policies, industrial development, and societal impacts. In this kind of context, it is essential to know "how things work."

Indeed, even the definitions of environmental science, environmental health, and environmental policy have changed over time. For example, the National Environmental Health Association says: "Environmental health and protection refers to protection against environmental factors that may adversely impact human health or the ecological balances essential to long-term human health and environmental quality, whether in the natural or manmade environment. These factors include but are not limited to air, food and water contaminants, radiation, toxic chemicals, wastes, disease vectors, safety hazards and habitat alterations" (Gordon et al., 1993).

Similarly, the World Health Organization (WHO) has emphasized the importance of infectious diseases as an environmental concern (Anon., 1992). WHO has stated "The most immediate problems in the world are ill-health and premature death caused by biological agents in the human environment—in water, food, air, and soil." For other agencies, the "environment" should include such areas and considerations as ecology, social interactions, and lifestyles; along with the more traditional questions of climate change, chemical pollution of environmental media, and physical agents in the environment (e.g., noise).

At a local level, a survey of county health departments in the United States was conducted to ascertain their views on overall environmental health issues. The surveyors found the following list of concerns (in descending order of priority): groundwater contamination, illegal dumping of waste, private well contamination, recycling, hazardous materials spills, radon, illegal dumping of hazardous wastes, surface water contamination, asbestos, public water contamination, leaking underground storage tanks, and lead contamination (NACHO, 1992).

From the foregoing, it is apparent that the definition of what constitutes environmental hazards has broadened over time and very much represents the perspective of the performing agency or organization. As the nature of environmental concerns and policies has expanded, so have the nature and number of national and local laws, regulations, and international agreements that are intended to control environmental pollution and its effects on human health and the ecosystem.

This chapter provides an overview of major federal laws and international laws and agreements that impact environmental pollution, hazardous waste generation and management, and remediation of past damage to the environment from pollution and contamination. Because of the large number, complexity, and impact of the existing federal, state, and local statutes concerning hazardous materials, this chapter can at best be a guide to more authoritative works on the body of relevant law (e.g., Arbuckle et al., 1991; ELI, 1989, 1992; Miller, 1992). Moreover, the reader is cautioned to consult legal counsel and agencies with designated statutory authority in order to obtain detailed information about any specific law or regulation.

A primary aim of this chapter is to illustrate the breadth of U.S. federal statutes that pertain to hazardous wastes and toxic substances. In addition, state-based legislation that illustrates greater involvement by the public is described, as well as actions taken by the United Nations and European Community to protect the environment and minimize health and ecological impacts of environmental degradation.

The directions being taken by national, state, and international legislative bodies and councils toward controlling environmental degradation and waste generation and management will be described. In general, it will be argued that the trend is toward reducing waste generation, greater regulation of waste transportation and disposal, increasing collection and dissemination to the public of information from waste generators, and developing more international agreements on environmental protection.

Given that consequential economic and social costs come with mandated behaviors such as those expressed in law or regulation, there is always a natural tension between those who legislate and regulate and those who are affected by the laws and regulations. While arguments can be made for and against individual statutes, in part or whole, it is argued here that the increased complexity of international trade, global environmental concerns, and world political changes will make more, not less, likely the trend toward legal prescriptions to control pollution and environmental hazards.

FICTIONAL CASE STUDY: GOODEWORKS, INCORPORATED

In order to illustrate particular points about relevant U.S. federal statutes, this chapter presents the story of the creation and growth of a fictitious company. The company, GoodeWorks, Incorporated, is founded by an entrepreneur who develops a new, improved laundry detergent. We will follow the company's encounters with various U.S. federal laws that pertain to the manufacture of the product. As GoodeWorks, Inc., prospers because of its detergent, the company expands its product line and increases production. These developments have consequences in terms of testing its products for toxicity, managing wastes, interacting with the public, and international marketing.

Awaiting our entrepreneur is a considerable and formidable body of federal and state statutes concerning the creation and operation of commercial enterprises. Not the least in terms of import and difficulty are laws and regulations meant to protect the environment, the consumers of commercial products, and public health.

Commencing in the 1940s, national legislation bearing on environmental protection and reduction of health risks has been enacted in the United States. More recently, some states have enacted legislation that significantly shapes actions on hazardous waste and toxic substances in the environment. Additionally, recent international conventions have been negotiated that deal with global environmental concerns, including hazardous wastes. All of this has potential bearing on GoodeWorks.

THE IDEA AND START OF A NEW PRODUCT

As background, an inventor spent long years of experimentation to develop an environmentally benign laundry detergent that would be less costly than competing detergents. A successful detergent was made by combining some readily available industrial solvents thought to be of low toxicity and naturally occurring materials extracted from common, indigenous plants. The detergent was named SolShine by its developer. Sales of the detergent at two small, local hardware stores were encouraging. Moreover, limited product testing of the detergent found it to be a marked improvement over commercially available competitors.

During the course of developing SolShine, the developer obtained information from the local environmental protection agency that limited what could be discharged into the sewer system. Other wastes were collected and disposed of through a commercial waste management company.

Given the initial success of SolShine, the GoodeWorks company was formed and attention given to production of the detergent and the development of a marketing and distribution system. It is a family-owned business. With this direction of the company came the need to know about and adhere to applicable local, state, and federal laws and regulations on toxic substances and waste management.

Table 1 Major U.S. Federal Environmental Laws

Federal Insecticide, Fungicide, and Rodenticide Act governs the sale and use of pesticide products. 7 U.S.C. 136 *et seq.* (1947)[1991]

Clean Water Act established the sewage treatment construction grant programs, and a regulatory and enforcement program for discharges into U.S. waters. 33 U.S.C. 1251 *et seq.* (1948)[1987]

Clean Air Act requires EPA to set mobile sources limits, ambient air quality standards, emission standards, standards for new sources, air quality deterioration requirements, and focus on areas that do not attain standards. 42 U.S.C. 7401 *et seq.* (1955)[1991]

Resource Conservation and Recovery Act provides cradle-to-grave regulation for solid and hazardous waste. 42 U.S.C. 6901 *et seq.* (1965)[1984]

National Environmental Policy Act established U.S. environmental policy and requires federal agencies to assess the environmental impact of major federal actions. 42 U.S.C. 4321–4370b (1969)

Occupational Safety and Health Act requires the Department of Labor to develop and enforce workplace standards. 29 U.S.C. 651 *et seq.* (1970)

Ocean Dumping Act regulates the intentional disposing of materials into ocean waters and establishes research on effects of, and alternatives to, ocean disposal. 33 U.S.C. 1414b, 1414c (1972)[1990]

Safe Drinking Water Act established primary drinking water standards, regulates underground injection practices, and establishes a groundwater control program. 42 U.S.C. 300f *et seq.* (1974)[1986]

Toxic Substances Control Act regulates the testing of chemicals and their use in commerce. 15 U.S.C. 2601 *et seq.* (1976)[1988]

Environmental Research, Development, and Demonstration Authorization Act provides authority for EPA research programs. Generally, 42 U.S.C. 4361, 4363, 4365, 4366 (1976)[1980]

Comprehensive Environmental Response, Compensation, and Liability Act, or Superfund, establishes a fee-maintained fund to clean up abandoned hazardous waste sites. 42 U.S.C. 9601 *et seq.* (1980)[1986]

Oil Pollution Act expands oil spill prevention, preparedness, and response capabilities of the federal government and industry. 33 U.S.C. 2701–2761 (1990)

From CRS (1991) and Miller (1992).

GoodeWorks, Inc., bravely entered the world of complex environmental laws and sometimes byzantine regulations.

*Initial Legal Considerations**

Focusing on the federal statutes, several of those listed in Table 1 (and which are summarized in the Appendix) will impact how GoodeWorks, Inc., does its business. Given that this book is concerned primarily with biotechnology and hazardous waste issues, several of the federal statutes in Table 1 will be discussed with regard to their impact on GoodeWorks' waste management responsibilities and practices.

As previously noted, the company's founder formulated a new, improved detergent by combining readily available industrial solvents and several natural ingredients derived from indigenous plants. One of the first considerations to face GoodeWorks, Inc., was whether its cleaning product possesses toxic properties. To

* The reader is again cautioned that the laws and regulations that are described in this case study are subject to change by legislative bodies and interpretation by regulatory agencies.

ascertain any toxic properties of a substance or product requires testing it for toxicity. Only four statutes give the federal government authority to require chemical manufactures to test their products (Miller, 1991a). They are the Toxic Substances Control Act (new or existing industrial chemicals); the Federal Insecticide, Fungicide and Rodenticide Act (pesticides testing); the Federal Food, Drug and Cosmetic Act (drugs and food additives); and the Clean Air Act (testing of fuel additives). Because the GoodeWorks product is not a food additive, drug, medical device, pesticide, or fuel additive, only the Toxic Substances Control Act (TSCA) might apply to SolShine.

Concerned about any possible toxicity of SolShine, GoodeWorks' founder contracted with a private toxicology company to test the product for toxicity, particularly any dermatologic effects, given the intended use of the detergent. Included in this testing was a determination of the product's breakdown products, that is, the substances released into wastewater as a consequence of using the product. GoodeWorks was encouraged to find that SolShine was found nontoxic under the conditions used by the testing laboratory and that the breakdown products were innocuous. Armed with this gratifying outcome, GoodeWorks applied for a patent for its new product. In time, a patent pending status was granted by the U.S. Patent Office.

Because GoodeWorks' detergent was a new formulation of chemical substances, the company, as a matter of precaution, supplied EPA with a Premanufacture Notification under TSCA provisions, even though all of the industrial solvents in SolShine were already listed on the EPA TSCA compilation of "existing chemicals." The company determined that the substances being used in its detergent might be subject to TSCA "significant new use" coverage where "existing chemicals" were being proposed for use in a new way that might increase human or environmental exposure. The toxicological data and other data provided to EPA were provided as confidential business information, given GoodeWorks' concern for protecting the formulation of its product. EPA's review of the company's data led to a determination of no further action by EPA, thus freeing the company to commence production of SolShine. A limited production of the detergent was undertaken by the entrepreneur and family members. It was, in the truest sense, a basement and garage operation.

From some cursory market research, GoodeWorks feels that SolShine could be marketed to upscale, environmentally conscious customers. Therefore, it is decided that not only should the product and its wastes be environmentally benign, so should the manufacturing process and packaging. Ideally, GoodeWorks would like to consider everything from the recyclability, degradability and recycled content of the packaging, to the degradability of the product, sources of the production components, and the integrity of the manufacturing process. The company is comfortable with the product itself, and believes that commercial packaging material can be obtained to meet environmental protection goals set by the company.

GoodeWorks is primarily concerned with the development of a good manufacturing process and successful navigation of the relevant environmental laws and regulations. The company contracts with an environmental management consultant to help with environmental compliance and for advice on implementing environmentally sound manufacturing processes. Of course, GoodeWorks also retains an attorney.

The Investment Loan and Start-up

After 2 years of marketing the detergent to a small circle of family-run stores, GoodeWorks turns its first profit. The company's founder decides to go into major production of SolShine and to market it on a regional basis. Unfortunately, despite its generally sound financial status, investment capital could not be obtained from local lending institutions. Without capital, GoodeWorks' growth would not be possible. The company's founder decides to outreach to a local, private investor known for providing venture capital. The result is a multimillion-dollar loan, but two points above the prime rate and with the investor getting an option to buy 25% of GoodeWorks within 5 years. This arrangement turns out to be a major, fateful decision for GoodeWorks, Inc.

The Facility Issues

With the necessary development capital in hand, GoodeWorks considers its three manufacturing options to be: (1) purchase an existing facility and use it substantially as is, (2) significantly modify and retrofit the equipment in an existing facility, or (3) construct a new facility. Each possibility must be carefully analyzed for cost and ease of compliance with environmental regulations.

Existing Facility — GoodeWorks learns from its environmental consultant that less stringent standards for emissions of pollutants may apply to existing industrial facilities. However, if an old production facility is purchased, there is the possibility that the site may have been polluted by the previous owners or operators, potentially exposing GoodeWorks to cleanup liability under the Comprehensive Environmental Response, Compensation, and Liability Act (CERCLA, or Superfund) or similar state statutes. An extensive inspection prior to purchase would be required to identify problems such as prior waste disposal in the ground, in lagoons, or in old containers, and underground storage tanks.

New Facility — If GoodeWorks constructs a new plant for the manufacture of SolShine, stringent New Source Performance Standards under the Clean Water Act and the Clean Air Act may apply to the facility. However, because using the best technology possible is GoodeWorks' intention, the company does not feel this would present a problem. Given the company's long-term environmental goals, GoodeWorks decides to build a new facility and consults with architects and engineers to design a cost-effective manufacturing process with minimal environmental impact for which environmental compliance would not be problematic. It is key that environmental considerations are integrated into the design phase of the project rather than imposed onto a completed plan. In discussions, however, most of the process options lead to the development of significant amounts of waste products. The variations merely alter which environmental medium (air, water, land, or another solid waste disposal site) would be contaminated.

GoodeWorks determines it advantageous to incorporate low-technology waste reduction methods such as choosing nontoxic components at the front end, and using treatments within the manufacturing process's loop, such as bioremediation, and settling tanks to minimize the end-product wastes. Further, GoodeWorks replaces one of the original solvents, which tended to create more toxic waste products, for a less toxic, water-based solvent, thus eliminating some of the waste output. By altering the design of the production inputs and processes, GoodeWorks is able to significantly affect the resulting amount of waste generated each day.

Concerned about its relationship with the local community, GoodeWorks is aware that a large waste management company had tried to site a medical waste incinerator in the area 2 years ago. Although the proposal was readily approved by local government officials, it was met quickly with organized public opposition. Eventually, caught in extensive legal challenges, the company canceled the project. Therefore, in planning the facility, GoodeWorks contacts local officials and other public leaders and offers to enter into a "Good Neighbor" agreement. The company pledges to use clean technologies to the fullest extent possible, reduce the production of hazardous substances as much as practicable, make information available to the community as to GoodeWorks' activities, and be involved in community service projects. The community does not voice any opposition to the facility, indeed there was considerable positive sentiment that jobs were being created for the community.

Having factored waste minimization and community concerns into the design of its new production facility, the plant is constructed and placed in full operation within 12 months. GoodeWork' future, though not secured, looks promising.

Regulatory Concerns

GoodeWorks realizes that in addition to complying with the Clean Air Act and Clean Water Act with regard to emissions to environmental media covered by these acts, the company will be responsible for storage, disposal, and minimization of any hazardous waste under the Resource Conservation and Recovery Act (RCRA), the reporting requirements of the Emergency Planning and Community Right to Know Act (also known as Title III of the Superfund Amendments and Reauthorization Act of 1986), and spill reporting requirements under CERCLA, Title III, the Clean Water Act, Oil Pollution Act, Clean Air Act, and possibly TSCA (for spills of polychlorinated biphenyls [PCBs]). GoodeWorks, in anticipation of a large effort to comply with waste and toxic substance regulations, decides to hire a full-time environmental manager and staff.

Waste Production — It appears to company management that GoodeWorks will be most impacted by the Clean Water Act, both in the design of the facility and by permitting requirements for releases to water. The plant must be designed pursuant to industry-specific requirements set by EPA that reflect the Best Available Demonstrated Control Technology for that industry. Under the new design the plant will have some discharge into the sewer system that goes to a Privately Owned Treatment Works for which pretreatment standards for soap manufacturers must be met.

In addition, GoodeWorks finds that low-volume, nontoxic discharges will need to be made into a stream running through the property. This discharge will require a permit under the National Pollutant Discharge Elimination System under the Clean Water Act. Such a permit reflects various requirements and limitations imposed on the facility by the Clean Water Act. This includes emissions limits established by EPA for specific pollutants and state-established limitations that are imposed on a facility and pollutant-specific basis, depending on the quality and use of the body of water for which the pollutant is intended. Pollutants are not limited to toxic or chemical pollutants, but include conventional pollutants such as oil and fecal matter, and nonconventional, nontoxic pollutants such as ammonia and phosphorus.

In GoodeWorks' case, the stream empties into a lake that provides drinking water for the surrounding area. The limitations imposed on the GoodeWorks facility will reflect specific limits on emissions in order to preserve water quality.

The manufacturing process does not result in releases to the air of substances covered under the Clean Air Act, so that law is not applicable. However, under RCRA, it is the responsibility of the operator to determine if solid waste produced by the facility is a hazardous waste under the act and thus subject to regulation. The GoodeWorks process will create some residual waste that the company must analyze to determine if it is considered a hazardous waste under RCRA. GoodeWorks would have to obtain a generator identification number from EPA (or a state, if the state is implementing the program), if the company is found to be a generator of hazardous waste. In addition, the volume of waste production must be estimated to determine what category of generator the facility would be in. The generator category will determine which regulations apply and how much waste can be stored on site and for how long. GoodeWorks' environmental manager, after consulting with the state's Department of Environmental Protection, secures the necessary permits.

Waste Transportation — For hazardous waste that is to be transported and disposed of off-site, GoodeWorks must prepare manifests (Uniform Hazardous Waste Manifest) that follow the waste stream from transport to disposal to ensure appropriate handling along the way, and protect the generator, who could be held responsible under CERCLA for improper disposal. In addition to EPA regulations concerning the transport of hazardous waste, the U.S. Department of Transportation has regulations specifying methods of transportation, labeling, and other requirements.

The manifests also serve to promote waste minimization by requiring a statement such as the following:

> I have a program in place to reduce the volume and toxicity of waste generated to the degree I have determined to be economically practicable, and I have selected the method of treatment, storage, or disposal currently available to me that minimizes the present and future threat to human health and the environment.

GoodeWorks feels that its manufacturing and waste management processes are consistent with this declaration.

The small amount of residual waste produced by the manufacturing process, but not discharged under the Clean Water Act, is not listed by EPA as a hazardous waste and does not display any of the four characteristics of a hazardous substance: toxicity, corrosivity, ignitibility, and reactivity, or meet other criteria for hazardous waste. However, the machinery itself requires solvents for cleaning, and refrigerants that must be disposed of pursuant to RCRA. Used oil from lubrication applications is not technically considered a hazardous waste.

Community and Worker Right-to-Know — GoodeWorks also must comply with the provisions of Title III of the Superfund Amendments and Reauthorization Act of 1986, which are primarily intended to make information on hazardous substances available to the community. This involves working with state and local emergency response groups to provide information on the substances found on plant premises and participating in the emergency response planning activities. Initially, GoodeWorks must determine if it is using, storing, or disposing of substances subject to the various reporting requirements under Title III. Material Safety Data Sheets or other information must be available for substances designated as "extremely hazardous" that are on site, and made available to emergency responders. In addition, regular or accidental releases of a wider universe of hazardous substances may need to be reported to the EPA's Toxics Release Inventory (TRI). GoodeWorks realizes that one impact of the TRI is to give the surrounding community and customers a profile of the company's waste production, which could impinge on the positive corporate image being sought.

In addition to complying with community right-to-know requirements, GoodeWorks must comply with the Hazard Communication Standard (HCS) promulgated by the Occupational Safety and Health Administration (OSHA). The HCS is known as the workers' "Right-to Know" standard and its concept has been widely adopted by numerous states and some local governments. The HCS requires GoodeWorks to develop and provide its workers with information about hazardous chemicals in the workplace. GoodeWorks' environmental consultant develops a Material Safety Data Sheet (MSDS) for each hazardous chemical on the premises, trains and educates its employees about chemical hazards, and places warning labels in the workplace. GoodeWorks' founder and plant supervisors participate in all training and education programs required of other employees.

In marketing SolShine, GoodeWorks consults with Federal Trade Commission (FTC) guidelines for environmental claims to ensure that claims regarding SolShine and its packaging are appropriate. The guidelines cover such claims as environmental benefit, degradability, recyclability, and compostability of the product and the packaging and other potential environment-related claims. GoodeWorks concludes that its detergent is being marketed in accord with FTC guidelines, and national marketing of SolShine commences.

The Spill

About 20 months after full operations commenced in its new production facility, GoodeWorks is notified early one morning that two 55-gallon barrels of solvents were accidentally spilled on the property. Upon arrival, the plant manager finds that the spill is substantially contained by concrete barriers constructed for that purpose, though some of the liquid has spilled out of the containment area and flowed toward the small stream that traverses the plant property.

Potentially, a spill of a hazardous substance is subject to multiple reporting requirements. GoodeWorks' environmental manager calls the corporate lawyer. The company is advised that it must determine if the chemical released is covered by the Reportable Quantity (RQ) provisions of CERCLA. If so, and the quantity released exceeds the threshold for reporting, the National Response Center (NRC) must be notified. (Similarly, releases into the navigable waters of the United States must be reported to the NRC under the Clean Water Act.) Additional Title III reporting requirements may also apply for spills of listed substances that exceed threshold quantities and move beyond the facility premises. Under Title III, such spills must be reported to state and local emergency responders. Where multiple requirements apply, a call to one does not necessarily satisfy the others. To be conservative, GoodeWorks reports the spill to the NRC, which finds no basis for action, and to the local authorities who also see no need to respond.

However, some of the employees involved in the cleanup of the spill become concerned that the plant was not being operated properly and that they were in danger from chemical exposures. Although GoodeWorks does, in fact, have the appropriate paperwork on chemicals used in the plant, particularly Material Safety Data Sheets, the information had not, through oversight, been made available to the concerned employees. The employees do not raise the issue internally, but complain to OSHA instead.

Under authorities in the Occupational Safety and Health Act, OSHA decides to inspect the plant. GoodeWorks, trying to be cooperative, voluntarily allows entry to the inspectors, and does not request that they obtain a warrant. The OSHA investigators look for compliance with chemical specific exposure limits, proper labeling of chemicals, appropriate use of protective equipment, control procedures, monitoring of employee exposures to chemicals, appropriate medical examinations, and access to records regarding exposure. Except for minor, technical infractions, GoodeWorks passes inspection and no citations are issued. However, plant management recognizes its failure to adequately provide training to its new employees, and corrective measures are taken.

THE NEW PRODUCT!

Over the 5 years after construction of its production plant, GoodeWorks' SolShine detergent becomes increasingly popular, and sells well. Customers report that it does wonders as a pre-soak for grease and oil stains, particularly motor oil on clothes. GoodeWorks, ever the innovative entrepreneurial company, wonders if SolShine might be useful for oil cleanup in other settings and begins to test it. The product is found effective in small-scale laboratory tests for dispersing and degrading oil and petroleum products. Given these encouraging results, GoodeWorks develops and markets as a degreaser a more concentrated version of the product, which is named Sol-E-Oily.

The company also explores marketing the product on a large scale. Under the National Oil Spill and Pollution Contingency Plan, specific products are listed as approved for use on oil spills. To have a product included on the list, the manufacturer must compile specified data, such as chemical and efficacy information, and submit it to EPA. GoodeWorks arranges to have the data developed.

Another customer reported that she had spilled a bottle of Sol-E-Oily on a stack of newspapers waiting in her basement to be recycled. Two weeks later she noticed that where the detergent had been spilled, the ink, which is petroleum-based, had totally dissolved off the paper, but when she washed an ink-stained shirt with Sol-E-Oily it had not come clean. GoodeWorks experimented under many different conditions to duplicate the reported results. The company's chief chemist found that in dry, contained, sterile environments there was no effect of Sol-E-Oily on the ink. However, in one experiment the chemist had applied the detergent to damp newspaper and let it sit outside. The stack was kept damp and after 3 weeks the ink was gone. The stack was sent to a laboratory for further examination to determine which agent of the product was the miracle worker.

It turned out that the agent responsible was a common bacterium that thrived in the damp environment and might possibly have been encouraged to grow by the GoodeWorks compound. This finding led GoodeWorks to investigate if Sol-E-Oily and the bacterium could be marketed for purposes of environmental cleanup of oil spills or petroleum-based substances. GoodeWorks does not require formal permission to pursue large-scale testing of the product or the bacterium, which is naturally occurring, however, the company cannot create the conditions to test it on a large scale in the environment without violating various environmental laws.

Further, it is difficult to use the product in an actual contamination scenario without data as to its effectiveness and safety. GoodeWorks is not the first to encounter this dilemma and is referred to the EPA Superfund Innovative Technology Evaluation program for assistance. This program is designed to promote the development and implementation of new cleanup technologies for use in the Superfund program.

GoodeWorks also considers contracting with a bioengineering firm to somehow improve the bacterium, realizing that actually bringing such a product to market is difficult because of financial costs and potentially adverse reaction from the public. Such engineered products can be used in a contained environment, such as a bioreactor, without formal permission, although EPA requests voluntary notification of research and development activities using engineered organisms. Releasing such organisms to the environment for use in environmental cleanup would require approval from EPA under TSCA. No such approval has yet been requested, according to the company's attorney. GoodeWorks shelves the Sol-E-Oily bioremediation project for now.

GoodeWorks does start using Sol-E-Oily and the bacterium to treat its own oily solvent waste. However, the company's attorney advises that this might put the facility in violation of RCRA by functioning as a treatment facility without a permit. As a generator of 100–1000 kg of waste per month, the company is allowed to store and treat waste for 180 days. To fully degrade the waste using this new method takes up to 200 days. Strict requirements apply to the plant as a treatment facility that GoodeWorks is not willing to implement at this time, so the company ceases the treatment activity.

GoodeWorks discovers, however, that certain exceptions in RCRA allow for the treatment of waste within a closed loop system, or effectively treating the waste within the production process before it actually becomes waste. The company's director of engineering modifies the plant to accommodate some bioremediation treatment within the loop. This significantly reduces the amount of waste that must be handled under RCRA, saving GoodeWorks money. Given the projected cost savings for waste management and a generally strong economic future, the person who had provided investment capital 5 years earlier, exercised his option to buy 25% of GoodeWorks and become a limited partner. The operation of what was once a family-owned enterprise becomes more complex. More business consultants and attorneys representing the new partner begin to appear.

GOODEWORKS GROWS!

With its business expanding rapidly, GoodeWorks finds that additional storage space is needed for the increasing inventory of packaging material and other items, and decides to build a storage building on an unused section of the property near the stream. However, the corporate environmental manager discovers that the area in mind could be considered a wetland under the Clean Water Act, which is currently defined as;

> Wetlands inundated or saturated by surface or groundwater at a frequency and duration sufficient to support, and that under normal circumstances do support a prevalence of vegetation typically adapted for life in saturated soil conditions.

In order to proceed as planned, GoodeWorks would have to apply for a permit from the Army Corps of Engineers. The company may also risk some adverse publicity and public opinion. Instead, it is decided to purchase a nearby warehouse that is unused. The owner, a produce distributor who is a friend of GoodeWorks' founder, asserts that the building was used only to store fruits and vegetables during 10 years of ownership. GoodeWorks gets a great deal on the property and the company's owner is quite pleased.

Three months later, one of the warehouse employees reports that a black, viscous material is oozing from the ground under the north side of the warehouse. GoodeWorks, to its dismay, finds from laboratory analysis that the substance is a noxious soup of various hazardous materials that are apparently buried under the building. Further sampling and extensive inquiries uncover that the site was indeed used as a toxic waste dump in the 1960s and early 1970s. GoodeWorks' owner is mortified by this bad luck, and poor judgment in not fully investigating the property prior to acquisition.

The warehouse site is quickly nominated by the state for listing on EPA's Comprehensive Environmental Response, Compensation, and Liability Information System (CERCLIS). After investigation, 2 years later, the site scores sufficiently high on EPA's Hazard Ranking System (HRS) for placement on the EPA Superfund National Priorities List (NPL). By this time, GoodeWorks' successful SolShine and Sol-E-Oily detergents have made the company somewhat of a "deep pocket." EPA names the company as a Potentially Responsible Party (PRP), along with the former produce vendor, and does not attempt to discover the identities of the multiple parties responsible for the original dumping.

After protracted negotiations, GoodeWorks and EPA enter into a settlement agreement, whereby GoodeWorks agrees to undertake much of the work involved in cleaning up its warehouse site. The former site owner promptly declared bankruptcy after being named as a PRP. GoodeWorks then enters the legal morass of identifying other responsible parties and suing them for their contribution to the cleanup effort. GoodeWorks also sues its insurance company, which has refused to indemnify GoodeWorks for pollution-related costs. Adding to the difficult state of financial affairs, GoodeWorks' minority share owner decides to sell his holdings to an investment company, which expects to later try to assume total ownership of GoodeWorks.

GoodeWorks is closely involved in the consultants' preparation of the Remedial Investigation/ Feasibility Study for the site cleanup and urges the consideration of a bioremediation technique. A consultant finds that the waste is quite heterogeneous and that bioremediation techniques, though possible, would require a longer time to be effective than other techniques. In the Record of Decision (ROD), on-site incineration is chosen as the cleanup method. When presented by EPA to the public, there is considerable negative reaction to incineration because of health concerns about emissions, and the community undertakes legal action to force a different remediation decision. Following protracted negotiation, bioremediation technology is agreed to, rather than incineration, and the residual contaminated soil is disposed of in a licensed facility.

Legal and site cleanup costs force GoodeWorks into its first year of red ink. Given these costs, GoodeWorks decides to convert to a publicly held corporation, offering stock to the public in order to fund another plant expansion and a new warehouse.

Toxic Chemical Release Inventory

During this same time period, a local environmental group decides to prepare a report on local industry emissions of toxic substances into the environment, and it accesses through EPA all available Toxic Chemical Release Inventory (TRI) data. This includes the data for the reporting year that was GoodeWorks' first year in operation. Subsequent changes and refinements in the process allowed GoodeWorks to further reduce emissions in the following years. The first year, in addition to the routine releases reported by GoodeWorks, the solvent spill was also reported, although arguably not required. The group obtains a printout of releases from the GoodeWorks' ZIP code listing, which contains a number of industrial operations.

The group then publishes a "most wanted for crimes against the environment" poster listing the polluting industries in that region, including GoodeWorks. The company undertakes a public relations campaign and renews regular meetings with the local organization in order to enhance communications. In time, an honest exchange of information with the community pays off, and positive relations are restored. In fact, the local environmental group is able to provide GoodeWorks with some new information on solid waste recycling, which saves GoodeWorks additional money on disposal costs.

WRAP-UP OF THE GOODEWORKS CASE STUDY

As we leave GoodeWorks, the company is fending off an attempted leveraged buyout on the part of its minority stockholder. The expenses associated with its defense are high and postpone for 2 years its planned expansion of the company's products into international markets. During this period, GoodeWorks retains a consultant knowledgeable of European Union directives on hazardous substances and waste. The company has come a long way from its days of basement production of SolShine, and its future, though not assured, looks promising.

The foregoing narrative is obviously rather simplistic and not intended as a guide to the intricacies of U.S. environmental control. Rather, it serves to illustrate several salient points about the nature of U.S. environmental control. First, the approach is primarily focused on controlling end-of-the pipe waste into specific environmental media. There are exceptions, such as the the TSCA, National Environmental Policy Act (NEPA), and the Federal Insecticide, Fungicide, and Rodenticide Act (FIFRA), that operate prior to contaminants entering the environment, but media specific and general waste management statutes like RCRA and CERCLA have the principal impact on standard industrial activities in this country. Progress is, however, being made in promoting the concepts of pollution prevention and minimization, due to growing evidence of both the ecological benefits and potential cost savings as the price of disposal and cleanup of waste rises. There is also a growing recognition that traditional cleanup methodologies for improperly released wastes sometimes do little more than simply move the waste from one environmental medium to another.

Second, the laws do not always serve to promote maximum environmental control and protection. As illustrated, industrial plants already in existence when new laws and regulations go into effect are often allowed by legal exemption to operate with less stringent controls. Although this is done to prevent facility closings and the attendant economic losses, it perversely can provide disincentives for facilities to upgrade or open new plants that would come under greater regulatory control.

Third, obstacles exist for the implementation of innovative technologies in some contexts, due to technology-based controls and impediments to testing and gaining approval for the use of new technologies.

These points are quite relevant to a consideration of the legal and regulatory implications of bioremediation techniques for waste treatment and the future of these techniques. EPA is taking steps, such as the SITE program and other initiatives, to encourage and facilitate the use of new technologies like bioremediation that can be more cost-effective and complete, and where the waste is neutralized rather than simply moved. It is anticipated that as the focus shifts from the end-of-the-pipeline control to pollution prevention and minimization, the regulatory scheme will have to be adjusted to more easily accommodate new approaches. Further, in the context of the ethical considerations raised in Chapter 1, bioremediation can be used to limit current exposures to humans and the environment and at the same time not result in a legacy of waste to be dealt with by future generations.

U.S. FEDERAL LEGISLATION

FEDERAL STATUTES

The previous sections have illustrated how various U.S. federal statutes can potentially affect a company that produces materials that may impact the environment. What was not discussed in the fictitious example of GoodeWorks are various state laws and international agreements that could affect how the company goes about its business. The sections that follow give some detail about other legal and regulatory requirements that relate to hazardous wastes and biotechnology.

National governments in the industrialized countries have come to realize the importance of controlling the generation, transportation, storage, and destruction of hazardous wastes. As noted previously in Chapter 1, the public's concern for the effects of hazardous wastes on public health has been a central expression in national legislation and international treaties.

There are 12 statutes in the United States that form the major portion of the legal basis for federal programs administered by the EPA and other federal agencies (CRS, 1991; Miller, 1992). They are listed in Table 1, along with the date of initial enactment of the statute or its antecedent, the year of the most recent amendments [shown in brackets], and a brief statement of purpose for each statute. Summaries of each statute are found in the Appendix. Each federal statute has attendant regulations developed by the EPA or other regulatory agencies, but these are not described here. The reader is cautioned that the cited statutes are complex and their implementation is controlled by regulations developed by particular government agencies authorized under specific laws. Amendments to the statutes change periodically and regulations are revised; hence, caution should be exercised when attempting to comply with provisions under specific statutes. Consultation with the applicable federal, state, or local authority is advised on any matter of regulatory compliance.

U.S. FEDERAL BIOTECHNOLOGY REGULATIONS

In the brave new world of the 1990s, biotechnology* is extolled as holding the promise for breakthroughs in the fields of medicine, foods, agricultural products, energy, and environmental cleanup (Anon., 1991). These same products have also been described as a "low probability, high consequence risk"; where, "while there is only a small possibility that damage could occur, the damage that could occur is great"(SOI, 1984). Since the mid-1980s, policy-makers and the courts have been grappling with the issues of regulating these products under the existing legal/regulatory framework and also developing new requirements for bringing such products to market.**

The products of biotechnology fall within the scope of relevant laws primarily by their intended use (OSTP, 1985). The TSCA regulates the introduction of new chemical substances for commercial purposes. The EPA established in 1984 that microorganisms were considered chemical substances under the act. However, naturally occurring microorganisms were not considered "new" for purposes of the TSCA inventory and are treated as if they were listed. (Generally, substances that are not listed on the inventory are considered new and therefore subject to premanufacture notice requirements.) Intergeneric, engineered microorganisms used for commercial purposes are considered new and subject to the premanufacture notice requirements for new chemical substances.***

---

* Biotechnology in the regulatory arena is generally defined as

> Any technique that uses living organisms or substances from those organisms to make or modify a product and perform services, to improve plants or animals, or to develop microorganisms for specific uses. These techniques include the use of novel technologies such as recombinant DNA, cell fusion, and other new bioprocesses. (Anon., 1991)

** *Diamond v. Chakrobatry*, 447 U.S. 303, established that genetically engineered organisms could be patented.

*** EPA has published a Proposed Rule on Microbial Products of Biotechnology: Proposed Regulation Under the Toxic Substances Control Act (59 Fed. Regist. 45526-45585, September 1, 1994). The rule is proposed under Section 5 of TSCA, and proposes procedures to screen new microorganisms intended for commercial use based on an unreasonable risk standard. The rule also proposes notification procedures and microorganisms exempt from notification.

EPA has also asked for voluntary notification of environmental releases for research and development. Additional regulations have been proposed or considered, including a significant new use rule for pathogenic microorganisms, a rule exempting small-scale research and development, and other reporting provisions (EPA, 1989). In the interim, EPA continues to operate under the 1986 Policy statement, published in conjunction with the Federal Coordinated Framework for Regulation of Biotechnology (OSTP, 1986).

Microorganisms intended for use as pesticides are regulated by the FIFRA. Other major pollution control statutes may be relevant in two contexts. First, products of biotechnology could be considered pollutants under these statutes, and thus subject to emissions limitations, handling, and cleanup requirements. Second, as addressed in this chapter, microorganisms may be used as pollution control techniques, and must meet the various technology-based or performance-based requirements under relevant statutes.*

Releases to the environment for any use by the federal government or with federal money are potentially covered by NEPA, and may require preparation of Environmental Assessments, and Environmental Impact Statements.** Also of importance in this area are guidelines from the National Institutes of Health (NIH, 1986). These guidelines set out specific practices for constructing and handling recombinant DNA molecules and organisms and viruses containing recombinant DNA molecules, and apply to all such research carried out with federal money or at an institution that receives any NIH support for such research.

The advancement of biotechnology products has challenged the existing regulatory framework. While agencies and courts are interpreting and redefining their statutory tools, policy-makers have sought to overlay some degree of order, consistency and coordination. Through the "Coordinated Framework for Regulation of Biotechnology," the Office of Science and Technology Policy (OSTP) of the Executive Office of the President has set out relevant existing laws and regulations, stated the policies of the major federal agencies reviewing the research and products of biotechnology, and established a mechanism for interagency coordination through the creation of the Biotechnology Science Coordinating Committee (OSTP, 1986).

In 1992 the OSTP published an additional document concerning the federal government's approach to the regulation of biotechnology, which states:

> In 1986 the "Coordinated Framework" was issued to explain the proper allocation and coordination of oversight responsibilities under the several relevant statutes and among the several relevant federal agencies, The Coordinated Framework thus addressed who shall have oversight authority in each instance, but did not address how that authority should be exercised in the frequent situation in which a statute leaves the implementing agency latitude for discretion. (OSTP, 1992)

* Primarily, the Clean Air Act, Clean Water Act, Oil Pollution Act, CERCLA, and RCRA.

** In the 1985 case of *Foundation on Economic Trends v. Heckler,* several organizations and individuals sued the National Institutes of Health (NIH), which had approved the release of a biotechnology product by the University of California. The Court of Appeals upheld the lower court's injunction against the release, finding that the NIH had not adequately assessed the environmental implications through an Environmental Assessment under NEPA. 756 F.2d 143 (Ct. App. D.D.C. 1985).

This "Scope" document sets out a "risk-based" approach to the regulation of biotechnology, where: "Within the scope of authority provided by statute, federal agencies shall exercise oversight of planned introduction of biotechnology products into the environment only upon evidence that the risk posed by the introduction is unreasonable." Further, "the Federal government regulatory oversight should focus on the characteristics and risks of the biotechnology product—not the process by which it is created." This risk-based approach was strongly advocated by the President's Council on Competitiveness in its February 1991 Report on National Biotechnology Policy. This report also recommended that the administration oppose legislative efforts to create new or modify existing regulatory structures for biotechnology.

It remains to be seen how such competing interests as the development and marketing of new and useful products by a lucrative and burgeoning industry versus concern about known and unknown human health and ecological risks from the release of such products will be accommodated into a regulatory regime.

ILLUSTRATIVE STATE LEGISLATION

State and local governments have enacted various laws and regulations bearing on toxic substances in the environment. These statutes are intended to meet specific needs or to extend or fill gaps in federal statutes. It is not possible to catalog all the state laws pertinent to this chapter. However, two state laws illustrate the important trend toward making information about toxic substances more readily available to the public.

MASSACHUSETTS HAZARDOUS WASTE MANAGEMENT ACT

The Massachusetts Hazardous Waste Management Act established a comprehensive framework for protecting public health and the environment from the generation, transportation, and disposal of hazardous wastes. The law provides for the state to develop and enforce standards for the identification and management of hazardous wastes. Of particular note is the requirement that the state conduct programs of public participation on matters of policy and programmatic significance. Generators of hazardous wastes are required to maintain records and report to the state the quantity and type of hazardous wastes generated, stored, treated, or disposed of. In turn, the state is required to annually furnish information identifying the hazardous wastes within their communities to local boards of health (Massachusetts Legislature, 1979).

CALIFORNIA SAFE DRINKING WATER AND TOXIC ENFORCEMENT ACT

A second state statute, which has attracted a great amount of attention from environmentalists, the lay public, private industry, and government officials is California's Safe Drinking Water and Toxic Enforcement Act, which is more commonly known as Proposition 65 (which was the number of the state referendum that led to the Act) (Jennings, 1986). Proposition 65, according to news releases of the day, was passed as a referendum on the belief that industry needed to assume a greater role in warning the public about hazardous chemicals and to reduce discharges of toxics into the environment. The Act shifts a burden of proof to industry for knowing the health effects of carcinogens and reproductive toxics. It creates a policy that for known carcinogens and reproductive toxicants, any detectable release into the environment shall be presumed as hazardous, unless evidence can demonstrate otherwise.

There are several key provisions of Proposition 65 that are germane to this chapter. They include: (1) the requirement that the Governor of California publish a list of carcinogens and reproductive toxic chemicals, (2) the prohibition of discharges of listed chemicals that might enter sources of drinking water, (3) warnings regarding public exposure to listed chemicals, (4) increased criminal penalties for violations of the state's Hazardous Waste Law, and (5) allowing citizens to initiate suits and collect a portion of fines for violations of Proposition 65. These provisions, and others in the Act, have been controversial since the Act's passage into law. Without passing judgment on the merits of the Act, it seems appropriate to observe that placing a greater burden on industry for toxicological testing, information gathering, and dissemination to the public is a theme common to several federal and state statutes of recent years. Moreover, the notion of citizens' "right-to-know" about toxic releases into the environment is being promoted by international environmental and health bodies, as will be discussed in the next section.

INTERNATIONAL DIRECTIVES AND AGREEMENTS

UNITED NATIONS ACTIONS

The United Nations, through its environmental program (UNEP), has drafted a declaration of 27 principles for encouraging environmentally responsible development (N.Y. Times, 1992a). Two of the principles seem especially appropriate for the subject of this chapter. Principle 10 states (in part), "Environmental issues are best handled with the participation of all concerned citizens, at the relevant level. At the national level, each individual shall have appropriate access to information concerning the environment that is held by public authorities, including information on hazardous materials and activities in their communities, and the opportunity to participate in decision-making processes." This principle reinforces the importance of the Toxic Chemical Release Inventory section of the Superfund Amendments and Reauthorization Act of 1986. The move is clearly to require the reporting, collation, and dissemination to the public of information about the release of hazardous substances into the environment.

Principle 14 asserts, "States should effectively cooperate to discourage or prevent the relocation and transfer to other states of any activities and substances that cause severe environmental degradation or are found to be harmful to human health." International cooperation to effectively implement this principle will be needed to prevent waste generation and environmental degradation in developing countries as the result of industries being relocated from industrialized countries (LaDou, 1991).

WHO, which is an agency of the United Nations, has issued the *European Charter on Environment and Health* (WHO, 1989). The charter is the product of the First European Conference on Environment and Health and was developed by ministers and other senior representatives from the environment and health administrations of 29 European countries and from the Commission of the European Communities. The charter establishes a series of entitlements and responsibilities of governments and individuals, articulates principles for public policy, sets strategic elements in support of public policies, and states priorities for actions necessary to protect the environment and health (WHO, 1989).

A number of recommendations in the European Charter on Environment and Health have direct bearing on hazardous and nonhazardous wastes. One of the entitlements and responsibilities states, "Every government and public authority has the responsibility to protect the environment and to promote human health within the area under its jurisdiction, and to ensure that activities under its jurisdiction or control do not cause damage to human health in other states. Furthermore, each shares the common responsibility for safeguarding the global environment." Under principles for public policy, three principles are particularly germane to waste issues. Principle 8 states, "The entire flow of chemicals, materials, products and waste should be managed in such a way as to achieve optimal use of natural resources and to cause minimal contamination." Principle 9 states, "Governments, public authorities and private bodies should aim at both preventing and reducing adverse effects caused by potentially hazardous agents and degraded urban and rural environments." And Principle 11 states, "The principle should be applied whereby every public and private body that causes or may cause damage to the environment is made financially responsible (the polluter pays principle)." (The last-stated principle parallels the statutory philosophy evident in the U.S. federal CERCLA of 1980.) These kinds of statements from the WHO are important in that they can be adopted by countries as principles that shape national legislation.

Basel Convention

Another factor that could potentially affect waste management issues on an international scale is the Basel Convention (ASIL, 1989). Negotiated under the auspices of UNEP, the Basel Convention was completed in March 1989. Over 50 countries have currently signed the Convention, including the United States, Mexico, and Canada. The Convention entered into force 90 days after 20 signatory countries ratified it, which occurred on May 5, 1992, following ratification by Australia. Ratification signals a party's readiness to fully implement the Convention, including having the necessary legislative authorities to enforce its terms.

The Convention covers transboundary movements of hazardous waste and household waste ash, and it prohibits movement between nonparties, unless these countries have a separate agreement that ensures sound waste management. The Basel Convention specifies that government-to-government notice must be given and consent obtained before waste exports proceed. It also sets an "environmentally sound management" standard as the basis for all movements. If the exporting government has reason to believe waste will not be managed in the importing country in an environmentally sound manner, then the exporting country must not allow a shipment to proceed, even if the importing country has agreed to accept it.

On August 11, 1992, the U.S. Senate ratified the Basel Convention, thereby committing the United States to the terms of the agreement. Congress will need to pass implementing legislation before the United States officially adheres to the Convention (BNA, 1992b). The impact that the Basel Convention will have on the United States and other industrialized countries is currently unclear, in part because the amount of waste exported is uncertain. (According to EPA estimates, less than 1% of the hazardous waste generated in the United States and less that 10% of that in European countries is exported [Johnson, 1991]). It is likely that legislation in support of implementing the conditions of the Basel Convention, together with further amendments to the Resource Conservation and Recovery Act, will give EPA additional statutory authority to regulate the export of nonhazardous solid waste, to limit exports of solid waste based on the management of the exported waste in the receiving country, to limit certain imports of solid waste, and to administer some kind of registration or permit program for waste exports and imports (Johnson, 1991).

Rio Conference on the Environment

The member states of the United Nations met in Rio de Janeiro, Brazil, in 1992 to conduct the United Nations Conference on Environment and Development (UNCED). On June 14, 1992, the conference approved Agenda 21, which is a comprehensive statement of concerns, findings, and recommended actions on specific environmental areas (UNEP, 1992). Agenda 21 is a document of 580 pages that is organized into 40 chapters, commencing with a Preamble, which states "Humanity stands at a defining moment in history. We are confronted with a perpetuation of disparities between and within nations, a worsening of poverty, hunger, ill health and illiteracy, and the continuing *deterioration of the ecosystems* (emphasis added) on which we depend for our well-being." Agenda 21 is a significant agreement among the nations of the world, which reflects a global consensus and political commitment at the highest level for economic development and environmental cooperation.

Several chapters of UNCED Agenda 21 relate to the purposes of this book; however, two are particularly relevant. They are Chapter 20, "Environmentally Sound Management of Hazardous Wastes Including Prevention of Illegal International Traffic in Hazardous Wastes," and Chapter 16, "Environmentally Sound Management of Biotechnology." Chapter 20 opens by stating, "Effective control of the generation, storage, treatment, recycling and reuse, transport, recovery, and disposal of hazardous wastes is of paramount importance for proper health, environmental protection and natural resource management, and sustainable development. This will require the active cooperation and participation of the international community." Further, "Prevention of the generation of hazardous wastes and the rehabilitation of contaminated sites are the key elements, and both require knowledge, experienced people, facilities, financial resources and technical and scientific capacities." In a similar voice of concern, Chapter 16 (p. 12) contains the following statement, "Despite increasing effort to prevent waste accumulation and to promote recycling, the amount of environmental damage caused by overconsumption, the quantities of waste generated and the degree of unsustainable land use appear likely to continue growing." These statements from Agenda 21 clearly commit the United

Nations members to a cradle-to-grave approach of reducing the volume of hazardous wastes, and the need to take an ecologic perspective and sustainable development approach toward reducing the impact of hazardous wastes.

Chapter 20 (p. 3) establishes four overall targets: (1) preventing or minimizing the generation of hazardous wastes as part of an overall integrated production approach; eliminating, or reducing to a minimum, transboundary movements of hazardous wastes, consistent with the environmentally sound and efficient management of those wastes; and ensuring that environmentally sound hazardous waste management options are pursued to the maximum extent possible within the country of origin; (2) ratification of the Basel Convention on the Control of Transboundary Movements of Hazardous Wastes and Their Disposal and the expeditious elaboration of related protocols; (3) ratification and full implementation of regional agreements bearing on transboundary shipments of hazardous wastes; and 4) elimination of the export of hazardous wastes to countries that prohibit the import of such wastes. Chapter 20 details a series of recommended policies and actions that support each overall target.

Chapter 16 of Agenda 21 pertains to biotechnology. Agenda 21 defines biotechnology as "a set of enabling techniques for bringing about specific man-made changes in deoxyribonucleic acid (DNA), or genetic material, in plants, animals and microbial systems, leading to useful products and technologies" (p. 2). Specific mention is made of the use of biotechnology in program areas to increase the availability of food, feed and renewable raw materials; improve human health; and enhance protection of the environment. Within the area of enforcing protection of the environment, is the objective "to prevent, halt, and reverse environmental degradation through the appropriate use of biotechnology in conjunction with other technologies, while supporting safety procedures as an integral component of the program" (p. 12). This objective clearly indicates international support for bioremediation as a means to reduce hazardous wastes and as a method for waste site remediation.

The implementation of Agenda 21, and other agreements from the Rio conference, by the nations of the world will determine the future fate of ecological systems, environmental quality, and human health of peoples at risk of environmental hazards. However ambitious and promising these agreements such as Agenda 21 seem, little will be achieved if national governments do not commit to actions to support individual recommendations in the agreements and to efforts to collaborate across national borders.

EUROPEAN COMMUNITY REGULATIONS AND DIRECTIVES

The following sections will highlight regulations and directives developed by the European Community (EC). Because the member nations of the EC continue to develop an integrated legislative and economic structure, some background is important. In particular, it is important to distinguish between the EC and the European Union (EU). The EC was created in 1957 by the Treaty of Rome. The EU was created by the Treaty of Maastricht and came into existence as an umbrella organization in November 1993. The EU now consists of membership from 15 nations: Austria, Belgium, Denmark, Finland, France, Germany, Greece, Italy,

Luxembourg, the Netherlands, Portugal, Republic of Ireland, Spain, Sweden, and the United Kingdom. As an umbrella, the EU functions as a means for its member countries to cooperate in areas that include a single market, foreign policy, mutual recognition of national credentials, and exchange of information bases. This cooperation is in various forms, officially referred to as three pillars: the three European Communities (European Economic Community, European Community for Coal and Steel, European Atomic Energy Community). The European Economic Community became the European Community. The EC is therefore a structural member of the EU.

The EC has four principal institutions: the Commission, the Parliament, the Council of Ministers, and the Court of Justice. The organization and functions of these institutions is beyond the purposes of this chapter, but can be found elsewhere (Smith and Hunter, 1992). Over the past 20 years, the EC and some member states have generated a substantial amount of environmental legislation. Although the EC legislation on environmental protection has been substantial, its impact on member states has been limited, according to one source, though the European legal environment is changing and may make the EC legislation of greater impact in the future (Smith and Hunter, 1992).

The two most common forms of EC legislation are regulations and directives. Regulations become law throughout the EC as of their effective date, generally enforceable in each member state. Directives, in general, are not directly and generally enforceable in member states (Smith and Hunter, 1992). Directives set out goals for the member states to achieve through national legislation. The EC has adopted a substantial body of environmental legislation, covering a broad range of environmental concerns and issues. Enacted legislation includes conducting environmental impact statements of projects, air and water pollution, solid and hazardous wastes, noise pollution, radioactive waste, conservation of wild fauna and flora, urban waste treatment, freedom of information on the environment, expanded waste regulations, and the establishment of the European Environment Agency (Smith and Hunter, 1992). Pending EC legislation relevant to waste problems includes (1) a packaging waste proposal and a plastic waste proposal, (2) regulation of the shipment of waste that would require advance government consent prior to shipment, (3) restructuring of air and water pollution control frameworks to achieve integrated pollution control, and (4) civil liability proposals for environmental pollution analogous to those in CERCLA (Smith and Hunter, 1992).

In 1990 the EC created the European Environment Agency. It does not currently have any regulatory authority; its purpose is to collect and disseminate information on the environment.

EC Waste Directives

A series of EC directives pertain to waste management, hazardous substances, and biotechnology. Eight directives are specific to rules on controlling and managing wastes. Two directives are on biotechnology (ELI, 1992). The directives on wastes, and the year of EC enactment, are given in Table 2.

Each directive contains specific policy statements and administrative details, which are beyond the scope of this chapter (ELI, 1992).

Table 2 EC Directives on Wastes

Year	Coverage	Content
1975	On waste	Rules to member states to develop authorities, plans, and procedures to control and manage waste
1975	Waste oils	Rules for disposal of waste oils, giving priority to the processing of waste oils by regeneration
1976	PCBs and polychlorinated terphenyls	Rules for prohibiting PCB discharges into the environment and disposal of PCB-containing waste
1978	Titanium dioxide	Rules for the disposal of waste from the titanium dioxide industry
1978	Toxic and dangerous waste	Established EC rules on dangerous waste, listing specific substances and wastes
1982	Titanium dioxide	Established rules for surveillance and environmental monitoring of environments polluted by titanium oxide
1984	Transfrontier shipments	Established EC rules for transfrontier shipments of hazardous waste, including rules on notification, labeling, and transport requirements
1991	Hazardous waste	Established rules for the controlled management of hazardous waste, including lists of categories of hazardous wastes and specific substances considered hazardous

EC Biotechnology Directives

Two EC directives, both issued in 1990, are specific to biotechnology. One directive addresses the contained use of genetically modified microorganisms (GMOs). The other addresses their deliberate release into the environment. At the heart of the directive on the contained use of GMOs, are requirements that the user (of GMOs) carry out a prior assessment of the contained uses to determine the risks to human health and the environment, and the assessment shall take into account specific parameters set out in the directive (which include extensive characterizations of the donor, the modified microorganism, health considerations, and environmental impact). Additionally, the user of GMOs must keep records and provide them to the appropriate authorities. The EC directive on deliberate release of GMOs into the environment centers on the development of a technical dossier necessary for a competent authority (i.e., national body) to evaluate any effects of the release on human health or the environment. The information to be provided in the dossier consists of data on: (1) the notifier, (2) specifics of the GMO, (3) conditions relating to the release and the receiving environment, (4) interactions between the GMOs and the environment, and (5) monitoring, control, waste treatment, and emergency response plans (ELI, 1992).

The effects of these EC directives on waste control and management and on biotechnology are currently unclear, but as member states adopt them in part or whole, their adoption will undoubtedly lead to greater regulatory action within and between member states. In particular, the biotechnology directive will lead to testing protocols and information gathering by industry and governments in support of approvals to release GMOs into the environment.

SUMMARY AND CONCLUSIONS

This chapter has outlined key federal statutes in the United States, illustrative state laws, and international policies and directives that are intended to protect human health and the environment from the effects of pollutants and other environmental hazards. Reflection on the evolution of environmental laws and regulations permits some general observations: Over time, environmental regulations have become more comprehensive and more complex. Regulatory agencies have increasingly been given more enforcement authority by legislative bodies. The economic costs of environmental laws and regulations are enormous, especially when dealing with the remediation of environmental problems created in the past. The public's right to know information about toxic releases and environmental contamination is increasingly evident in legislation. An attitude that the polluter should pay the costs of environmental damage and associated consequences is ingrained in legislation on waste control and management. To what extent these trends in legislation become more globally pervasive is a matter yet to be decided.

APPENDIX

PROFILES OF MAJOR U.S. ENVIRONMENTAL LEGISLATION*

Toxic Substances Control Act (TSCA)

TSCA was enacted in 1976. In a public health sense, the act was intended to identify toxic (and therefore presumed deleterious to human health) substances prior to their entry into commerce, or substances already in commerce found or suspected to have toxic properties. TSCA gave EPA broad authorities to regulate toxic substances. The key provisions of the act are as follows (Miller, 1991a):

Section 5 — Premanufacture Notice (PMN) — Under Section 5 (a) a manufacturer must notify EPA 90 days before producing a new chemical substance, defined as any chemical not listed on a specially compiled inventory list under Section 8(b). Notification is also necessary for older chemicals, already on that list, if EPA concludes that there is a significant new use which increases human or environmental exposure.

Section 8(b) — Inventory List — TSCA required EPA to compile a list of those substances produced within the 3-year period preceding the promulgation of applicable regulations, which turned out to be January 1, 1975. Any substance not reported for the inventory by August 30, 1980, or subsequently added through the PMN process, must undergo premanufacture review before it may be manufactured or imported for a commercial purpose.

*Profiles are based on federal statutes as in effect December 1994.

Section 8(a) — Reporting Requirements — EPA is given authority to require information about use, production volume, and other aspects of chemicals that may be of potential concern because of possible toxicity or harm to the environment.

Section 8(d) — Health and Safety Studies — This section requires EPA to promulgate rules requiring any person who manufactures, processes, or distributes in commerce any chemical substance or mixture lists of health and safety studies and copies of the studies. EPA's rules exempted distributors and certain kinds of studies not useful to EPA. The final rule, while complicated, has at its heart the requirement that chemical manufacturers and processors must provide EPA with copies of health and safety studies pertinent to the toxicity of a substance.

Section 4(a) — EPA is authorized to require the testing of any chemicals, both old and new, if an unreasonable risk to health or the environment is suspected. Testing may also be required if a chemical will be produced in such quantities that significant human or environmental exposure could occur. Mixtures are also subject to the above rules, but only when the effects cannot be determined or predicted by testing the individual chemical substances that comprise the mixture.

Other Provisions — Coverage under TSCA, or amendments, includes controls, including restricted use or bans of PCBs, chlorofluorocarbons, and asbestos.

It is apparent that TSCA gives EPA broad authority to (a) induce testing of existing chemicals, currently in widespread commercial production or use, (b) prevent future chemical risks through premarket screening and regulatory tracking of new chemicals, (c) control unreasonable risk of chemicals, and (d) gather and disseminate information about chemical production, use, and possible adverse effects to human health and the environment (CRS, 1991). TSCA has been amended several times to include legislative directives about how to control specific toxic substances, e.g., asbestos and radon. Given that a large number of substances considered or classified by law or regulation as being hazardous, TSCA provides statutory authority to require industry to develop toxicological and other information about a substance's hazard to health and the environment.

Federal Insecticide, Fungicide, and Rodenticide Act

The Federal Insecticide, Fungicide, and Rodenticide Act (FIFRA) is the federal government's principal statute on the regulation and control of chemical pesticides. The act is a descendent of one of the earliest environmental statutes, the Insecticide Act of 1910, which was an act primarily concerned with protecting consumers from ineffective products or deceptive labeling (Miller, 1991b). FIFRA, in its current construction, is the result of several congressional attempts to increase federal regulation of pesticides. FIFRA, both in law and in regulatory actions, is a complicated statute, deriving in part from the several amendments that changed key features of antecedent versions. In its current construction, FIFRA has the following major functions:

1. Pesticide Registration — All new pesticide products used in the United States must first be registered with EPA. To register a new pesticide requires

the submission to EPA of the product's complete chemical formula, a proposed label, and a full description of the tests made of the product and the results upon which the claims are based. Manufacturers can ask for trade secret protection in order to protect information claimed to be vital to commercial propriety.

2. Control Over Pesticide Usage — EPA has authority to restrict use of pesticides. FIFRA permits the classification of pesticides into general and restricted categories, with the latter category available only to certified applicators. Certification standards are developed by EPA to regulate how certified applicators apply restricted pesticides.
3. Removal of Pesticides from the Market — FIFRA mandates EPA to take action against those pesticide products considered to pose a risk to public health and the environment. EPA's actions can take the form of a cancellation order (which is used to initiate review of the substance, during which the product can continue to be manufactured and placed in commerce), or a suspension order (which is an immediate ban on the production and distribution of a pesticide product). There are also different administrative procedures attending a cancellation order or a suspension order that would determine how quickly EPA's action would take effect (Miller, 1991b).
4. Imports and Exports — Section 17 of FIFRA provides that imports of pesticide products shall be subject to the same requirements of testing and registration as domestic products. However, FIFRA excludes U.S. exports from coverage under the Act, other than for certain record keeping provisions (Miller, 1991b).

FIFRA has a number of implications for hazardous waste generation and management, primarily through linkage to other federal statutes. The Resource Conservation and Recovery Act of 1976 gives EPA the authority to regulate the disposal of generated hazardous wastes, including the disposal of pesticides from manufacturers. The Federal Water Pollution Control Act of 1972, under Section 301, requires all industrial enterprises, including pesticide manufacturers and formulators, to apply to EPA for discharge permits if they release effluent into any body of water. The same statute, Section 307 permits pesticides to be controlled as "toxic substances," thereby leading to the development of special discharge standards. The Comprehensive Environmental Response, Compensation, and Liability Act, as amended, directs the Agency for Toxic Substances and Disease Registry (ATSDR), in consultation with EPA and the National Toxicology Program, to initiate a program of research to fill gaps in scientific knowledge for prioritized CERCLA hazardous substances. The program of research is, by statute, to be coordinated with EPA's authorities under FIFRA and TSCA, in both instances possibly leading to EPA rule-making requiring manufacturers of a particular hazardous substances to provide the research findings identified by ATSDR.

Resource Conservation and Recovery Act (RCRA)

The Resource Conservation and Recovery Act (RCRA) of 1976 is the principal federal statute bearing on the management of wastes. The Act states "The Congress hereby declares it to be the national policy of the United States that, wherever feasible, the generation of hazardous waste is to be reduced or eliminated as expeditiously as possible. Waste that is nevertheless generated should be treated, stored, or disposed of so as to minimize the present and future threat to human health and the environment." RCRA deals both with hazardous waste and nonhazardous waste, though the main emphasis in the Act is on the former. Over 500,000 companies and individuals in the United States who generate over 172 million metric tons of hazardous waste each year are covered under RCRA regulatory programs (Case, 1991).

RCRA is the Congressional prescription that wastes be treated, stored, and disposed of so as to minimize the current and future threat to human health and the environment (ELI, 1989; Case, 1991; Miller, 1992). RCRA is a regulatory statute designed to provide cradle-to-grave control of hazardous waste by imposing management requirements on generators of hazardous waste and transporters, and upon owners and operators of treatment, storage, and disposal facilities.

The statute applies principally to active facilities, whereas CERCLA, as amended, applies mainly to inactive sites. RCRA, as amended, represents a significant challenge to the regulated community. In particular, industry is challenged to find new ways to minimize, treat, and dispose of hazardous waste. The use of innovative technologies, like bioremediation, to reduce waste is the subject of active development.

Wastes covered under RCRA are defined in the statute. RCRA defines "solid waste" as being "any garbage, refuse, sludge from a waste treatment plant, water supply treatment plant, or air pollution control facility and other discarded material, including solid, liquid, semisolid, or contained gaseous material resulting from industrial, commercial, mining, and agricultural operations, and from community activities, but does not include solid or dissolved material in domestic sewage, or solid or dissolved materials in irrigation return flows or industrial discharges which are point sources subject to permits under section 402 of the Federal Water Pollution Control Act, as amended, or source, special nuclear, or byproduct material defined by the Atomic Energy Act of 1954, as amended" (ELI, 1989). RCRA therefore applies to almost any waste regardless of its physical form. EPA regulations have further clarified the definition of waste coverable under RCRA.

RCRA is a complex and comprehensive statute, implemented by EPA by a set of extensive regulations. The major provisions of RCRA are as follows:

Subchapter II — Office of Solid Waste; Authorities of the Administrator — Establishes the EPA Office of Solid Waste and the authorities of the EPA Administrator in carrying out the provisions of the Act.

Subchapter III — Hazardous Waste Management — This section of the Act is specific to the management of hazardous waste. EPA has defined solid waste, under RCRA, to include any discarded material, provided a regulatory exclusion or specific variance granted by EPA or an authorized state does not apply. Excluded from EPA's definition of solid waste are certain solid wastes: (1) household waste, (2) agricultural wastes which are returned to the ground as fertilizer, (3) mining overburden returned to the mine site, (4) utility wastes from coal combustion, (5) oil and natural gas exploration drilling waste, (6) wastes from extraction, beneficiation, and processing of ores and minerals, including coal; (7) cement kiln dust wastes, (8) arsenical-treated wood wastes generated by end users of such wood; and (9) certain chromium-bearing wastes (Case, 1991). EPA has also provided some additional limited exemption under narrowly defined circumstances of use of the generated waste.

Subchapter III requires EPA to compile listings of hazardous wastes and to develop standards applicable to generators of hazardous waste; transporters of hazardous waste; and owners and operators of hazardous waste treatment, storage, and disposal facilities. Permits for treatment, storage, and disposal of hazardous waste are required. EPA is given authority to inspect hazardous waste facilities coverable under RCRA and is given enforcement powers to ensure compliance with federal RCRA requirements. Each state is required to submit to EPA a continuing inventory that describes the location of each site at which hazardous waste has at any time been stored or disposed of. Similarly, each federal agency must provide the same kind of information to EPA. Subchapter III also sets forth requirements on the export of hazardous waste to other countries. In general, the exporter must furnish EPA information about the nature and amount of the waste, the country of destination, the ports of entry, the manner of transport, and the name and address of the ultimate treatment, storage, or disposal facility. Following receipt of the export information, EPA must request the Secretary of State to contact the receiving country to obtain that country's willingness to receive the exported hazardous waste.

Among the most significant provisions of the 1984 amendments to RCRA are the prohibitions on disposal of hazardous wastes on land (Case, 1991). In a phased approach, the Act bans the disposal in landfills of bulk or non-containerized liquid hazardous wastes, and hazardous wastes containing free liquids. The Act then required EPA to determine whether to ban, in whole or in part, the disposal of all RCRA hazardous wastes in land disposal facilities. At the same time, EPA must establish treatment standards for each restricted waste based on the Best Demonstrated Available Technologies (BDAT). If the restricted waste is first treated to BDAT levels, the treated waste or residue can then be placed in land disposal facilities.

Subchapter IV — State or Regional Solid Waste Plans — Regulation of nonhazardous waste, under RCRA, is the responsibility of the states. The federal involvement is limited to establishing minimum criteria that prescribe the best practicable controls and monitoring requirements for solid waste facilities. Disposal of solid waste in open dumps is prohibited, but RCRA provides EPA with no enforcement authority for banning open dumps. (EPA's enforcement authority under RCRA only covers hazardous waste.)

Subchapter VI — Federal Responsibilities — Federal statutes sometimes exempt federal agencies from an act's coverage. This may be for reasons of national security, economic factors, or political reasons. RCRA holds each department, agency, and instrumentality of the executive, legislative, and judicial branches of federal government having jurisdiction over any solid waste management facility or disposal site, or engaged in any activity resulting, or which may result, in the disposal or management of solid waste or hazardous waste to the same expectations and requirements "as any person is subject to such requirements." Only the President can exempt a department's solid waste management facility if it is in the nation's paramount interest.

Subchapter VIII — Research, Development, Demonstration, and Information — EPA is given authority to conduct research and studies on a range of areas that include adverse health and welfare effects of solid waste releases; resource conservation systems; production of usable forms of recovered resources; identification of solid waste components; improvements in land disposal practices; methods for disposal of, or recovery of resources;

methods of hazardous waste management; and adverse effects on air quality. EPA is also directed to develop, collect, evaluate, and coordinate information on subjects pertaining to solid waste.

Subchapter IX — Regulation of Underground Storage Tanks — EPA is directed to develop a comprehensive regulatory program for "underground storage tanks" (Bosco and Randle, 1991). RCRA directs EPA to promulgate release detection, prevention, and correction regulation applicable to all owners and operators of underground storage tanks (UST), as may be necessary to protect human health and the environment. EPA estimates there are 700,000 UST facilities with about 2,000,000 tanks covered by this regulation.

Subchapter X — Demonstration Medical Waste Tracking Program — A series of reports of medical waste being washed up on Atlantic Coast beaches in the eastern United States led to passage of the Medical Waste Tracking Act of 1988. Congress added Subchapter X to RCRA to set up a demonstration program for tracking the shipment and disposal of medical wastes in a selected number of states. The states covered by this subchapter are New Jersey, New York, Connecticut, states contiguous to the Great Lakes, and any state that petitions EPA to be a party to the demonstration project, which is currently ongoing.

Comprehensive Environmental Response, Compensation, and Liability Act

The Comprehensive Environmental Response, Compensation, and Liability Act (CERCLA, or Superfund), was enacted in 1980 and reauthorized by the U.S. Congress in 1986 (Stoll, 1991). The intent of the law is "To provide for liability, compensation, and emergency response for hazardous substances released into the environment and the cleanup of inactive hazardous waste disposal sites." CERCLA's basic purposes are to provide funding and enforcement authority for cleaning up waste sites created in the past, and for responding to hazardous substance spills. The statute includes provisions for remediating waste sites, responding to public health concerns, enforcement authorities to identify "potential responsible parties," and emergency removal of chemical spills. The Superfund trust fund is funded by an "environmental tax" on select industries, primarily the petrochemical industry in 1980, but since 1986 more broadly based.

CERCLA was enacted in 1980 following the discovery of releases of hazardous substances from abandoned landfills into community residences, in particular, the community of Love Canal, New York, which was evacuated following the discovery that it overlay an abandoned chemical dump. CERCLA's principal purposes are to provide funding and enforcement authority to EPA for cleaning up (i.e., remediating) the very large number of abandoned waste sites in the United States. The Act's central philosophy is to require the parties, called the Potentially Responsible Parties (PRPs), to bear the costs of cleanup of sites to which the parties have contributed wastes and for costs attending environmental and health problems created by releases of substances from the waste sites. In addition, CERCLA gives EPA broad legal authority to identity PRPs and require them to pay for the costs of a site's remediation and other actions related to the effects of hazardous substances released from the site. PRPs include the past and current owners and/

or operators of a site, those who arranged for the transportation of hazardous substances to the site, and those who arranged for the treatment or disposal of the substances, and they are subject to retroactive liability. This is established under the legal concept of "retroactive joint and several liability" for parties whose wastes had contributed to environmental degradation at the waste sites. The concept means that a company or other accountable entity which long ago had disposed of wastes could now be held responsible for all of a site's remediation costs unless other responsible parties could be identified and costs shared. Needless to say, retroactive joint and several liability has often led to litigation as to who pays what amount of a site's remedial costs.

Under CERCLA, EPA has identified approximately 40,000 waste sites that potentially require remediation. The annual cost of CERCLA programs is currently approximately $1.6 billion. EPA has prioritized the large number of CERCLA sites into a ranked number, called the National Priorities List (NPL), which currently numbers about 1300 sites. NPL sites are eligible for site remediation funds out of the trust fund, called the Hazardous Substances Superfund, which was established by CERCLA. The current average cost of remediating an NPL site is approximately $30 million; though the remediation of large federal agency sites, e.g., weapons facilities, will be much more costly. The 1990 Five-Year Plan of the Department of Energy calls for expenditures totaling more than $30 billion on environmental restoration and waste management activities (OTA, 1991, p. 5).

CERCLA authorizes EPA to pay for site cleanups out of the Superfund trust fund and, where possible, out of the liability scheme of the Act, i.e., from costs recovered from the PRPs. Responsible parties' liabilities cover not only the actual costs of remediation, but also the site investigation, feasibility study, design costs, and cost of health studies. The types of parties who may be liable for site-associated costs are specified by CERCLA to be: (1) current and past "owners or operators" of the site; (2) parties who transported wastes to the site; and (3) parties (usually referred to as "generators") who arranged for wastes to be disposed or treated, either directly with an owner/operator or indirectly with a transporter (Stoll, 1991). It is common to have multiple Potentially Responsible Parties associated with a particular CERCLA site.

In addition to the identification, ranking, remediation, and cost recovery provisions of CERCLA that pertain to uncontrolled releases of hazardous substances from sites, CERCLA contains other provisions of note as pertains to hazardous waste:

- "Removal" actions are effected by EPA in instances where a short-term, limited response to a manageable environmental release is indicated (e.g., a spill of hazardous substances from a transportation mishap), rather than a long-term "remedy (i.e., remediation)" is indicated, e.g., for an NPL site.
- CERCLA of 1980 created the Agency for Toxic Substances and Disease Registry (ATSDR) within the U.S. Public Health Service for the purpose of examining public health implications of hazardous substances in the community environment, with emphasis on those released from CERCLA sites. ATSDR conducts public health assessments of all NPL sites, develops toxicological profiles on priority hazardous substances, conducts epidemiologic and other applied research, provides medical education for physicians and other health professionals, responds to emergency releases of

hazardous substances (e.g., through transportation spills), and maintains a national registry of persons exposed to select hazardous substances (Johnson, 1990).

The Emergency Planning and Community Right-to-Know Act of 1986 was enacted as a freestanding provision of the Superfund Amendments and Reauthorization Act of 1986 (Arbuckle et al., 1991a). This Title III provision of CERCLA, as amended, is a significant statement from Congress concerning the obligations of government and private industry to protect the public from releases of hazardous substances. Under Title III, state and local governments are required to develop emergency plans for responding to unanticipated environmental releases of a number of acutely toxic materials (Arbuckle et al., 1991a). (Title III requires different kinds of notifications according to groups of chemical substances.) Additionally, businesses covered by Title III must notify state and local emergency planning entities of the presence and amounts in inventory of hazardous materials on their premises and to notify federal, state, and local authorities of planned and unplanned environmental releases of those substances. Regulatory agencies, in turn, are required to make available to the public the data on releases of substances in the environment. EPA has developed a reporting system for businesses to use when providing information for placement in the Toxics Release Inventory (TRI). Through administrative arrangement with the National Library of Medicine (NLM), National Institutes of Health, the TRI database can be accessed through the NLM. Citizen and environmental groups have used the TRI data to bring attention to the amounts of hazardous substances released by industry within geographic areas of concern. The impact has been to draw attention to alleged excess amounts of pollution, thereby bringing public pressure to reduce releases.

Clean Air Act

The federal Clean Air Act (CAA) was enacted by Congress in 1970, heavily amended in 1977, and again substantively amended in 1990. The Act, as amended, is a comprehensive, complex statute that controls air pollution emissions and regulates government, business, and community lifestyles that affect the releases of air toxics into the environment. The 1970 statute federalized air pollution control regulation, made public health protection the basis for much of that regulation, required automotive and other industries to meet emission standards, and created an extensive regulatory system to effect the Act's goals. The 1977 amendments added special provisions for geographic areas with air cleaner than national standards to prevent their deterioration in air quality, and special provisions were added pertaining to nonattainment areas, i.e., geographic areas that had failed to meet national air quality standards (Randle and Bosco, 1991). The 1990 amendments substantively revised the earlier version of the CAA. These amendments, which are still being developed into federal regulations, added comprehensive provisions to regulate emissions of air toxics, acid rain, and substances thought to be a threat to the ozone layer. In addition, the 1990 amendments added an elaborate permit program and markedly strengthened enforcement provisions and requirements for nonattainment areas, mobile source emissions, and automobile fuels (Randle and Bosco, 1991).

The Clean Air Act adopted the strategy of developing National Ambient Air Quality Standards for individual air toxics, then places most of the responsibility on the states to achieve compliance with the standards. (EPA has established air quality standards for relatively few pollutants—ozone, carbon dioxide, particulates, and lead, given limitations in scientific data to support the development of standards.) The development of individual State Implementation Plans (SIPs) begins by dividing each state into geographic air quality control regions, and ascertaining whether and how much air pollution in each region exceeds the limits allowed by the air quality standards. Control requirements are then imposed to reduce emissions from the various sources. Under the 1977 amendments to the CAA, states were required to develop SIPs that would meet the air quality standards by 1982, except for ozone, for which the deadline was 1987 (Quarles and Lewis, 1991). The CAA, and its 1977 amendments, placed heavy emphasis on reducing emissions from automobiles and other vehicles (i.e., mobile sources of air pollutants).

The 1990 amendments to the CAA contain several new provisions concerning standards for air pollutants, enforcement, requirements of states, and other actions. According to one source (Spensley, 1992), the major changes are as follows:

Title I — Attainment and Maintenance of National Ambient Air Quality Standards — Requirements for SIPs were expanded. The new requirements address the three criteria pollutants that have been the primary cause of urban air pollution problems: ozone, carbon monoxide, and particulate matter. The new requirements for SIPs range from improved emission inventories to mandatory transportation control measures and specified reduction in ozone levels. Reclassification of nonattainment air quality control regions for these three pollutants is based upon the severity of the pollution. Nonattainment regions that fail to meet statutory deadlines are subject to increased sanctions.

Title II — Mobile Sources — Three new requirements were added pertaining to reduction of emissions from mobile sources. More stringent emission standards for tailpipe emissions of hydrocarbons, carbon monoxide, and nitrogen oxides. The stricter standards are to be phased in beginning with vehicles produced for car model year 1994. Second, the control of emissions from gasoline evaporation during refueling of vehicles is required. Third, cleaner (reformulated) gasoline will be required for use in cities with the worst ozone problems, starting in 1995.

Title III — Hazardous Air Pollutants — A new program is created for regulating hazardous air pollutants and a significantly expanded list of 189 new hazardous air pollutants. The new amendments replace an older EPA risk-based approach to setting standards for hazardous air pollutants with a technology-based control approach.

Title IV — Acid Deposition Control — In order to reduce acid rain, reductions in sulfur dioxide and nitrogen oxide are mandated. The goal of Title IV is to reduce sulfur oxides emissions from coal-fired electric utilities by 50% by the year 2000.

Title V — Permits — A new federal permit program to be administered by the states will require operating permits for the most significant sources of regulated emissions. Prior to the 1990 amendments, individual federal permits were required only for new or modified sources, and existing sources were governed only by general requirements embodied in SIPs (Spensley, 1992). EPA has issued a final rule setting up the permitting program; approximately 34,000 major sources of air toxics releases are estimated to require permits (BNA, 1992a).

Title VI — Stratospheric Ozone Protection — Production phaseout of chemicals harmful to the Earth's ozone barrier is mandated. EPA has published an initial list of two classes of substances known to harm the ozone layer: Class I substances that are primary causes of ozone depletion (all fully halogenated chlorofluorocarbons, halons, carbon tetrachloride, and methyl chloride) and Class II substances that are a threat to the ozone layer (hydrochlorofluorocarbons).

Title VII — Enforcement — EPA is given significant new enforcement authorities, including an increase of civil and criminal penalties for violations of the Act.

It is clear that the 1990 amendments to the CAA will have a significant impact on the American public, though the extent of the impact is yet to be evident as regulations to implement the amendments are yet to be released. The economic impact will be considerable, though the approximate cost is not agreed on. Industry has estimated the cost of the new amendments to the CAA will cost between $25 to $100 billion annually. Estimates from the President's Council of Economic Advisors are closer to $25 billion annually, beginning in the year 2005 (Spensley, 1992).

Federal Water Pollution Control Act

The principal federal law governing pollution of the Nation's waterways is the Federal Water Pollution Control Act, more commonly called the Clean Water Act (CWA). Although originally enacted in 1948, the current Act was completely revised by amendments in 1972, giving the CWA most of its current shape. The CWA regulates pollutants discharged to surface waters through "point sources" (i.e., pipes, ditches, and other discrete conveyances of pollutants). Pollution sources are required by the CWA to treat their wastes to meet the most stringent of two sets of requirements, based on either technologic feasibility or attainment of desired levels of water quality (Miller, 1992). The 1987 amendments to the CWA authorized measures to address nonpoint source pollution (e.g., stormwater runoff from agricultural lands, forests, and urban areas) by directing states to develop and implement nonpoint pollution management programs.

The Act states, "The objective of this Act is to restore and maintain the chemical, physical, and biological integrity of the Nation's waters. In order to achieve this objective it is hereby declared that, consistent with the provisions of the Act, (1) it is the national goal that the discharge of pollutants into the navigable waters be eliminated by 1985, (2) it is the national goal that wherever attainable, an interim goal of water quality which provides for the protection and propagation of fish, shellfish, and wildlife and provides for recreation in and on the water be achieved by July 1, 1983, (3) it is the national policy that the discharge of toxic pollutants in toxic amounts be prohibited, (4) it is the national policy that Federal financial assistance be provided to construct publicly owned waste treatment works, (5) it is the national policy that aerate waste treatment management planning processes be developed and implemented to assure adequate control of sources of pollutants in each State, (6) it is the national policy that a major research and demonstration effort be made to develop technology necessary to eliminate the discharge of pollutants into the navigable waters, waters of the contiguous zone, and the oceans; and (7) it is the national policy that programs for the control of nonpoint sources of pollution be developed and implemented in an expeditious manner so as to enable the goals of this Act to be met through the control of both point and nonpoint sources of pollution" (ELI, 1989).

Given the Act's stated goal and policies, the current Act could be said to consist of three major parts, one part being the Title II and Title VI provisions that authorize federal financial assistance for municipal sewage treatment plant construction. A second major part is the regulatory requirements, found throughout the Act, that apply to industrial and municipal discharges of pollutants. The third major part are the provisions in Title I that establish research, investigations, training, and information authorities (CRS, 1991; ELI, 1989).

The Act has forced the development and use of technologies to reduce the quantities of pollutants released into waterways. The CWA gave industries until 1977 to install "best practicable control technology" to clean up waste discharges. Later amendments to the CWA required a greater level of pollutant cleanup, generally requiring that industry use by 1989 the "best available technology" that is economically achievable. Failure to meet statutory deadlines could lead to enforcement action.

The process of deriving treatment requirements to attain given water quality is a complicated four-step process: (1) the state first establishes the desired highest use for a surface water, (2) the state adopts scientific criteria (made available by EPA) for water quality to support the designated use, (3) the state must determine how much of a pollutant may be discharged into a body of water without violating the criteria, and (4) the state must determine how much of the pollutant may be discharged by a given point source (Miller, 1992).

Under Titles II and VI, Congress has authorized grants for planning, design, and construction of municipal sewage treatment facilities since 1956 (CRS, 1991). Since that time, over $57 billion has been appropriated for grants to aid wastewater treatment plant construction (CRS, 1991). Federal grants are made for several kinds of water treatment projects, based on a priority list established by the states. Grants are generally available for up to 55% of total project costs (CRS, 1991).

Title I establishes an important research, investigation, training, and information authority for EPA under the CWA. In general terms, the Act mandates EPA to: (1) conduct and promote the coordination and acceleration of research, investigation, experiments, training, demonstration, surveys, and studies relating to the causes, effects, extent, prevention, reduction, and elimination of pollution; (2) encourage, cooperate with, and render technical services to pollution control agencies, and public and private agencies; (3) conduct, in cooperation with state water pollution control agencies and other interested agencies, public investigations concerning the pollution of any navigable waters; (4) establish advisory committees of recognized experts; (5) establish, equip, and maintain a water quality surveillance system; and (6) initiate and promote research for measuring the most effective practicable tools and techniques for measuring the social and economic costs and benefits of activities that are subject to regulation under the CWA (ELI, 1989).

Safe Drinking Water Act

The Safe Drinking Water Act (SDWA) was enacted by Congress in 1974 and extensively amended in 1986. The 1974 Act was in response to congressional finding that chlorinated organic chemicals were contaminating major surface and underground water supplies, that widespread underground injection operations were a threat to aquifers, and that the infrastructures of public water supply systems were increasingly inadequate to protect the public health (Randle, 1991). The 1986 amendments gave EPA additional regulatory and enforcement authorities and mandated that EPA issue safe drinking water standards within 3 years for 83 specified contaminants, with standards for 25 additional contaminants to be issued every 3 years thereafter.

There are six parts to SDWA, as amended. Parts A and E consist of definitions used throughout the statute and general provisions, respectively. The other parts establish various responsibilities and authorities for EPA and states.

Part B covers public water systems. It requires EPA to set minimum national standards to protect public health from drinking water contaminants. EPA was directed to develop, publish, and promulgate national primary drinking water regulations on 83 substances within 36 months following enactment of the 1986 amendments. In addition, the 1986 amendments require EPA to devise a priority list of additional contaminants for regulation and, every 3 years thereafter, issues Maximum Contaminant Levels (MCLs) and Maximum Contaminant Level Goals (MCLGs) for another 25 contaminants. The MCL is the National Primary Drinking Water Standard; the MCLG is the desired standard if technology permits its achievement. For carcinogens, the MCLG is zero. Part B also authorizes EPA to grant a state primary enforcement responsibility for public water systems under conditions specified in the statute.

Part C governs protection of underground sources of drinking water. Part C directs EPA to regulate underground injection, which is the subsurface emplacement of fluid through a well or dug-hole (Randle, 1991). Emplacement of fluid could occur, for instance, through a septic tank, a cesspool, a dry well, or a fissure. EPA is directed by the SDWA to regulate state programs, that in turn, were to regulate underground injection. Part B also requires states to develop their plans to protect wellhead areas from contaminants.

Part D provides the EPA Administrator with emergency powers to take such actions as deemed necessary in order to protect the health of persons due to an *imminent* and substantial *endangerment* from contamination of a public water system or an underground source of drinking water. Part D also specifies *criminal and civil penalties* for persons convicted of tampering with public water systems.

Part F is concerned with prohibiting the use of lead solder, pipes, or flux in drinking water systems and with removal from schools of water coolers with lead-lined tanks. It requires EPA to identify such coolers by brand name and provide this information to states, which in turn, are to develop programs of public education, water testing, and removal actions to replace lead-lined water coolers in schools.

The Safe Drinking Water Act and its amendments are important not only for their establishment of standards for contaminants in drinking water systems, but for the use of these standards in other federal regulatory contexts. For instance, drinking water standards are used in some CERCLA sites to establish groundwater cleanup levels. Regarding hazardous waste issues, the 1984 amendments to the Resource Conservation and Recovery Act contain a number of provisions directly applicable to hazardous waste injection wells. In particular, under these provisions, EPA must review all RCRA listed hazardous wastes to determine whether injection or other land disposal of those wastes may continue (Randle, 1991).

Oil Pollution Act

The Oil Pollution Act (OPA) of 1990 is legislation enacted by the U.S. Congress in recognition of the large number of oil spills occurring annually in the nation, and in particular, the legislation was markedly influenced by the 1989 oil spill in Alaska from the vessel *Exxon Valdez*. The OPA is a comprehensive amendment to the Clean Water Act, Section 311. The OPA is designed to expand oil spill prevention, preparedness, and response capabilities and authorities of government (OSWER, 1991). The Act establishes a new liability and compensation regime for oil pollution incidents in the aquatic environment and provides the resources necessary for the removal of discharged oil. The Oil Spill Liability Trust Fund, which is similar in statutory concept to the CERCLA trust fund, created a $1 billion fund to be used to respond to, and provide compensation for damages caused by, discharge of oil. (Similar to authorities in CERCLA, the OPA provides that the responsible party for a vessel or a facility from which oil is discharged, or which poses a substantial threat of a discharge, is liable for certain damages and costs of removal of oil.) In addition, the OPA provides new requirements for contingency planning both by government and industry and establishes requirements on construction (e.g., Section 4115 of the Act mandates newly constructed tank vessels must be equipped with double hulls, with the exception of vessels used only to respond to discharges of oil or hazardous substances), manning, and licensing for tank vessels (OSWER, 1991).

Several U.S. federal agencies are responsible for implementing the OPA. In general, EPA is responsible for oil spill prevention, preparedness, and response activities associated with nontransportation-related onshore activities. The U.S. Department of Transportation is generally responsible for oil spill planning and response activities for tank vessels, transportation-related onshore facilities, and deepwater ports. The U.S. Department of Interior is responsible for oil spill planning and response activities for offshore facilities except deepwater ports. States that have laws governing oil spill prevention and response are covered under the OPA. For example, Section 1019 of the OPA provides states the authority to enforce, on the navigable waters of the state, OPA requirements for evidence of financial responsibility.

Ocean Dumping Act

The Ocean Dumping Act, the first two titles of the Marine Protection, Research, and Sanctuaries Act of 1972, regulates intentional ocean disposal of materials and authorizes related research. The statute contains three titles (CRS, 1991). Title I of the Act prohibits all ocean dumping, except that allowed by permits, in any ocean waters under U.S. jurisdiction, by any U.S. vessel, or by any vessel sailing from a U.S. port. Any dumping is banned of radiological, chemical, and biological warfare agents and any high-level radioactive waste. Permits for dumping any other materials can be issued by EPA where it is determined that the dumping will "not unreasonably degrade or endanger human health, welfare, or amenities, or the marine environment, ecological systems, or economic potentialities." Permits issued under the Ocean Dumping Act must specify the kind of material to be disposed, the amount to be transported for dumping, the location of the dumpsite, the length of time the permit is valid, and special provisions for surveillance (CRS, 1991). Subsequent amendments to the Ocean Dumping Act prohibited the dumping of municipal sewage sludge or industrial waste.

Four federal agencies have responsibilities under the Ocean Dumping Act (CRS, 1991). EPA has primary authority for regulating ocean disposal of all substances except dredged spoils, for which the Army Corps of Engineers has authority. Under Title II of the Act, the National Oceanic and Atmospheric Administration has authority for researching the effects of human-induced changes to the marine environment, while EPA is authorized to carry out research and demonstration activities pertaining to phasing out sewage sludge and industrial waste dumping. The U.S. Coast Guard is charged with maintaining surveillance of ocean dumping. Title III authorizes the establishment of marine sanctuaries.

To what extent the Ocean Dumping Act, and international agreements intended to reduce ocean pollution, have been effective is unknown. However, it is apparent that large amounts of ocean pollutants exist, as measured by what washes up on beaches. In 1991 the Center for Marine Conservation organized 118,200 volunteers to clean up 3800 miles of U.S. coastline in one day. They picked up 2.8 million pounds of trash (N.Y. Times, 1992). That amount obviously only represents a tiny fraction of what is in the oceans and the animals that live there.

National Environmental Policy Act

The National Environmental Policy Act of 1969 (NEPA) is a short, concise statute that states U.S. national policy on the environment and establishes goals for federal programs in terms of their impact on the environment. NEPA is divided into two titles. Title I declares congressional direction for national environmental policy and was the first such congressional policy declaration (Vanderver et al., 1991). Title II creates the Council on Environmental Quality, which is placed in the Executive Office of the President, and defines its responsibilities. Once source considers NEPA to have had a profound impact in the sense that the Act focused the federal government's attention on environmental consequences and forced government agencies to assess the impact of their actions on the environment (Miller, 1992). NEPA's concepts have been emulated by many state statutes and other national governments.

Title I of NEPA states "... it is the continuing policy of the federal government, in cooperation with state and local governments, and other concerned public and private organizations, to use all practicable means and measures, including financial and technical assistance, in a manner calculated to foster and promote the general welfare, to create and maintain conditions under which man and nature can exist in productive harmony, and fulfill the social, economic, and other requirements of present and future generations of Americans." In support of the state policy, NEPA states six goals for the Nation:

1. fulfill the responsibilities of each generation as trustee of the environment for succeeding generations;
2. assure for all Americans safe, healthful, productive, and aesthetically and culturally pleasing surroundings;
3. attain the widest range of beneficial uses of the environment without degradation, risk to health and safety, or other undesirable and unintended consequences;
4. preserve important historic, cultural, and natural aspects of our national heritage, and maintain, wherever possible, an environment that supports diversity and variety of individual choice;
5. achieve a balance between population and resource use that will permit high standards of living and a wide sharing of life's amenities; and
6. enhance the quality of renewable resources and approach the maximum attainable recycling of depletable resources.

Title I requires that an environmental impact statement (EIS) shall be prepared by all agencies of the federal government to "include in every recommendation or report on proposals for legislation and other major federal actions significantly affecting the quality of the human environment." The EIS is to include: "(a) the environmental impact of the proposed action, (b) any adverse environmental effects which cannot be avoided should the proposal be implemented, (c) alternatives to the proposed action, (d) the relationship between local short-term uses of man's environment and the maintenance and enhancement of long-term productivity; and (e) any irreversible and irretrievable commitments of resources which would be involved in the proposed actions should it be implemented" (ELI, 1989). The EIS process involves participation and review at all levels of government and allows interested public groups to be involved.

The Council on Environmental Quality (CEQ) was created by Title II of NEPA. The CEQ is to assist and advise the President in the preparation of an annual Environmental Quality Report to Congress. The report must include: (1) the status and condition of the major natural, manmade, or altered environmental classes of the nation; (2) current and foreseeable trends in the quality, management, and utilization of such environments; (3) the adequacy of available natural resources; (4) a review of government and nongovernment programs as to their effect on the environment and on the conservation, development, and utilization of natural resources; and (5) a program for remedying the deficiencies of programs, together with recommendations for legislation (ELI, 1989). In addition, CEQ advises the President on policies and legislation, gathers timely information concerning trends in environmental quality, and to conduct studies related to environmental quality and ecological systems.

Environmental Research and Development Demonstration Act

Many federal statutes that pertain to environmental protection and concern for effects of environmental hazards on public health contain provisions for research and development. The intent of Congress has been to encourage research in order to advance scientific knowledge about specific environmental hazards and their consequences. According to an analysis by the Congressional Research Service, EPA's statutory mandate for research and development grew piecemeal from parts of many environmental protection laws, enacted and amended over the years, currently involving at least 13 separate statutes (CRS, 1991).

The Environmental Research, Development, and Demonstration Authorization Act (ERDDAA) of 1976 was intended to codify EPA's research and development programs under one authorization statute. ERDDAA was enacted annually through 1980, which was the last year. ERDDAA contains no language specific to research on hazardous wastes, though other statutes, e.g., the Comprehensive Environmental Response, Compensation, and Liability Act, as amended, do contain directives to conduct research and demonstration programs on waste generation, management, and toxicity of hazardous substances.

Occupational Safety and Health Act

The Occupational Safety and Health Act of 1970 (OSHAct) was the first federal legislation to deal comprehensively with health and safety problems in the workplace. The declared purpose and policy of the statute is "to assure so far as possible every working man and woman in the Nation safe and healthful working conditions and to preserve our human resources" (Moran, 1991). To effectuate this goal, the OSHAct created three new government entities. The Occupational Safety and Health Administration (OSHA), which is located in the Department of Labor, is directed to developed workplace safety and health standards, conduct workplace inspections, enforce regulatory actions developed by the agency; helps set up state occupational safety and health programs and monitors their effectiveness; and it conducts education and training programs for safety and health professionals. The National Institute for Occupational Safety and Health (NIOSH), which is placed in the Department of Health and Human Services, is tasked with conducting scientific research, performing workplace health evaluations, and developing criteria documents that contain recommendations for safety and health exposure conditions. The Occupational Safety and Health Review Commission was established to adjudicate disputes arising from enforcement of the Act.

The most consequential OSHA responsibility under the OSHAct is to develop and promulgate safety and health standards. This is done through regulatory action, taking into account all relevant scientific information and public concerns. OSHA standards affect most workplaces, large and small. Employers convicted of willful violation of OSHA standards may face civil or criminal penalties, depending on the seriousness of the infraction. OSHA inspectors conduct inspections of workplaces to assess compliance with OSHA safety and health standards. Since 1977, OSHA has attempted to direct most of its inspections at high-hazard industries — construction, petrochemicals, and manufacturing (Bingham, 1992). Where no specific standards exist for a workplace condition, OSHAct directs each employer to provide "a place of employment which is free from recognized hazards that are causing harm to employees." This "general duty" clause is used by OSHA to control workplace hazards that are obvious and for which no specific standard exists (Bingham, 1992; Moran, 1991).

An OSHA standard of particular importance is the Hazard Communication Standard (HCS). One source considers it the most significant job safety and health regulatory action ever adopted (Moran, 1991). The HCS is known as the "Right-to-Know" standard and its concept has been widely adopted by numerous states and some local governments. The HCS requires the employer to develop and provide to employees information about hazardous chemicals in the workplace. It requires each employer to adopt its own material safety data sheet (MSDS) for each product containing a hazardous chemical on the premises, train and educate its employees on those chemical hazards, and enforce the placement of warning labels in the workplace (Moran, 1991).

Language related to waste management is found in the 1986 amendments to the Comprehensive Environmental Response, Compensation, and Liability Act (CERCLA) that directs OSHA to promulgate standards for the health and safety protection of employees engaged in hazardous waste operations. The CERCLA amendments stipulate that the OSHA regulations pertaining to hazardous waste workers shall include the following: (1) site analysis (each hazardous waste site is to have a specific plan for worker protection), (2) training requirements for contractors to provide workers with training in hazardous waste operations, (3) medical surveillance of workers, (4) protective equipment requirements, (5) engineering controls concerning the use of equipment, (6) maximum exposure limits for workers, (7) information programs informing workers of hazards, 8) handling of hazardous waste, (9) new technology that would maintain worker protections, (10) decontamination procedures, and (11) emergency response requirements (ELI, 1989, p. 245). The subject OSHA regulations were promulgated in final form in 1990 (OSHA, 1989).

REFERENCES

American Society of International Law (ASIL), International legal materials, *Am. Soc. Int. Law,* 28, 31, 1989.

Anon., Report on National Biotechnology Policy, White House Council on Competitiveness, Washington, D.C, 1991.

Arbuckle, J. G., Bosco, M. E., Case, D. R., Laws, E. P., Martin, J. C., Miller, M. E., et al., Eds., *Environmental Law Handbook,* Government Institutes, Inc., Rockville, MD, 1991.

Arbuckle, J. G., Randle, R. V., and Wilson, P. A. J., Emergency Planning and Community Right-to-Know Act (EPCRA), in *Environmental Law Handbook,* Arbuckle, J. G., Bosco, M. E., Case, D. R., Laws, E. P., Martin, J. C., Miller, M. E., et al., Eds., Government Institutes, Inc., Rockville, MD, 1991a, 190.

Arbuckle, J. G., Vanderver, T. A., Jr., and Randle, R. V., Water pollution control, in *Environmental Law Handbook,* Arbuckle, J. G., Bosco, M. E., Case, D. R., Laws, E. P., Martin, J. C., Miller, M. E., et al., Eds., Government Institutes, Inc., Rockville, MD, 1991b, 65.

Bingham, E., The Occupational Safety and Health Act, in *Environmental and Occupational Medicine,* 2nd ed., Rom, W. N., Ed., Little, Brown, Boston, 1992, 1325.
Bosco, M. E. and Randle, R. V., Underground storage tanks, in *Environmental Law Handbook*, Arbuckle, J. G., Bosco, M. E., Case, D. R., Laws, E. P., Martin, J. C., Miller, M. E., et al., Eds., Government Institutes, Inc., Rockville, MD, 1991, 443.
Bureau of National Affairs (BNA), Current developments, *Environ. Rep.,* July 3, 715, 1992a.
Bureau of National Affairs (BNA), Senate ratifies Basel Convention on transboundary shipments of waste, *Environ. Rep.,* August 21, 1255, 1992b.
Case, D. R., Resource Conservation and Recovery Act (RCRA), in *Environmental Law Handbook,* Arbuckle, J. G., Bosco, M. E., Case, D. R., Laws, E. P., Martin, J. C., Miller, M. E., et al., Eds., Government Institutes, Inc., Rockville, MD, 1991, 406.
Committee on the Future of Environmental Health, The Future of Environmental Health, National Environmental Health Association, Denver, 1992.
Congressional Research Service (CRS), Summaries of Environmental Laws Administered by the Environmental Protection Agency, The Library of Congress, Washington, D.C., 1991.
Environmental Law Institute (ELI), *Environmental Law Deskbook,* Environmental Law Institute, Washington, D.C., 1989.
Environmental Law Institute (ELI), *European Community Deskbook,* Environmental Law Institute, Washington, D.C., 1992.
Environmental Protection Agency (EPA), Request for comment on regulatory approach, *Fed. Regist.,* 54, 7027, 1989.
Gordon, L., Davis, T., Powitz, R., Stern, B., Treser, C., and Wiant, C., The future of environmental health, Part I, *J. Environ. Health,* 55, 28, 1993.
Jennings, B., Analysis of Proposition 65—Safe Drinking Water and Toxic Enforcement Act, Senate Office of Research, State of California, Sacramento, 1986.
Johnson, B.L., Implementation of Superfund's health-related provisions by ATSDR, *Environ. Law Reporter,* 20, 10277, 1990.
Johnson, S., The Basel Convention: The Shape of Things to Come for United States Waste Exports. Environmental Law 1991, Northwestern School of Law, Lewis and Clark College, Chicago, 1991.
LaDou, J., Deadly migration, *Technol. Rev.*, July, 47, 1991.
Massachusetts Legislature, Hazardous Waste Management Act, Chapter 21C, Department of Environmental Protection, Boston, MA, 1979.
Miller, J. G., Environmental law and the science of ecology, in *Environmental and Occupational Medicine,* 2nd ed., Rom, W. M., Ed., Little, Brown, Boston, 1992, 1307.
Miller, M. L., Toxic substances control act, in *Environmental Law Handbook*, Arbuckle, J. G., Bosco, M. E., Case, D. R., Laws, E. P., Martin, J. C., Miller, M. E., et al., Eds., Government Institutes, Inc., Rockville, MD, 1991a, 280.
Miller, M. L., Federal regulation of pesticides, in *Environmental Law Handbook,* Arbuckle, J. G., Bosco, M. E., Case, D. R., Laws, E. P., Martin, J. C., Miller, M. E., et al., Eds., Government Institutes, Inc., Rockville, MD, 1991b, 328.
Moran, R. D., Occupational Safety and Health Act, in *Environmental Law Handbook,* Arbuckle, J. G., Bosco, M. E., Case, D. R., Laws, E. P., Martin, J. C., Miller, M. E., et al., Eds., Government Institutes, Inc., Rockville, MD, 1991, 370.
National Association of County Health Officials (NACHO), Current Roles and Future Challenges of Local Health Departments in Environmental Health, National Association of County Health Officials, Washington, D.C., 1992.

National Institutes of Health (NIH), Guidelines for research involving recombinant DNA molecules, *Fed. Regist.,* 51, 16958, 1986.

New York Times, World's oceans are sending an S.O.S., *New York Times Magazine,* May 3, 1992a.

New York Times, Draft of environmental rules: Global partnership, *New York Times,* April 4, 1992b.

Occupational Safety and Health Administration (OSHA), Hazardous waste operations and emergency response, *Fed. Regist.*, 54, 9292, 1989.

Office of Science and Technology Policy (OSTP), Regulatory matrix for biotechnology products, *Fed. Regist.,* 50, 47174, 1985.

Office of Science and Technology Policy (OSTP), Federal coordinated framework for regulation of biotechnology, *Fed. Regist.,* 51, 23323, 1986.

Office of Science and Technology Policy (OSTP), Exercise of federal oversight within scope of statutory authority: Planned introductions of biotechnology products into the environment, *Fed. Regist.,* 57, 6753, 1992.

Office of Solid Waste and Emergency Response (OSWER), OPA Q's & A's: Overview of the Oil Pollution Act of 1990, Environmental Protection Agency, Washington, D.C., 1991.

Office of Technology Assessment (OTA), Complex Cleanup: The Environmental Legacy of Nuclear Weapons Production, Congress of the United States, Washington, D.C., 1991.

Quarles, J. and Lewis, W., Jr., Mopping up after the Clean Air Act; navigating the rules and fog, *Legal Times,* February 11, 1991.

Randle, R. V., Safe Drinking Water Act, in *Environmental Law Handbook,* Arbuckle, J. G., Bosco, M. E., Case, D. R., Laws, E. P., Martin, J. C., Miller, M. E., et al., Eds., Government Institutes, Inc., Rockville, MD, 1991, 149.

Randle, R. V. and Bosco, M.E., Air pollution control, in *Environmental Law Handbook,* Arbuckle, J. G., Bosco, M. E., Case, D. R., Laws, E.P., Martin, J. C., Miller, M. E., et al., Eds., Government Institutes, Inc., Rockville, MD, 1991, 524.

Smith, T. T., Jr. and Hunter, R. D., The European Community environmental legal system, *European Community Deskbook,* Environmental Law Institute, Washington, D.C., 1992, 3.

Spensley, J. W., An overview of the Clean Air Act Amendments of 1990, *Shephard's Clean Air Act Rep.,* 1, 3, 1992.

Stoll, R. G., Comprehensive Environmental Response, Compensation, and Liability Act, in *Environmental Law Handbook,* Arbuckle, J. G., Bosco, M. E., Case, D. R., Laws, E. P., Martin, J. C., Miller, M. E., et al., Eds., Government Institutes, Inc., Rockville, MD, 1991, 471.

Subcommittee on Investigations and Oversight (SOI), The Environmental Implications of Genetic Engineering, Committee on Science and Technology, House of Representatives, U.S. Congress, Washington, D.C., 1984.

United Nations Environment Program (UNEP), United Nations Conference on Environment and Development, United Nations, New York, 1992.

World Health Organization (WHO), *European Charter on Environment and Health,* WHO Regional Office for Europe, Copenhagen, 1989.

Index